Sugarcane

Second Edition

Edited by

Glyn James

Blackwell
Science

© 2004 by Blackwell Science Ltd
a Blackwell Publishing Company

Editorial Offices:
Blackwell Science Ltd, 9600 Garsington Road, Oxford
OX4 2DQ, UK
   Tel: +44 (0) 1865 776868
Blackwell Publishing Professional, 2121 State Avenue,
Ames, Iowa 50014-8300, USA
   Tel: +1 515 292 0140
Blackwell Science Asia Pty Ltd, 550 Swanston Street,
Carlton, Victoria 3053, Australia
   Tel: +61 (0)3 8359 1011

First edition published 1984 by Longman Group Ltd
Second edition published 2004 by Blackwell Science Ltd

Library of Congress
Cataloging-in-Publication Data
is available

ISBN 0-632-05476-X

A catalogue record for this title is available from the
British Library

Set in 10.5/12 pt Ehrhardt
by Sparks, Oxford – www.sparks.co.uk
Printed and bound in India by
Replika Press PVT. Ltd, Kundli 131028

The publisher's policy is to use permanent paper from
mills that operate a sustainable forestry policy, and
which has been manufactured from pulp processed
using acid-free and elementary chlorine-free practices.
Furthermore, the publisher ensures that the text paper
and cover board used have met acceptable environmental
accreditation standards.

For further information on Blackwell Publishing,
visit our website:
www.blackwellpublishing.com

Contents

Foreword   vi

Preface   vii

1 An Introduction to Sugarcane   1
The origins and spread of sugarcane   2
Movement and development of the noble
  canes   3
The ancestry of cultivated sugarcane   3
The early commercial cane varieties   4
  Saccharum officinarum or the noble canes   4
  The end of the noble cane era   5
  Saccharum spontaneum   6
  Saccharum barberi   7
  Saccharum sinense   7
  Saccharum robustum   7
  Saccharum edule   7
The botany of sugarcane   8
  The stem   8
  The leaf   10
  The roots   12
  The inflorescence   13
  The seed   15
Cultivation of sugarcane   16
  Sugarcane cultivation and the slave trade   17
  The introduction of indentured labour   17
The production of sugar   18
  Sugar bounties   18
Research   18
Developments in the twentieth century   18
References   18

2 Plant Improvement of Sugarcane   20
Introduction   20
Germplasm   20
  Introduction   20
  Basic germplasm   20

Commercial (*Saccharum* spp. hybrid)
  germplasm   26
  Germplasm conservation   26
Cross-pollination   27
  Introduction   27
  Flower induction   27
  Cross-pollination   28
  Parental selection   29
  Seed storage   30
Selection   30
  Introduction   30
  Selection in original seedlings   32
  Selection of clones   33
  Genotype × environment interaction   33
  Competition and plot technique   34
  Experimental design   35
  Automation   35
Breeding objectives   36
  Productivity   36
  Disease resistance   37
  Pest resistance   39
  Milling characteristics   40
  Sugar quality   40
  Crop ideotype   40
  Row spacing   41
  Achievements   41
Future options   43
  Molecular-assisted selection   43
  Genetic transformation   44
  Functional genomics   45
Conclusions   46
Acknowledgements   47
References   47

3 Diseases   54
Introduction   54
Principles of disease control in sugarcane   55

Variety resistance	56
Seedcane quality	56
Field control practices	56
Selection of varieties for disease resistance	57
Fungal diseases	57
Pineapple disease	57
Pokkah boeng	58
Red rot	59
Common rust *Puccinia melanocephala* Sydow and orange rust *P. kuehnii* Butler	60
Smut *Ustilago scitaminea* Sydow	61
Miscellaneous fungal diseases	63
Bacterial diseases	63
Gumming disease *Xanthomonas axonopodis* pv *vasculorum* (Cobb) Vauterin	63
Leaf scald *Xanthomonas albilineans* (Ashby) Dowson	64
Ratoon stunting disease *Leifsonia xyli* subsp. *xyli*	66
Other bacterial diseases	69
Viral diseases	69
Mosaic, sugarcane mosaic virus and sorghum mosaic virus	70
Yellow leaf syndrome, sugarcane yellow leaf virus; sugarcane yellows phytoplasma	72
Miscellaneous viral diseases	73
Phytoplasmal diseases	73
References	74

4 Pests of Sugarcane — **78**

Introduction	78
Stalk borers	78
Biology	78
Damage	79
Distribution	80
Control	80
Soil pests	85
Damage	85
Biology	85
Distribution	86
Control	87
Termites	88
The biology of other soil insect pests	89
Sap feeders	90
Biology	90

Damage	90
Distribution and control	91
Leaf feeders	93
Damage	94
Distribution	94
Biology and control	94
Vertebrate pests of sugarcane	95
Rodents	95
Other vertebrate pests	96
References	96

5 Sugarcane Agriculture — **101**

Introduction	101
Systems of cultivation	101
Soil conservation and field layouts	101
Row cropping	102
Cambered beds	103
The system in Guyana	104
Florida and Mozambique	106
Louisiana banks	106
Cultivation on sloping land	106
Yield	108
Ratoons and ratooning	108
Land preparation	110
Seedcane production	112
Planting	113
Ratoon management	114
Weed control in sugarcane	116
Chemical weed control	117
Irrigation and drainage	122
Appropriate irrigation systems	123
Furrow irrigation	123
Overhead irrigation	127
Drip irrigation	133
Soil moisture instrumentation	136
Drainage	137
World Wide Web information sources	140
References	141

6 Sugarcane Agronomy — **143**

Soil management problems	143
Site selection	144
Clearing	145
Planning and layout	145
Field design	145
Land preparation	146
Nurseries	147
Commercial planting	148

Crop management	150
Crop control	151
Harvest	152
Field factors and cane quality	153
Ratooning	155
Monoculture	156
Specialty crops	157
Organic sugarcane	157
Transgenic sugarcane	157
Inter-cropping	157
Basic economics	158
References	158

7 Harvest Management   **160**

Pre-harvest burning	160
Green cane harvesting	161
To burn or not to burn	162
Unplanned cane fires	163
Fire control measures	163
Fire control equipment	163
Reaping and transport	164
General	164
Manual and mechanised harvesting	164
Manual cane cutting	165
Cane loading	165
Mechanised cutting	169
Harvesting aids	171
Yield losses from mechanised harvesting	171
Cane transport	172
Field-edge to mill transport	175
Offloading systems	176
Cane storage	178
Co-ordination and control of harvesting	179
References	180

8 Cane Payment Systems   **181**

Introduction	181
Different types of cane payment systems	181
Fixed cane price	181

Fixed revenue sharing	182
Variable revenue sharing	182
Payment for which sugar products?	183
Incentives to improve technical performance	183
Sampling and testing of cane	183
Payment systems	184
Systems based on the average quality of growers' cane	184
Systems based on the quality of the individual grower's cane	184
Incentives to improve the cane quality and sucrose recovery	185
Limitations	186
Incentives to expand production	187
Cane payment systems and the length of the campaign	187
Relative payment schemes	189
Grower/miller incentives to expand	189
Incentives to produce refined sugar	190
Impact on incentives for different socio-economic groups	190
Integrated operations	190
Co-operative mills	190
Growers on leased or tribal land	190
Smallholders	191
Recent performance of cane payment systems	191
Conclusion	193

9 Project Planning   **195**

Introduction	195
The role of the consultant	195
Pre-feasibility study	197
Feasibility study	198
Further reading	201

Author Index   **203**

Subject Index   **211**

Foreword

By way of introduction as the editor, I am an Honours graduate of the University of Wales and was awarded my Doctorate by the University of London and conferred a Diploma of Membership of the Imperial College in 1966. Following 30 years experience in sugarcane, I was awarded a Fellowship of the Institute of Biology in 1996.

After 3 years research experience into the diseases and nematode problems of the coffee in Kenya, I have over 36 years experience in sugarcane in several countries. My major involvement in sugarcane has been in applied research to ensure optimum and sustainable cane and sugar production without detriment to the environment. I have published over 50 technical papers, and have held appointments as Director of Research and Agriculture Manager, and acted as Consultant in several countries in the world's sugarcane agro-industry.

Glyn James

Preface

This is an update of Frank Blackburn's comprehensive book *Sugar-cane*, which was published by Longman in 1984. I was asked to act as the editor in April 1998, and undertook the task of approaching several possible collaborators for producing this edition. By the beginning of the following year I was fortunate enough to have selected a team to write about the topics on which they are the recognised authorities in the world's sugarcane agro-industry. The list of authors is as follows:

- **Rod Ellis, Glyn James, Bob Merry, David Weekes** and **Ben Yates**, Booker Tate Ltd, England.
- **Philip Digges, Gareth Forber** and **Martin Todd**, LMC International Ltd, England
- **Nils Berding, Mike Cox** and **Mac Hogarth**, Bureau of Sugar Experiment Stations, Australia.
- **Roger Bailey** and **Graeme Leslie**, South African Sugar Experiment Station.
- **James Irvine**, Texas A&M University, United States.

As all of the authors are very busy specialists, their main programmes of work delayed the intended publication date of mid 2000. This was not made any easier because of the difficulties that I had in communicating with them; I was working in a relatively remote area in Sumatra, Indonesia, until the end of 2001 when I retired. The final chapter arrived at the beginning of 2003. Consequently, I am extremely grateful to my colleagues for their work and tolerance with my incessant requests, and for the publisher's patience.

Blackburn's book dealt with the history and botany of the crop through to its milling and the processing of sugar. However, as there are currently a number of excellent publications covering the factory side, this revised edition mainly deals with the agricultural aspects. The intended audience is students, agriculturalists and others who have an association with the crop. Consequently, the book is not unnecessarily technical, and any further reading is listed in the comprehensive reference sections added to most of the chapters.

Glyn James
March 2003

Chapter 1
An Introduction to Sugarcane

Glyn L. James

Worldwide, man is known for his 'sweet tooth'; however, it is not known whether this delight in sweet things is inherent or acquired. Nevertheless, it has been with us for a very long time. Sugar not only enhances the flavours of food and intensifies its colour, but it also has other properties, e.g. it can be used as a preservative or as a substrate for fermentation, and it is also a source of energy. Consequently, it is a very useful commodity.

In Europe especially, sugar was a luxury up until the early eighteenth century. It is only since the Caribbean islands and tropical north and south Americas were colonised by Europeans, that sugar became available on the world market in a large enough quantity and at acceptable prices for everyday use. In England, the yearly per capita consumption increased 20-fold from the 1660s to 1775[1], and a further five times from 1835 to 1935[2]. The rise in prosperity over almost three centuries in Britain alone allowed more people to purchase more sugar and, with the increasing popularity of tea, gave them more opportunity to use it. The industrial processing of food and its requirement for sugar was associated with the Industrial Revolution in Britain.

It is difficult to determine when cane sugar first became the principal sweetener, but relatively accurate assumptions can be made. Cane sugar first achieved dominance on the subcontinent of India more than 2500 years ago[3], and it was in that country and China that commercial sugar was first produced from sugarcane. It was not until the early eighteenth century, however, that sugar began to become widely used in western Europe. Sugarcane was unknown in the New World until Columbus introduced it on his second voyage in 1493. Returning Crusaders brought news of cane sugar to the nobility of northern Europe, and they began to import it as a medicine and as a rare and costly additive to food and drink. Gradually sugar moved from the medicine chest and guarded storehouse to the kitchen. Sugar was sold in loaves weighing up to 40 pounds. During the sixteenth century, white sugar was a new way of displaying wealth, especially when served from a silver caster.

The consumption of sugar continued to increase as its price fell in Europe during the seventeenth and eighteenth centuries. A very significant boost in its use came with the introduction of new drinks, e.g. lemonade was invented in Paris during 1630[4]. Chocolate, made from the beans of the tropical southern American plant cacao, and coffee and tea from Africa and the East were all normally taken with sugar. During the eighteenth century, the decline in the price of tea made it cheaper than chocolate or coffee, and readily accessible even to the poor. In England, tea became the drink of the masses, replacing gin, which taxes and the rising price of grain were making a luxury. It even threatened the consumption of beer. Home brewing had once been widespread, but by 1800 was reducing. While the East India Company encouraged the consumption of tea in Britain, tea drinking did not develop in European society to the same extent; especially in the south where wine maintained its hold on people's habits. By the end of the eighteenth century, sugar approached the status of a staple food in the English diet[5]. Cane sugar remained unchallenged until beet-sugar came onto the market in the early nineteenth century. Now, refined, pure sugar can be produced from both sources, and there is no difference between them.

The Industrial Revolution in Britain created a new working class, and gave it a sugar-laden diet as the price of the commodity declined. Because of the increased money that could be made, industrialisation attracted people from the countryside to the back-to-back houses and tenements of the developing industrial towns and cities. However, in these dwellings the occupants had to buy what they ate instead of growing their own requirements as before. The long hours men and women spent tending looms, mills, and other machinery meant there was less time spent preparing meals at home. Consequently, one of the British working class's responses was to incorporate many cups of sweet tea into its diet – 'a calorie-laden stimulant that revives the spirit, blunts the pangs of hunger, but does not nourish'[5]. The second response was to abandon the careful cooking of traditional dishes in favour of cold or quickly heated, shop-bought food. For example, the 'jam-buttie', which is a sugar-rich, high calorie 'convenience food' which could be quickly prepared and quickly eaten, is nothing more than factory-made jam spread on a slice of factory-made bread. The consumers of that time therefore could be said to have become another exploited group associated with sugar as the slaves were earlier[6].

The key variable in accounting for the differences in sugar consumption between societies is wealth, but culture, fashion, and availability of the commodity are also significant. The populations of richer industrial countries consume more sugar per capita than those of poorer countries. But the influence of other variables within a rich country come into play, as the richer, better-educated, diet-conscious inhabitants consume less than the poorer and less well-educated ones. Nevertheless, the greatest consumers of sugar are, not surprisingly, some of the producers, and in Australia, New Zealand, and northern Europe the per capita consumption was above 40 kg per year in the 1980s. Consumption was somewhat less in the USA, however, where dietary concerns were prominent. In parts of Africa and the Far East, the annual per capita consumption is still only around 5 kg.

THE ORIGINS AND SPREAD OF SUGARCANE

There are records of the use of sugarcane in India and China as far back as chronicled history goes, and there can be little doubt that primitive man cultivated it long before then, albeit not for the manufacture of commercial sugar. During the 1920s, Brandes[7] described and illustrated the swidden agriculture or garden culture of sugarcane for chewing by primitive tribes in Papua New Guinea with a whole range of other crops, e.g. bananas, betel nuts, breadfruit, sago, sweet potatoes, and yams. In addition, the sugarcane existed in a wide range of types or varieties that differed extensively in colour and form.

There can hardly be any doubt that the improvement of sugarcane varieties has a history as long as its cultivation. According to Stevenson[8], this historic trail fades back 'into the realms of conjecture'. It is only necessary to credit primitive man with the powers of observation and sufficient intelligence to realise the value of natural variants of the canes in his garden. As these canes were used for chewing, improved forms would be the ones that were sweeter and/or less fibrous. In addition, bright and unusual colours would be of interest, and would account for the presence of the brilliant coloured and striped types that are currently present in the island of New Guinea (i.e. Irian Jaya and Papua New Guinea).

Artschwager & Brandes[9] hypothesise that a common ancestor of sugarcane originated in southern Asia and spread southeastwards across an ancient land mass that extended from southern Asia to Australia during the early Cretaceous period, some 60 million years ago. The origin of the early Cretaceous *Saccharum* from a sub tribe Andropogoneae in Asia is conjectural; but Celarier[10] suggests that the ancestral genus evolved from *Erianthus* with contributions from *Miscanthus*. Artschwager & Brandes[9] state that there were two periods of sugarcane movement. The earliest period dating back to the early Cretaceous, when the enormous Asiatic-Australian continent during that period provided a continual land bridge over which primitive Asiatic canes possibly moved unaided by man to the region now known as Melanesia. They suggested the existence of a common wild ancestor throughout that entire region. Later, the advent of flood conditions during the late Cretaceous period over the Asiatic-Australian continent posed enormous restrictions to the movement of flora. Many plant species evolved for growth in and around brackish water tracts, and had some advantages in that their seeds or fruit were able to survive long periods drifting in ocean currents. The seeds produced by early *Saccharum* forms, however, could not survive prolonged immersion in saline water. Wild *Saccharum* types preferred freshwater habitats along flood plains and river banks, as they do today. Furthermore, the seed or fuzz was not suited for dissemination by birds or wind over broad expanses of water. It is therefore reasonable to conclude that distinct wild forms of sugarcane evolved in isolated spots, e.g. Asia, New Guinea, and other Melanesian islands. Considerable modification must have taken place by natural selection and hybridisation. Primitive man also took part in the selection and dissemination process at some later period. The present advanced 'chewing' canes, cultivated by the indigenous people in countries such as New Guinea, reflect prolonged selection.

MOVEMENT AND DEVELOPMENT OF THE NOBLE CANES

The Dutch research workers in Java first applied the term 'noble' to forms of *S. officinarum* during the early 1920s. It well describes the tall, thick, sometimes brightly coloured canes of the species.

The most prized canes first accompanied man eastward in ancient times on his early travels, which eventually covered half the circumference of the world. According to Brandes[11], there were the following three main movements from New Guinea.

(1) The first movement brought the introduction of *S. officinarum* to the Solomon Islands, Vanuatu, and New Caledonia at around 8000 BC.

(2) The next movement started around 6000 BC and was in a westerly direction to Indonesia, the Philippines, and ultimately to the Indian subcontinent via the Malayan Peninsula and Burma.

(3) The last movement is thought to have occurred around AD 600 to 1100, eventually reaching the

various island groups east of Vanuatu, e.g. Fiji, Tonga, Samoa, the Cook Islands, the Society and Marquesa Islands, Easter Island, and northward to Hawaii. Hawaiian legends refer to sugarcane being brought around AD 750 to 1000[12].

Each of these areas became important centres of diversification. To summarise, *S. officinarum*, or 'noble canes', originated around 5°S of the equator, were first moved into territories within 21°S, and later moved into territories bounded by 21°N.

THE ANCESTRY OF CULTIVATED SUGARCANE

From the following discussion it will be evident that the systemic grouping of the species within the genus *Saccharum* is still inconclusive. Barnes[12] maintained that India was the home of sugarcane, basing this conclusion on recorded ancient Hindu mythology but giving no other evidence. Some authors claim two separate origins of sugarcane, i.e. India and New Guinea. A more logical explanation, however, is development in the Cretaceous of a common wild ancestor in distinct Asiatic and Melanesian evolution centres, which were under different selection pressures. Parthasarathy[13] defined two broad groups: (1) northern India canes having *Saccharum barberi* and *S. sinense* parentage that are characterised by thin, hardy stalks, and (2) noble canes of *S. officinarum* parentage that are characterised by thick stalks. Northern Indian canes, including all those recorded in ancient Indian literature, were considered by Barber[14] and Jeswiet[15] to be indigenous, while the noble types were thought to have been introduced much later. Cytogenic and morphological evidence, however, indicate a later origin of the northern Indian canes than that of the noble ones. Parthasarathy[16] believed that the cultivated canes of northern India were derived from extensive hybridisation of *S. officinarum* with *S. spontaneum*, which grows wild over much of India from the Himalayas to the tropical south of the subcontinent. As well as morphological and cytological evidence supporting this conclusion, a common ancestry was put forward for *S. officinarum* and the genus *Sclero-stachya* from an unknown genus having a basic chromosome number of five. This further contributes to the opinion that *S. officinarum* was present on the Indian subcontinent over many years, and that it contributed to the later evolution of the northern Indian canes. In contrast, Mukherjee[17] placed the cultivated sugarcanes into the following three species:

- *S. officinarum*, not known in the wild state;
- *S. barberi* (the northern Indian canes), having thinner stalks and poorer juice quality than *S. officinarum*; and
- *S. sinense* group (i.e. the China canes and the Panache Indian canes), which is of a similar appearance to *S. barberi*.

On a worldwide basis, Mukherjee[17] regarded the cultivated canes to be a result of extensive hybridisation among *S. officinarum*, *S. spontaneum*, *S. barberi*, and *S. sinense*.

Grassl[18] stressed that *S. spontaneum* developed from an initially small botanic group to a huge one containing innumerable forms of varying rank and importance; he lists sixteen species that he has reduced to one, i.e. *S. spontaneum*.

Saccharum robustum is the second species of sugarcane native to New Guinea, New Britain, Vanuatu, and possibly also in Borneo and the Solomon Islands. Brandes[19] distinguishes *S. robustum* from *S. spontaneum* on the basis of its large vegetative structure and height. The wild species of *S. robustum* therefore evolved in the region of New Guinea, and possibly hybridised with other grasses. Primitive man then selected the sweetest, softest, and thickest canes from *S. robustum* for chewing. Eventually, such forms became dependent on man for their propagation as the species *S. officinarum*. Grassl[18] does not regard *S. edule* as an authentic *Saccharum* species.

Saccharum barberi and *S. sinense* are believed to have originated in India through natural hybridisation with *S. officinarum*. For years *S. barberi* was confined to India, while *S. sinense* was common in Burma, and China as well as India. Both eventually achieved commercial importance in Brazil, South Africa and the USA, as well as India and China. Nevertheless, both species have now been replaced by hybrid sugarcane varieties in all countries. To summarise, in Grassl's opinion there are four species in the genus, *Saccharum*:

- *S. officinarum* – the type cane of the genus;
- *S. robustum* – the 'wild ancestor' of *S. officinarum*;
- *S. spontaneum* – another 'wild ancestor' which is more primitive than *S. robustum*; and
- *S. barberi* – the origin of which is unclear.

Also in his view, there were two separate origins for sugarcane, i.e. India and New Guinea.

Note that Australia does not seem to have supported any wild *Saccharum* species, even though it is situated close to confirmed *Saccharum* habitats.

THE EARLY COMMERCIAL CANE VARIETIES

Alexander the Great took sugarcane from India in around 325 BC on his retreat to Europe. Later, sugarcane reached Spain, Madeira, the Canary Islands and São Tomé, off West Africa. While there is no definite evidence of the deliberate movement of the plant to those countries, Columbus certainly took sugarcane to the New World in 1493 on his second voyage from Spain. Attempts to grow the crop in Hispaniola (now Haiti and the Dominican Republic) failed at first, but success was achieved in 1506 in the western part of the island. In 1515 it was taken from Haiti to Puerto Rico, and was also introduced into Mexico in 1520. This was the start of the modern sugar industry in Mexico[12].

Saccharum officinarum or the noble canes

Otaheite, Bourbon or Creole cane

The original introductions into the Portuguese and Spanish colonies of the tropical regions of the New World came from sugarcane growing in Madeira and Spain. The variety became known as Creole, and was the basis for the development of the sugar industry in that region until the second half of the eighteenth century. Otaheite was the first of the noble canes to be cultivated on a worldwide scale. It was the only variety grown in Mauritius from the establishment of the industry there in the 1730s, and it was known as Lousier. Otaheite, renamed Bourbon, was taken from Mauritius to the French West Indies, and introduced into St Vincent in 1793. In 1795 one plant of Bourbon was obtained from Santo Domingo by the Government Botanist of Jamaica, and a further introduction was made to Jamaica by Captain Bligh one year later. Otaheite was also the standard variety of Brazil, British Guiana (now Guyana), Hawaii, India, Java, Mexico, and the other West Indies. It was grown in the West Indies as Bourbon, in Hawaii and Java as Lahaina, and in India as Vellai. Bourbon cane was the principal variety grown in the Caribbean until its collapse in 1890 to rind disease or red rot, caused by the fungi *Phaeocytostroma sacchari* and *Glomerella tucumanensis*, respectively. However, Otaheite was never outstanding as a breeding cane. Nevertheless, it was a remote ancestor of the remarkable variety POJ2878, and other famous Javanese hybrids.

Cheribon or transparent cane

Various forms of this cane were possibly the most important in the world in the late nineteenth and early twentieth centuries. Up to 1921 it had probably produced more sugar than all the other varieties combined[20]. Although POJ2878 and other varieties have surpassed its production since, Cheribon has had equal fame as a breeding cane from which many commercial hybrids have been produced. The Cheribon series of canes, which originated in Java during the first half of the nineteenth century, was a series of light, striped, and dark modifications. The Cheribon canes followed Otaheite (Bourbon) as the main commercial variety in most countries following the collapse of the latter to disease epidemics, i.e. around 1840 in Mauritius, during the 1860s in Brazil, about 1872 in Puerto Rico, from 1890 to 1895 in the other West Indies islands, and during the early twentieth century in Hawaii. The Cheribon series was also the most valuable among the noble canes for breeding.

Tanna or Caledonian cane

Originally from the island of Tanna in Vanuatu, these canes existed in a selection of striped, light,

and dark forms. The striped form was introduced into Mauritius during 1869 by Caldwell, and given the name Wopandon. White Tanna accounted for 63% of the area under cane in Mauritius during 1925, and arose as a self-coloured sport of Striped Tanna at Pamplemousses in north Mauritius during 1892. The success of White Tanna can be ascribed largely to its hardiness, its resistance to gumming disease or gummosis (caused by the bacterium *Xanthomonas axonopodis* pv. *vasculorum*), and its particularly good performance in the higher rainfall and cooler districts of the island. Tanna was taken as Yellow Caledonian to Hawaii in 1881, and was grown with success as the standard cane on the unirrigated plantations in that country up until 1925. It was also grown on a large scale commercially in Fiji. Tanna was unsuccessful as a breeding cane in Mauritius, however, because its flowers were almost completely sterile, and the few seedlings that were produced from it were inferior. In contrast, seedlings were obtained from Tanna in crosses with H109 (a variety of Otaheite or Lahaina lineage) in Hawaii. These were also successful as parents.

Seedling varieties of noble canes

Sugarcane fertility, which was discovered by Soltwedel in Java during the late 1880s, pointed out a new era in sugarcane culture[20]. Nevertheless, the first commercial varieties to be raised from seed were purely of noble cane origin. In Java, EK and SW11 originated from Otaheite and Cheribon respectively, while in Queensland, Q813 and HQ409 cane from Badilla, a form of *S. officinarum* that originally came from New Guinea. The most famous of the early Hawaiian seedlings was H109; the female parent was from an Otaheite arrow, and the male parent was Rose Bamboo – a light Cheribon[21]. The best of Harrison's early seedlings bred in Guyana were D109, D145, and D625, and these were all raised from the light, striped forms of Cheribon.

The end of the noble cane era

The noble cane varieties were succeeded in cultivation by hybrids having better agronomic characteristics. This was because experience led to the belief that, while noble varieties might be valuable in certain environments, they were generally limited in the scope of their usefulness under natural or induced poor soil conditions, or where diseases were common, e.g. sugarcane mosaic, root disease complexes or sereh. The change from depending on noble canes to hybrids was earliest in Java, and was most rapid there because sugarcane was grown in rotation with rice without ratooning the previous plant crop, i.e. allowing its regrowth after harvest. Hybrids appeared in cultivation there on a large scale in 1925, and by 1929 POJ2878, the 'wonder cane of Java', occupied around 90% of that country's sugarcane area. The spread of the sugarcane mosaic virus in the noble canes of Louisiana during the early 1920s resulted in a catastrophic drop in sugar production from around 200 000 tons per annum to 47 000 tons per annum by 1926. With the introduction of resistant hybrid varieties, however, production levels were quickly restored and improved. In Hawaii, the POJ varieties were first replaced by noble canes in around 1930, e.g. D1135, H109, and Yellow Caledonia (Tanna). Nevertheless, since 1940 the cane area has been dominated in that country by locally bred hybrids, and the first to be extensively grown was H32–8560.

Guyana replaced the noble cane varieties earlier than the rest of the West Indies, and by 1941 had almost 50 000 acres of POJ2878. In Barbados and Jamaica, the cultivation of noble cane varieties continued for longer, and hybrids did not make any significant inroads until around 1941. However, the varietal change was very rapid thereafter, and both countries have depended upon the production of hybrid varieties since 1948.

The original forms of *S. officinarum* possessed many desirable characteristics, and many were cultivated commercially up until the 1940s in some countries. The discovery of sugarcane fertility by Soltwedel in Java during the late 1880s[20], however, pointed out a new era in sugarcane culture. This enabled the production of more vigorous new progeny from inter- and intraspecific crosses. The key phenomenon of the interspecific cross (i.e. the crossing of *S. officinarum* with wild *S. spontaneum* or *S. robustum*) was the 'nobilisation' process in which the female parent, *S. officinarum*, contrib-

uted the quality or sugar content of the progeny, while the male parents contributed the growth vigour and disease resistance. Interspecific crosses are rare in other crops; but it has been commonplace in sugarcane and other allied genera of *Saccharum* for about 100 years. For the process of ennobling the wild *S. spontaneum* forms or other wild canes through their hybridisation with *S. officinarum*, Brandes[11] lists the desirable qualities of the 'ennoblised' ones as:

- high weight of the plants and the canes, and a high tonnage per ha;
- the ability of the stalks to drop old leaves or 'detrashing';
- disease and pest resistance; and
- a surprising adaptability to widely differing climates.

The highly specialised requirements of the modern sugar industry necessitated that *S. officinarum* should be hybridised with other species, most notably the mosaic disease resistant *S. spontaneum*. Nature is believed to have accomplished hybridisation alone, e.g. *Saccharum barberi* and *S. sinense*. In general, hybrid seedlings are more resistant to diseases, and are more adaptable to climatic variables than *S. officinarum*. Nevertheless, the desirable characteristics are still retained.

Saccharum spontaneum

The clones of *S. spontaneum* form a highly complex group, and types are known from north and east Africa, Anatolia and Turkmenistan, Borneo, Burma and the Indian subcontinent, China, Java, New Guinea, the Philippines, Sulawesi, and Taiwan. The fact that around 300 different types are known from India alone is an indication of the extent of the genetic pool that is available to sugarcane breeders.

Although the clones of *S. spontaneum* are so numerous and varied, they form a natural group which is easily distinguishable from *S. officinarum*. All are perennial grasses with slender culms, which are hard but very pithy, often with a hollow centre. These culms grow from clumps or stools, or can form continuous hedges or breaks with frequently aggressive rhizomatous tillering. *Saccha-*

rum spontaneum clones are not self-trashing, and their colour is pale green when young. As they age, they become white or yellowish, and are usually covered with a heavy wax bloom. The internodes of the culms are normally long, and the surface has no waxy markings or growth cracks. The nodes are always thicker than the internodes. Based on Panje's classification[22] on vegetative characters, there are two subspecies, *indicum*, which is tufted or prostrate (except a Javanese form known as 'Glagah'), and *aegyptiacum*, which is erect.

Saccharum barberi

Before the present-day commercial varieties, sugar production in northern India and what is now Pakistan was dependent for centuries on the indigenous, subtropical forms of cane which evolved there. Barber[23] differentiated five major types on the basis of vegetative characteristics of leaves, stems, and roots: Mungo, Nargori, Saretia, Sunnabile, and Pansahi. Most resemble the local wild grasses to such an extent that he thought that they were derived from *S. spontaneum*. The Pansahi clones were not exclusive to the Indian subcontinent, but were also found in south China, Indochina, and Taiwan. Accordingly, Jeswiet[24] included them in *S. sinense*, and left the other four to constitute *S. barberi*. The origin of the northern Indian canes is one for which innumerable theories have been put forward. Both Jeswiet and Barber believed them to have arisen from *S. spontaneum*. Parthasarathy[13] considered that *S. officinarum* and *S. spontaneum* formed the ancestral stock. While there are no records of the noble cane ancestors in India, both *S. barberi* and *S. sinense* have many characteristics similar to *S. officinarum* that do not occur in *S. spontaneum*.

Saccharum sinense

The name of this species denote 'Chinese cane'. Uba cane, which was the most typical of *S. sinense*, was widely grown commercially in many parts of the world. Deerr[25] associated it with the Brazilian for reed; however, it was known that *S. sinense* did not reach Brazil until the late 1860s, when it was sent at the time of the gummosis epidemic. Uba

was sent to Brazil from Mauritius in 1869, and shipments of it were sent to South Africa from both India and Mauritius in 1882. It was also sent to Louisiana and Puerto Rico in 1915, when it was found to be immune to mosaic virus disease[26]. Nevertheless, Uba was only an expedient variety until other hybrids were identified which were also resistant to mosaic and had a better milling quality. Notwithstanding, Uba was grown in South Africa to a greater extent than elsewhere, and even remained the principal variety when mosaic disease appeared there at the beginning of the twentieth century. As it was almost sterile, however, there was very little use for Uba as a breeding cane, and the few seedlings it did produce were poor and bore excessive leaves. Even so, some successful Barbados varieties and early Hawaiian canes have Uba in their ancestry.

Saccharum robustum

This *Saccharum* species also occurs in the wild similarly to *S. spontaneum*. It is only indigenous in New Guinea and the neighbouring islands. Brandes and Jeswiet collected 154 *S. robustum* clones during their expedition to New Guinea in 1928. They also collected an additional ten *Saccharum* species. In its natural habitat, *S. robustum* is often extremely vigorous, forming compact tufts or dense cane-breaks, which can grow up to 10 m high. *Saccharum robustum* is often used as fences in New Guinea. At one time, it was thought to offer new scope in sugarcane breeding; however, it was soon found that all of its clones were susceptible to the sugarcane mosaic virus. This is often quoted as evidence of its close relationship to *S. officinarum*[8].

Saccharum edule

This is a small group of clones that originates in New Guinea and its neighbouring islands. It is also closely related to *S. robustum*, so much so that certain authorities do not consider it to be a separate species[18]. *Saccharum edule* is characterised by unusual, swollen and aborted inflorescences, which are used as a source of food in Melanesia.

THE BOTANY OF SUGARCANE

Sugarcane is a tall perennial tropical grass, which tillers at the base to produce unbranched stems from 2 to 4 m or more tall, and to around 5 cm in diameter. It is cultivated for these thick stems, stalks or canes, from which the sugar is extracted.

Barber[23] in India and Jeswiet[24] in Java pioneered the study of sugarcane's morphology. Artschwager continued this work in the USA, and recommended the standardisation of taxonomic descriptions by describing the vegetative characteristics of wild forms of *Saccharum*[27,28]. Later he postulated the origin, and described the characteristics and descriptions of representative clones or varieties of *S. officinarum*, in collaboration with Galloway[5]. Meanwhile van Dillewijn[29] had written a definitive book on the botany of sugarcane.

The stem

The solid, unbranched stem, roughly circular in cross-section, is clearly differentiated into joints each comprising a node and an internode. The node consists of a lateral bud situated in the axil of the leaf, a band containing root primordia, and a growth ring (Fig. 1.1). The buds, which can be situated on, or just above, the leaf scar, may be round, small and adpressed to the stalk, or more prominent and pointed, depending on the variety. In certain varieties a bud groove or furrow can be found on the surface of the internode above the bud. Normally only one bud occurs at each node, and these buds are situated on alternate sides of the stalk. On occasion, however, more than one bud may be formed at a node. Each bud is an embryonic shoot consisting of a miniature stem with small leaves, the outermost ones having the form of scales. Several types of nodes and internodes are illustrated in Fig. 1.2.

Generally, the nodes are spaced at intervals of around 15 to 25 cm; but are much closer at the top of the stalk where elongation is taking place. The nodes are also much closer at the base (i.e. at or just below soil level) where new tillers are being produced. In commercial production, sugarcane is propagated from stem cuttings (i.e. seed-pieces or setts), each having two to four buds. The buds on

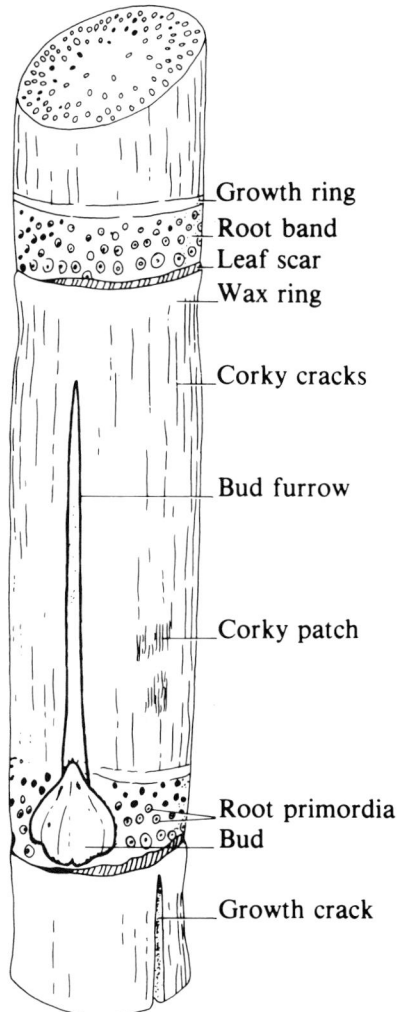

Fig. 1.1 Stem of sugarcane. (After Artschwager & Brandes 1958.)

the setts develop to give primary stems, the basal buds of which form secondary stems and so on (Fig. 1.3).

The colour and hardness of the stalks vary with the variety, and the stalks can range in diameter from around 2.5 cm to around 5.0 cm. Stalk hardness may also be influenced by the growing conditions. Each stem has a hard, wax-covered rind (epidermis) surrounding a mass of softer tissue (parenchyma) that is interspersed with fibres (vascular bundles). The wax layer prevents the loss of

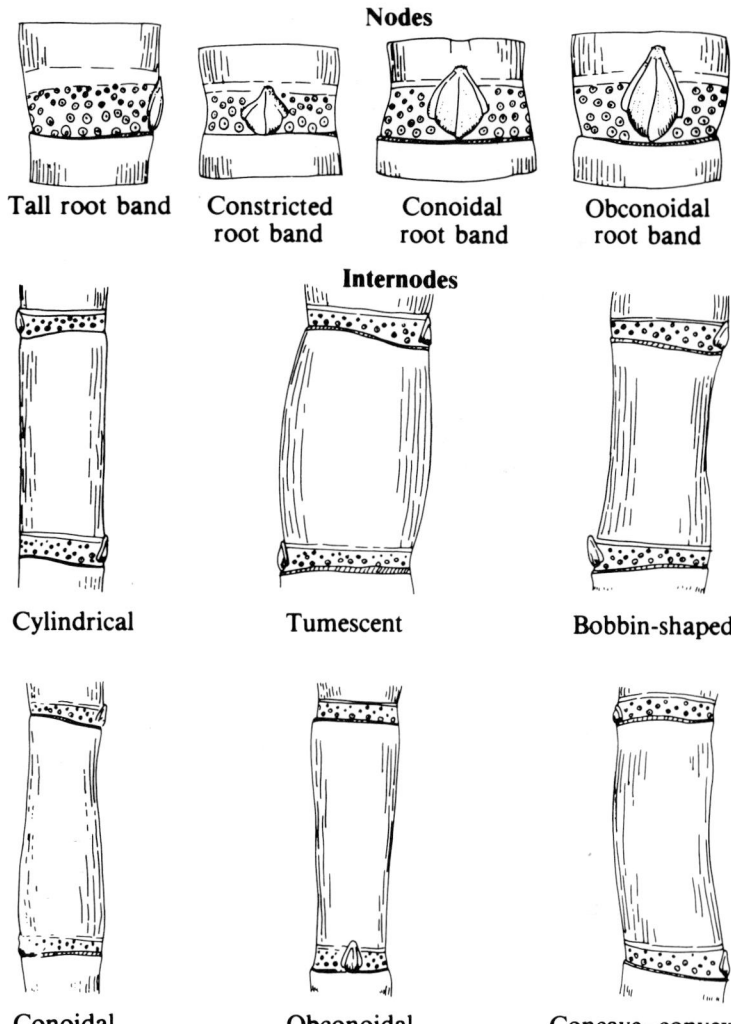

Nodes

Tall root band   Constricted root band   Conoidal root band   Obconoidal root band

Internodes

Cylindrical   Tumescent   Bobbin-shaped

Conoidal   Obconoidal   Concave–convex

Fig. 1.2 Types of nodes and internodes. (After Artscchwager & Brandes 1958.)

water from the stalk by evaporation, and the fibrous rind provides strength and rigidity. The fibres are more abundant towards the periphery of the stalk than in the centre, so the mechanical structure of the stalk is fundamentally tube-like. In the internode the fibres or vascular bundles run nearly parallel to each other, but in the node many of them branch or bend to supply the leaves, buds, or root primordia. In addition, as the ground tissue is often lignified, the nodes are much harder than the internodes. The juice containing sugar is stored in the thin-walled, parenchymatous tissue.

Varieties with thin canes have higher ratios of rind tissue to parenchyma compared to thicker stalked varieties. The hardness of the rind affects the milling quality or 'mill-ability' of the cane, and, where hand cutting is still practised, the thinner, harder canes are more difficult to harvest. In contrast, thicker, softer canes are easier to harvest, and are thus more acceptable to the harvesting gangs. Such soft rinded varieties are, however, also chosen for chewing.

The length and diameter of the internodes are also affected by other factors of which moisture, nutrition, and temperature are the most important. For example, a stalk might have short and thin internodes in its central portion, reflecting retarded growth during a period of moisture stress, with

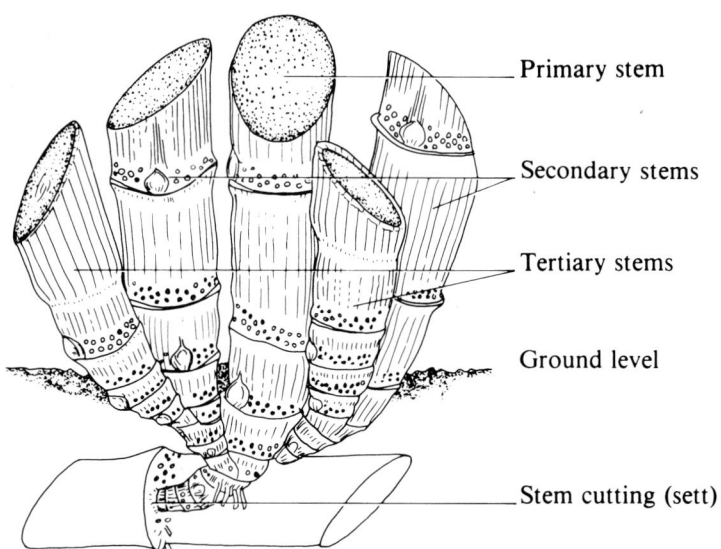

Fig. 1.3   The formation of tillers.

much longer and thicker ones both above and below which were developed during wetter periods.

There are large differences between varieties in the thickness of the wax coating. But, in general, wax is more heavily deposited immediately below the node than elsewhere.

The colour of the stem depends upon many factors. Two basic pigments are involved: red and blue anthocyanins in the epidermal cells and green chlorophyll in the deeper tissue. When both anthocyanin and chlorophyll are absent, the stem is yellow. The colour is usually subdued if an internode is covered by its leaf sheath; but it becomes distinct on exposure to sunlight. The immature top joints are pale yellow.

The leaf

The leaves are attached to the stem at the bases of the nodes, alternately in two rows on opposite sides of the stalk. Each leaf consists of two parts: the sheath and the blade or lamina (Fig. 1.4).

The sheath is tubular in shape and broader at the base than the top. It tightly encircles the stalk, and is separated from the long, tapering, pointed leaf blade by a ligule and one or two dewlaps dependent upon the variety. The leaf sheath is a thin structure that closely overlaps the stalk at the base, but tends to be less closely pressed against the stalk

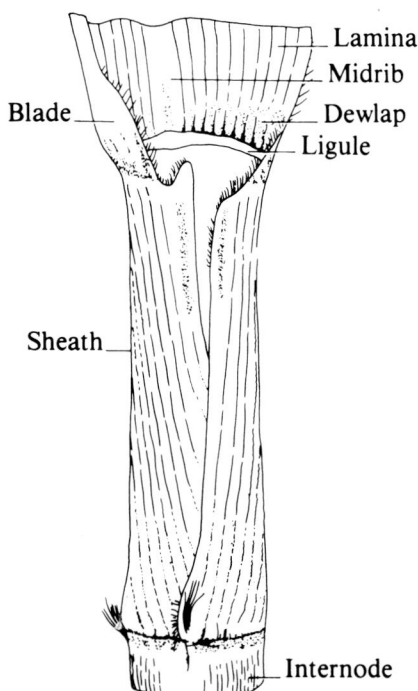

Fig. 1.4   The structure of a leaf.

towards the leaf lamina. The free margins of the sheath are on the opposite side of the stalk from the bud, which it surrounds and protects. The ligule, a membranous appendage to the sheath, is formed from elongated parenchyma cells without vascular bundles. It is translucent and hyaline when young, but with age dries, changes colour and becomes torn. Nevertheless, the ligule is an important diagnostic feature, and four types are recognised, i.e. linear or strap-shaped, deltoid or triangular, crescent-shaped, and arcuate or bow-shaped (Fig. 1.5). The dewlaps are wedge-shaped appendages made flexible by their collenchyma. They are also characteristic of each variety, and three main types are recognised, i.e. rectangular, deltoid, and ligular (Fig. 1.5). However, there are many intermediate forms of dewlap.

The leaf has a strong midrib, which is usually white and concave on the upper or adaxial surface, and pale green and convex on the abaxial side. The leaf blade broadens from the ligule to as much as 10 cm in width, and then narrows towards its pointed tip. Although there is considerable variation, leaves may be as much as 1 m in length. Motor or bulliform cells are situated along the midrib, which can cause the young leaves to in-roll in most varieties, i.e. involute curling. There are a few varieties, however, in which the leaves roll outwards, i.e. convolute curling. The reaction of the bulliform cells to cause leaf curling is a characteristic exhibited by most varieties during periods of moisture stress, i.e. when the rate of transpiration exceeds the rate of moisture uptake by the roots. Stomata occur on both sides of the leaf, but the stomatal density on the lower surface of the leaf is almost double that of the upper surface. It is curious that leaves usually curl involutedly to protect the upper surface when under moisture stress. As the transpiration rate is about equal from both surfaces, it immaterial which way the leaves curl[29]. It has been estimated that a fully expanded leaf has about 30 million stomata.

An undesirable characteristic in commercial canes is the development of siliceous cells to form hairs or spines on the leaves or leaf sheaths, or to give sharp cutting edges to the leaf blades. Harvest labour is reluctant to cut such varieties. In an industry where the cane tops are fed to cattle, the more coarsely toothed varieties are also unsuitable for forage.

The area of the individual leaf blades along the stalk is smallest at the base, and gradually increases towards the top until a maximum is reached. This is followed by a decrease near the top. When an adverse period of growth succeeds a favourable one, the surface area of newly produced leaves may be smaller. The subsequent leaf blades may have an increased area, however, if there is an improvement in the growing conditions. Thus, the normal appearance and shape of the leaves can be disturbed by environmental conditions in the same way as they affect internodes.

As the older leaves die, they may drop to the ground. Varieties that do this are known as 'free-

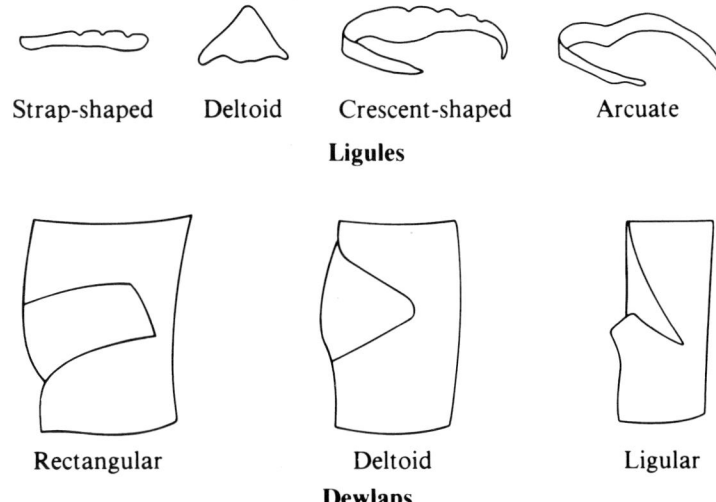

Strap-shaped　　Deltoid　　Crescent-shaped　　Arcuate

Ligules

Rectangular　　　　　　Deltoid　　　　　　Ligular

Dewlaps

Fig. 1.5 Types of ligules and dewlaps.

trashing'. This characteristic is of some practical importance, because the dead leaves (trash), if retained, impede harvesting and may shelter pests.

The roots

Soon after a seed-piece or sett is planted, two kinds of root will develop: first the sett roots and later the shoot roots. Sett roots develop from the initials on the root band (Fig. 1.1), and shoot roots develop from the root primordia on the new developing tillers. The sett roots are thin and branched, and those from the primordia of the tillers are thicker, fleshier, and much less branched. At first, the newly planted sett depends entirely on its own roots for the uptake of moisture and nutrients; but the

lifespan of these sett roots is limited. Later their function is taken over by the roots produced by the new tillers, i.e. the shoot roots. Each tiller develops its own root system (Fig. 1.6). The sett roots then die.

During subsequent growth, the shape of the root system is determined by the condition of the soil in which the cane is planted. The roots proliferate wherever conditions of available moisture, nutrients, and soil aeration are favourable. Furthermore, the depth of cultivation and the soil profile are of particular importance. The life of the shoot roots is also limited; but the root system as a whole is renewed as each new shoot produces its own roots. The rate of this process of rejuvenation is governed by the periodicity of tillering. This

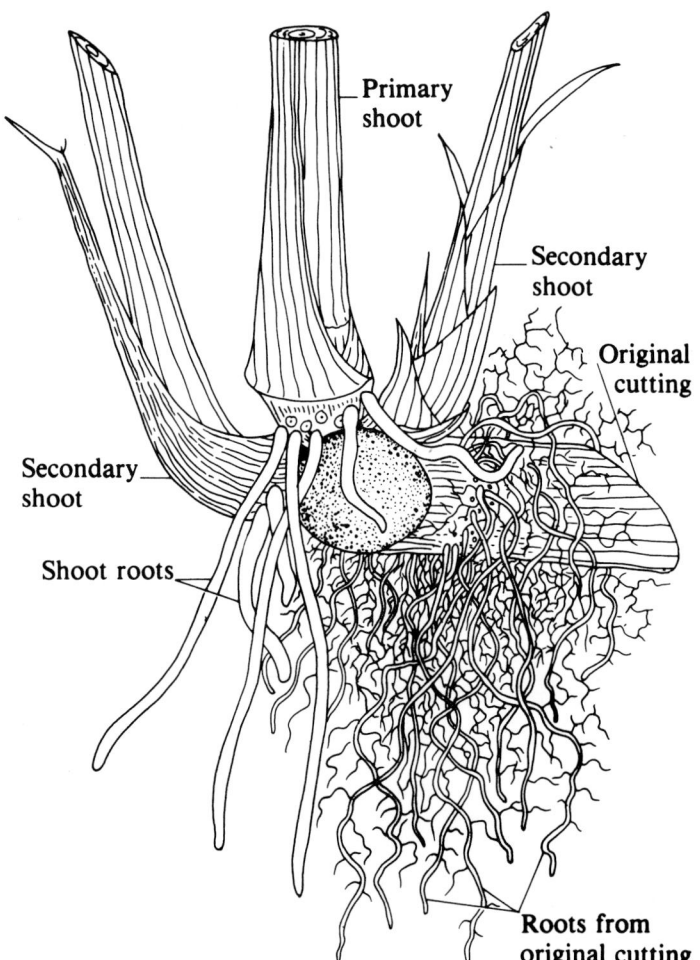

Fig. 1.6 Young cane plant showing two kinds of roots: sett roots from the root primordia of the cutting, and shoot roots originating from the root primordia of the shoots. (After Martin 1938.)

continuous production of new roots is of great importance, since it enables the plant to adjust itself to changing environmental conditions.

In Mauritius, Evans[30,31] examined the roots of several varieties in detail and divided them into three categories, i.e. superficial roots, buttress roots, and rope systems. The superficial roots absorb moisture and nutrients; the buttress roots provide stability; and the rope systems can penetrate to depths of 3–6 m, where the soil remains moist even during severe drought (Fig. 1.7). It should be emphasised, however, that the precise pattern of development is peculiar to local soil conditions. What is common in Mauritius may not be apparent elsewhere. In practice, it often difficult to distinguish between superficial and buttress roots, and rope systems are extremely rare. Nevertheless, whatever the pattern, approximately 50% by weight of the roots occur in the top 20 cm of soil, and 85% in the top 60 cm. Sugarcane roots can penetrate through soil with a water potential of < -15 to -20 bars, provided the main root mass has adequate water. Similarly, a few main roots can transport moisture to the plant through very dry soil. Root growth is affected by soil moisture and by soil temperature, as well as the volume of soil available for the roots to spread. Root growth is very slow when the soil temperature is below 18°C, but it increases progressively to an optimum at around 35°C. At increasingly higher soil temperatures root growth is also reduced[32].

The inflorescence

In 1965 Stevenson[8] described the factors that cause the apical meristem of a stalk to change to an intercalary one (i.e. to change from the vegetative to the reproductive phase). Flowering in sugarcane normally takes place when there is a slowing down of the growth due to shortening days and lower night temperatures. The main differentiating factor is undoubtedly length of day; but other conditions also affect the change from the vegetative to the reproductive state. These include temperature and altitude (these two are often associated), light and nutrition. Stevenson[8] states that if day length alone were decisive, sugarcane would flower twice a year in regions some distance from the equator, as there would be two times when the daylight hours would be favourable. In fact, flowering normally occurs after the summer solstice only, i.e. when the days are getting shorter. However, there are reports of occasional exceptions to this rule.

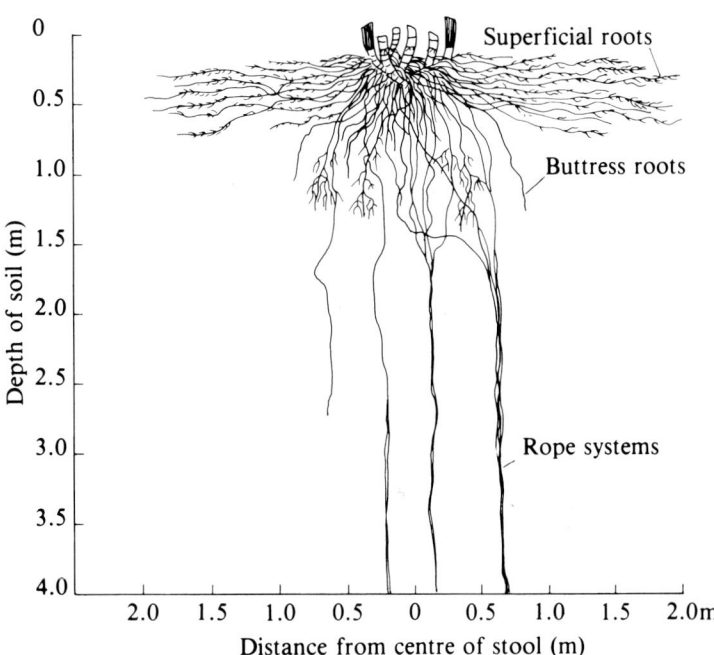

Fig. 1.7 The roots of an established cane stool. (After Evans 1936.)

The normal flowering period in the northern hemisphere is from late October to early December, and in the southern hemisphere from late April to early June. In regions some distance from the equator, if flowering occurs at all, emergence is over a short time only. The flowering season becomes longer as the equator is approached, until, in the equatorial regions, the season is so prolonged that the cane may flower almost all the year round. This is because day length and temperature never become limiting.

Generally, a 12.5 h day length and night temperatures above 18°C induce floral initiation if enough inductive cycles are given. Bull & Glasziou[33] state that the minimum number of cycles is at least ten. Coleman[34] has suggested that a quantitative amount of stimulus is accumulated to stimulate the differentiation of the floral primordia. In the vegetative state the unexpanded sheaths near the apex are shorter than the expanded ones below. The first sign of flowering is that successive sheaths become longer and the leaf blades become shorter. The apical meristem, which is surrounded by a leaf sheath, ceases to form leaves and develops into an inflorescence primordia about 3 months before the actual flower emerges. Then the distinctive flag leaf appears. Its sheath, which encloses the young panicle, is about 90 cm or more long, while the leaf blade is only about 15 cm long and shaped like a pennant. The stalk elongates and the panicle emerges (Fig. 1.8). The inflorescence, which is also known as the arrow, appears above the foliage to facilitate wind pollination. The loose, terminal panicle is 25–50 cm long with a silky appearance owing to rings of long hairs below each spikelet. The anatomy and morphology of the inflorescence have been described in detail by van Dillewijn[29] and Artschwager *et al.*[35].

The main axis of the panicle arises almost imperceptibly from the terminal internode, and gradually narrows until it merges into the terminal rachis of spikelets. The surface of the main axis is slightly furrowed and the lateral branches arise at the nodes, not always at the same level – some members of the whorl arising above and others below the main nodal region.

The bases of the panicle branches are swollen and thinly covered with short white hairs. At

Fig. 1.8   Flowering sugarcane. (Photograph: A.L. Down.)

the base of the panicle the primary branches are about 15 cm long, but shorter above. The secondary branches tend to arise in two rows, alternately along the primary branches, and may carry tertiary branches. The ultimate branches bear the spikelets, one of which is sessile and the other supported on a stiff pedicel (Fig. 1.9). At the base of each spikelet in the pair is a ring of silky white hairs that are longer than the spikelet. It is these hairs which give the arrows their characteristic appearance.

Both spikelets have two florets, the lower one of which is sterile and represented by a delicate, pointed lemma, which is shorter than the glumes. The structure of both spikelets is similar, with a pair of hard, boat-shaped glumes protecting the developing flowers. The upper floret of each spikelet is hermaphrodite, with no lemma except in *S. spontaneum* and some of its hybrids. When present, the lemma is a narrow scale with fine hairs at the top.

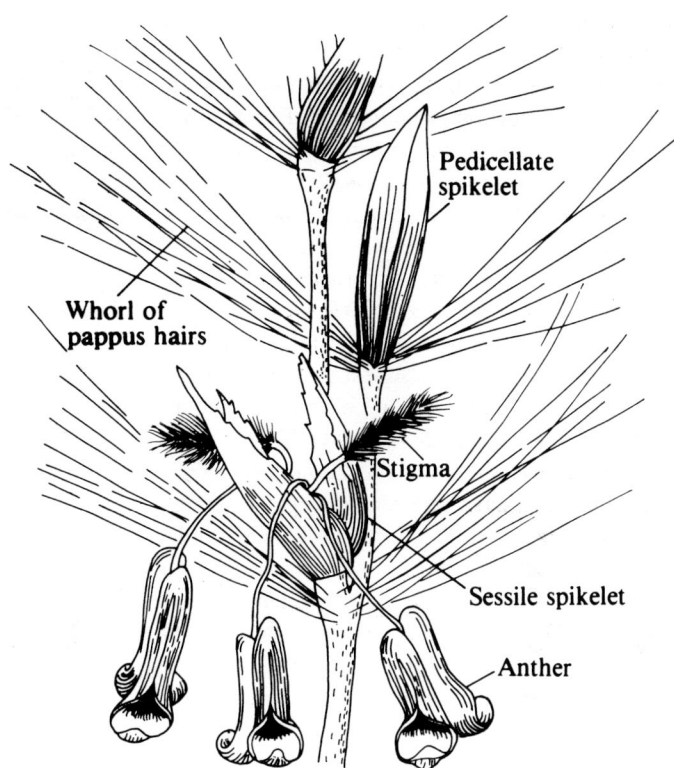

Fig. 1.9 Part of a lateral axis of an inflorescence of *S. robustum* (51 NG 71), showing the arrangement of spikelet. (From Stevenson 1965.)

At the base of the ovary are two short, wedge-shaped lodicules opposite the palea. The three stamens are in one whorl and have large lobed anthers. Indehiscent anthers are usually yellow or pale orange, while dehiscent anthers are brown or purple. The ovary is round, flattened on the ventral surface, and contains a single anatropous ovule. The pistil has two long terminal styles, each with a large brush-like feathery stigma, which is plum red. The spikelets open during the night or early morning, beginning at the top of the panicle and progressing downwards and inwards over 1 or 2 weeks. The lodicules swell and push the glumes apart, and the stigmas are extruded. The flowers are protogynous, the anthers appearing about three hours later when the filaments have lengthened so that the anthers hang down, well clear of the floret. High humidity delays anthesis. Natural pollination is by wind. The pollen grains remain viable for only a short time, and the anthers fall from the filaments soon after shedding the pollen. However, the stigmas persist. Most of the cultivated varieties exhibit considerable sterility of both pollen and ovules, so that flowers seen in commercial fields rarely set seed.

After pollination, it takes 21 to 25 days for the seed to fill and mature. Though the production of the inflorescence ends the production of leaves on that stem, lateral buds on the stem may develop, forming leafy side shoots. These have a lower sugar content than the main stem. As the production of the inflorescence also reduces the sucrose content of the stem, efforts are often made to delay flowering so as to maintain sucrose yield.

The seed

The seed is a dry, one-seeded fruit or caryopsis formed from a single carpel, the ovary wall (pericarp) being united with the seed coat or testa. The seeds are ovate, yellowish-brown, and very small (about 1 mm long). The withered stigma persists at the tip, and at the base are whorls of silky hairs for wind dispersal. The seeds are shed within the spikelet, individual florets breaking off at the nodes. Collectively the seeds are known as 'fuzz' or 'fluff'. The seeds soon lose their viability, but can retain their viability for a long time if freeze-dried. If the fresh seeds are to be planted within 2 weeks, they should be kept in a desiccator.

The complete inflorescences of commercial clones have been estimated to contain 25 000 florets; but the number of fertile florets is always much lower than this[36]. As many as 700 seedlings per g of fuzz have germinated, but most varieties produce much less. Mature, naked seeds weigh around 0.4–0.5 mg; hence at best only 35% by weight of the fuzz is true filled seed. Seed viability falls rapidly in ambient tropical conditions, but suitable drying by cold dehumidification and under deep-freeze conditions can maintain viability at a high level under reduced humidity conditions (see Chapter 2).

Germination of the seed in soil takes from 2 to 8 days at 35°C. Young seedlings are delicate up to the four-leaf stage, but thereafter grow rapidly. As the fuzz germinates better in light, it is placed on the surface of sterilised compost in shallow trays, and kept at high humidity. The seedlings can be transplanted about 6 weeks after germination.

CULTIVATION OF SUGARCANE

Sugarcane is grown commercially in the tropics and subtropics, and is known to be one of the oldest cultivated plants in the world. The boundaries of its cultivation are shown by van Dillewijn[37] to agree quite closely to the Palm Tree Line (Fig. 1.10). Nevertheless, it is important to note that the types of varieties that are suitable for cultivation in the subtropics are very different from the successful tropical ones.

The ideal environment for sugarcane is one in which rainfall (or irrigation) is well distributed during the growing season, but where the preharvest ripening period is relatively dry, and the sunshine hours are plentiful throughout the whole season. Mangelsdorf[38] has shown that there is a positive correlation between the number of sunshine hours per year and the yield of sugar.

Sugar is stored in the stalks, but conditions favourable for ripening are necessary if sugar production is to become economic. Ripening normally takes place during the cooler, drier times of the year. Where the crop is produced under irrigation, this can be induced by a reduction in the

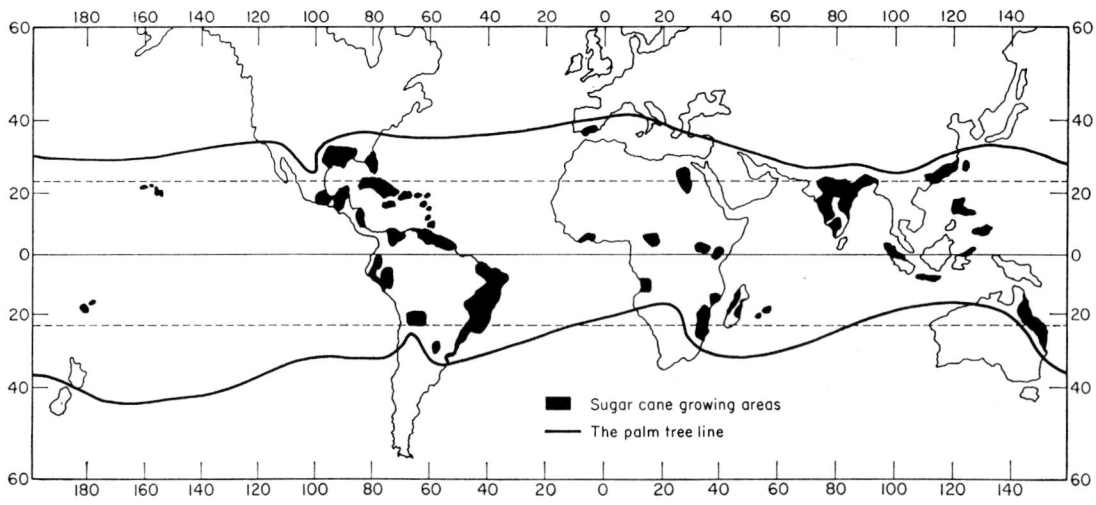

Fig. 1.10 World distribution of sugarcane cultivation and the Palm Tree Line.

water supply. Under the best ripening conditions, a tonne of sugar can be produced from 7 to 8 tonnes of cane, e.g. in Australia, Zambia, and Zimbabwe. However, in other countries (e.g. in Guyana and Sumatra), where the natural ripening conditions are not as good, the tonnes cane : tonnes sugar ratio is 12 : 1.

Varieties differ not only in their yield of cane but also in their juice quality. They also differ in the length of time required to reach maturity. There are also very marked responses to the environment, and even different ecological zones within a country. It is for this reason that, if the best selection is to be made, the final stages in a varietal selection programme must incorporate trials in a country that must be representative of all the main ecological zones.

After the plant crop has been harvested, it is normal to allow the crop to regrow once or several times so that two or more harvests are taken from the original planting, a procedure known as ratooning. At the end of the cycle, the crop is ploughed out and the field is replanted more or less immediately with sugarcane or another crop, or after a period of fallow.

Sugarcane cultivation and the slave trade

From the early sixteenth century, the development of the sugarcane industry in the Caribbean, and the tropical and subtropical regions of North, Central, and South America were dependent upon slaves and the slave trade in Africa. This continued until the slave system was abolished at various times in different countries. Between European countries and the sugar regions of the Western Hemisphere, much of the commerce in the eighteenth century was based on the outward shipment of slaves, and the homeward transport of sugar, molasses, and rum. Individual estates were small, and each had its own sugar mill, processing unit, and labour force of slaves. The whole estate was operated as a self-contained organisation, with cattle and other animals, food gardens and fruit orchards, pastures and woodlands, in addition to the sugarcane fields. Compared to modern standards, waste and inefficiency characterised the industry. However, labour was cheap, and the value of sugar and rum was high. The abolition of slavery between 1762 and 1865 had a profound effect upon the sugar industry in different countries. The action in Britain to end slavery extended over 70 years. The Act for the Abolition of Slavery came into effect in 1834, and a 6-year period of apprenticeship was then offered to the former slaves. This period was reduced to 4 years so that the status of slavery eventually ended in the British colonies in 1838.

In some of the British West Indian islands, notably Jamaica and Trinidad, a sharp decline in sugar production followed the abolition of slavery. This was caused by the unwillingness of the now free labour to carry out tasks they previously did as slaves, even though reasonable wages were offered. When any work was done, it was done indifferently. Consequently, sugar production in Jamaica fell from 72 730 imperial tons in 1832 to 28 037 tons in 1841. A similar decline happened in Guyana where sugar production was reduced by more than two-fifths, i.e. from 55 841 to 30 404 tons.

The introduction of indentured labour

Consequently, there was a need for more reliable labour. The first introduction of indentured or contract labour took place in 1838 to Guyana from the Indian subcontinent. Subsequently, various abuses and ill treatment were reported, and so the scheme was temporarily suspended until 1845 when these alleged bad practices were disproved. By 1848, more than 21 000 immigrants had arrived in Guyana, Jamaica, and Trinidad. However, they failed to solve the labour problem in Jamaica. Many died and some were repatriated, so by 1854 only a third of the original number remained in Jamaica. In contrast, the indentured labour system was the major cause of increased production in Guyana and Trinidad.

In later years, Fiji and South Africa (Natal) were among the countries where the indentured labour system was also adopted. In 1917, however, the system was abolished although it still prevailed in other British Colonies. Although there always had been the opportunity for repatriation, many of these initially indentured labourers chose to remain as settlers. This has resulted in the permanent establishment of Indians in some sugar-producing countries.

Controlled immigration of Chinese people also occurred in Guyana, Hawaii, Jamaica, and Mauritius, but on a much smaller scale. Chinese labourers were introduced into Hawaii in 1852 and continued arriving until 1898, but immigration ceased in the other three countries in 1879.

THE PRODUCTION OF SUGAR

The basic principles of making sugar from sugarcane are the extraction of the juice, and its concentration into a thick syrup from which the sugar separates by crystallisation. A type of refined sugar was first produced in the fourth century AD, but more advances in the process were made in the fifteenth century, when the industry first settled around Venice.

The introduction of a central factory system in Cuba, combined with the advantages of a very rich soil, favourable climate, and cheap labour, allowed sugar production to increase tenfold in the first half of the nineteenth century in that country. From 1851 to 1868 sugar output increased by 264 000 tonnes to 749 000 tonnes. Among other important events during that century was the resurgence of the industry in countries where it had been introduced much earlier, e.g. in the Philippines, Mexico, Java, and Peru. In addition, sugarcane industries started in Argentina, Australia (New South Wales and Queensland), Fiji, Hawaii, and South Africa (Natal). The combined production levels of all these countries exceeded 500 000 tonnes annually by 1900.

Sugar bounties

During the second half of the nineteenth century, the free-trade policy of the UK, combined with the sugar bounties (or subsidies) for beet-sugar production, seriously depressed the sugarcane industries in the British Colonies, especially in the West Indies. The beet-sugar bounties paid in Europe resulted in the production of more sugar than was needed for local consumption. The Napoleonic Wars deprived France of the supplies from her colonies, and the seizure of Mauritius by Britain in 1810 closed this source of supply. It was these events that led to the establishment of the beet-sugar industry, and by the middle of the nineteenth century France was again an exporter of refined sugar.

RESEARCH

Before the end of the nineteenth century, many sugarcane growing countries recognised that research was urgently needed to solve the problems of diseases. In addition, field practices were studied to improve yields. While numerous successful lines of inquiry had been carried out on extraction and refining, little work had been done on the cultural aspects of sugarcane production until field experiments were started in Guadeloupe, Hawaii, India, Java, Mauritius, and mainland USA. Research was also started in Guyana and the West Indies near the end of the nineteenth century.

DEVELOPMENTS IN THE TWENTIETH CENTURY

The period from around 1920 until the end of the twentieth century was perhaps the most important in the sugarcane industry, with large research stations becoming established in many countries, e.g. Australia, Brazil, Cuba, Hawaii, India, Indonesia, Mauritius, South Africa, and mainland USA. During this time the results of wide-scale research have been applied, agricultural and technical methods transformed, field mechanisation developed and improved, research and training extended and new methods established, the central and large factory systems expanded, automation introduced, and the methods for sugar handling and its transport on land and sea altered and developed. The agricultural results of this progress are covered in the subsequent chapters in this book.

REFERENCES

1 Sheridan, R.B. (1974) *Sugar and Slavery. An Economic History of the British West Indies.* Johns Hopkins University Press, Baltimore, MD.

2 Burnett, J. (1966) *Plenty and Want. A Social History of Diet in England From 1815 to Present Day.* Nelson, Edinburgh.

3 Balu, R. (1999) Sweet dreams. *Sugar y Azucar,* March.

4 McPhee, J. (1967) *Oranges.* Farrar, Straus and Giroux, New York.

5 Galloway, J.H. (1989) *The Sugar Cane Industry. An Historical Geography from its Origins to 1914.* Cambridge University Press, Cambridge.

6 Mintz, S.W. (1985) *Sweetness and Power. The Place of Sugar in Modern History.* Elizabeth Sifton Books, Viking, New York.

7 Brandes, E.W. (1929) Into primeval Papua by seaplane. *National Geographic Magazine.*

8 Stevenson, G.C. (1965) *Genetics and Breeding of Sugar Cane.* Longmans, Green and Co, London.

9 Artschwager, E. & Brandes, E. W. (1958) Sugarcane (*Saccharum officinarum* L). *Agriculture Handbook No. 122.* US Government Printing Office, Washington, DC.

10 Celarier, R.R. (1956) Cytotaxonomy of the Andropogoneae. 1 Sub tribes Dimeriinae and Saccharinae. *Cytologia.*

11 Brandes, E.W. (1956) Origin, dispersal and use in breeding of the Melanesian garden sugarcanes and their derivative, *Saccharum officinarum.* L. *Proceedings of the International Society of Sugar Cane Technologists,* 9.

12 Barnes, A.C. (1974) *The Sugar Cane.* Leonard Hill Books, London.

13 Parthasarathy, N. (1948) Origin of noble canes (*S. officinarum.* L). *Nature (London),* **161**.

14 Barber, C.A. (1920) The origin of sugarcane. *International Sugar Journal,* **22**.

15 Jeswiet, J. (1929) The development of selection and breeding of the sugarcane in Java. *Proceedings of the International Society of Sugar Cane Technologists,* 3.

16 Parthasarathy, N. (1946) The probable origin of North Indian sugarcanes. *Journal of the Indian Botanical Society,* **133**.

17 Mukherjee, S.K. (1957) Origin and distribution of *Saccharum. Botanical Gazette,* **119**.

18 Grassl, C. O. (1969) *Saccharum* names and their interpretation. *Proceedings of the International Society of Sugar Cane Technologists,* **13**.

19 Brandes, E. W. (1951) Changes in seasonal growth gradients in geographically displaced sugarcane. *Proceedings of the International Society of Sugar Cane Technologists,* 7.

20 Deerr, N. (1911) *Cane Sugar.* Norman Roger, Manchester, England.

21 Rosenfield, A. H. (1927) A monograph of sugar-cane varieties. *Journal of the Department of Agriculture Puerto Rico,* **11**.

22 Panje, R. R. (1933) *Saccharum spontaneum,* L. A comparative study of the forms grown at the Imperial Sugarcane Breeding Station, Coimbatore. *Indian Journal of Agricultural Science,* 3.

23 Barber, C.A. (1918) Studies in Indian sugarcanes, No. 3. The classification of Indian canes with special reference to the Saretha and Sunnabile groups. *Memoirs of the Department of Agriculture India, Botanical Series,* 9.

24 Jeswiet, J. (1925) Beschrijving der soorten van het suikeriet. Elfde Bijdrage. Bijdrage tot de systematiek van het geslacht *Saccharum.* Meded. Proefst. Java-Suikerina.

25 Deerr, N. (1949) *The History of Sugar,* Vol. I. Chapman & Hall, London.

26 Earle, F. S. (1919) The resistance of cane varieties to the yellow stripe or mosaic disease. *Insular Experimental Station Puerto Rico, Bulletin,* No. 19.

27 Artschwager, E. (1939) Illustrated outline for the use in taxonomic description of sugarcane varieties. *Proceedings of the International Society of Sugar Cane Technologists,* 6.

28 Artschwager, E. (1948) Vegetative characteristics of some wild forms of *Saccharum* and related grasses. *US Department of Agriculture Technical Bulletin.*

29 van Dillewijn, C. (1952) *Botany of Sugarcane.* Chronica Botanica, Waltham, MA.

30 Evans, H. (1935) The root system of the sugarcane. I. Method of study. *Empire Journal of Experimental Agriculture,* 3.

31 Evans, H. (1936) The root system of the sugarcane. II. Some typical root systems. *Empire Journal of Experimental Agriculture,* 4.

32 Blackburn, F. (1984) *Sugar-cane.* Longman, Harlow, Essex, England.

33 Bull, T.A. & Glasziou, K.T. (1975) Sugar cane. In: *Crop Physiology* (ed. L. T. Evans). Cambridge University Press, Cambridge.

34 Coleman, R. E. (1968) Physiology of cane flowering in sugar cane. *Proceedings of the International Society of Sugar Cane Technologists,* 13.

35 Artschwager, E., Brandes, E.W. & Starrett, R.C. (1929) Development of flower and seed of some varieties of sugar cane. *Journal of Agricultural Research,* 39.

36 Walker, D.I.T. (1984) In: *Sugar-cane* (ed. F. Blackburn). Longman, Harlow, Essex, England.

37 van Dillewijn, C. (1946) *Sugar-cane Breeding.* International Institute of Agriculture, Rome.

38 Mangelsdorf, A.J. (1939) Sugar-cane varieties in Hawaii. *Proceedings of the International Society of Sugar Cane Technologists,* 6.

Chapter 2
Plant Improvement of Sugarcane

Nils Berding, Mac Hogarth and Mike Cox

INTRODUCTION

Sugarcane breeding is widely acknowledged as the principal method for improving productivity in most sugarcane industries of the world. It contributes to increased productivity without an increase in growing costs, and pest-resistant cultivars may even reduce the cost of growing crops. Sugarcane breeding is also the principal method for controlling diseases such as smut *Ustilago scitaminea* Syd., common rust *Puccinia melanocephala* H. & P. Syd., sugarcane mosaic virus, Fiji disease virus, red rot *Glomerella tucumanensis* (Speg.) Arx and Mueller and leaf scald *Xanthomonas albilineans* (Ashby) Dowson. Breeding can also have a significant effect on the milling characteristics of cane and on the quality of the sugar produced.

Reviews of sugarcane breeding by Berding *et al.*[1] and Hogarth *et al.*[2] examined the role played by sugarcane breeding in improving productivity of sugarcane. The reviews examined the advances in technology that have been responsible for continuing gains in productivity from improved cultivars, and they speculated on the advances that are likely to be made in the future. Information from these reviews will be used in this chapter, which will also attempt to summarise more recent research findings.

Breeding programmes traditionally have three main components: collection and evaluation of germplasm, cross-pollination, and selection. Each of these components will be discussed in this chapter.

GERMPLASM

Introduction

The collection of germplasm with desired characteristics is the first step in any breeding programme. In sugarcane, the germplasm collection will typically include basic germplasm and commercial hybrids; both types of germplasm have important roles in a sugarcane improvement program. The uses of each type of germplasm will be discussed.

Conservation of germplasm resources is a major concern in all crops, and sugarcane is no exception. Methods used for germplasm conservation are discussed.

Basic germplasm

Contemporary commercial sugarcane clones are mostly interspecific hybrids between *Saccharum officinarum* L. and *S. spontaneum* L., although there has been limited infusion of *S. sinense* Roxb. and *S. robustum* Brandes and Jeswiet ex Grassl germplasm. However, sugarcane improvement is concerned with a broader range of germplasm. The remaining species in the genus *Saccharum* L. are *S. barberi* Jeswiet, and the sterile *S. edule* Hassk., which is of little interest in sugarcane improvement. Breeders are interested in the 'Saccharum complex'[3] which includes all genera involved in the origin of sugarcane. In addition to *Saccharum*, the complex includes the genera *Erianthus* Michx. sect. *Ripidium* Henrard, *Sclerostachya* (Hack.) A. Camus, and *Narenga* Bor. The *Saccharum* complex was revised by Daniels *et al.*[4] to include

Miscanthus Anderssen sect. *Diandra* Keng, as they considered that some botanical characteristics of *Saccharum* were not present in the taxa in the originally defined *Saccharum* complex.

Taxonomic considerations

Daniels & Roach[5] discussed the taxonomy and evolution of *Saccharum*. Their purpose was to describe the classical taxonomy of the complex to guide the collection and use of related germplasm. They also considered the germplasm resources of the complex in light of Harlan & de Wet's[6] general classification of crop germplasm into three pools for use. Daniels & Roach[5] considered that most sugarcane improvement operated within the primary gene pool (GP-1) and that most effort in GP-2 and GP-3 had been directed to obtain specific characteristics such as disease resistance and waterlogging tolerance.

Burner & Webster[7] thought that *Saccharum* was overclassified taxonomically, and that many of the divisions were artificial. The success of a wide range of interspecific and intergeneric crosses was considered testimony to this. However, this ignores the fact that unreduced gametes are functional in *Saccharum* and many of its relatives, enabling wide crosses to be made[5]. Webster & Shaw[8] reported that taxonomic relationships among the taxa of the *Saccharum* complex have not been studied carefully and have not been well defined, and considered that *Erianthus* Michx. should be a synonym of *Saccharum*. Consequently, they proposed that five species of *Saccharum* and one variety of *Erianthus* native to North America should be treated as *Saccharum*.

Collection

Sugarcane as a crop has a long and proud history of germplasm collection that has often had an internationally collaborative nature[9]. Early movement of sugarcane was driven by direct commercial necessity as commercial cultivation used *S. officinarum* clones, often lifted directly from swidden agriculture in New Guinea. The discovery of sexual fertility in sugarcane, initially in 1858, and again in 1885 and 1888, allowed commencement of

genetic improvement for *S. officinarum*, through intraspecific crossing[9]. In Java, to combat sereh disease in the 1880s, innovative Dutch breeders developed the concept of interspecific hybridisation as a means of seeking resistance from basic germplasm related to *S. officinarum*. They pursued a vigorous collection and introduction programme in search of suitable sources of resistance. We can thank innovative scientists such as Soltwedel, Wakker, Kobus, Wilbrink, Jeswiet, van Harreveld, and Bremer for their endeavours in the period 1885 to 1921 that laid the genetic foundations of contemporary sugarcane. Collectively, they developed the concept of nobilisation through interspecific hybridisation. Details of this important period have been summarised by Berding & Koike[10], Berding & Roach[9] and Berding & Skinner[11].

Characterisation and evaluation

Characterisation, as defined by the International Board of Plant Genetic Resources, 'consists of recording those characters which are highly heritable, can be easily seen by the eye, and are expressed in all environments'. Characterisation data are acquired after the origin or passport data, but precedes evaluation. Evaluation data facilitate use of germplasm in crop improvement. Indian researchers have made a significant contribution to characterisation of basic germplasm, i.e. *S. spontaneum*[12], *S. barberi*, *S. sinense*, *S. robustum* and *S. edule*[13], and *S. officinarum*[14]. Hawaiian researchers have characterised a large proportion of the population of *S. officinarum* clones present in the USDA-ARS World Sugarcane Collection maintained in Florida (P. H. Moore, pers. comm.). Although these data are unpublished, they are accessible through the USDA-ARS germplasm resource information network (GRIN), a resource which also contains limited descriptor data on some of the germplasm maintained at the USDA-ARS World Sugarcane Collection, Fort Pearce, Florida. More recently, Tai *et al.*[15] characterised a group of 125 *S. spontaneum* clones, being about one-third of the population present in the USDA-ARS World Collection, for four juice-quality and five morphological characters. Cluster and principal component analyses revealed considerable variation for all characters.

Despite the history of development and application of morphological descriptors to sugarcane germplasm, the major works referred to above used non-standard descriptor sets. Additionally, studies of germplasm groups using the descriptor set developed through collaborative discussion at the international level are non-existent[9].

An optimised morphological descriptor set was developed for *Saccharum* spp. hybrid germplasm by Gallacher[16]. He listed 24 vegetative macro-morphological descriptors (i.e. laminar of top visible dewlap leaf – 3; leaf sheath – 3; ligule – 4; auricle – 1; central internode – 4; wax and root band – 2; bud – 4; yield components – 3). These were selected by tree classification and a regularised discriminant function, and were considered to conform closely to an optimum descriptor set developed from the extensive data analysed. Many of these descriptors were not contained in the internationally accepted descriptor set. The use of morphological descriptors may seem dated in light of the development of biochemical and molecular markers. However, an optimised morphological descriptor set would have wide utility because of its ease of use.

Morphological descriptors have been overshadowed by developments in biochemical descriptors. Relative to morphological descriptors, these have intrinsic simplicity, as theoretically most can be regarded as single gene products, and have high heritability because of their closeness to the gene. Thus, they are reliable, and offer maximum discriminatory power. Examples are flavonoids[12], isozymes[17,18], restriction fragment length polymorphisms (RFLP)[19], and random amplified polymorphic DNA (RAPD) molecular markers[20] to discriminate among sugarcane clones. The use of molecular markers to discriminate among cultivars in other major crops (e.g. maize *Zea mays* L.) is well proven[21]. The potential power of molecular markers for discriminating among sugarcane clones is obvious. Not so obvious is the necessity for sugarcane improvement scientists to have such a tool available for use in germplasm collection management. Two examples are:

• A breeder may wish to periodically verify the trueness to label of clones in a germplasm collection. Appropriate molecular marker data could be accumulated when clones initially enter a collection. These data, accumulated in a database, could provide the basis for genetic distance analyses that may facilitate the subdivision of germplasm collections into homogeneous subsets, and consequently optimise cross-pollination combinations.

• Sugarcane improvement scientists may wish to set up a collaborative experiment, using a defined set of clones, across quarantine boundaries. Appropriate markers can verify that the clones are true to label.

To date, sugarcane breeders have not agreed on a standard marker system to use for management of germplasm collections. This is a high priority if molecular markers are to be used effectively on an international scale.

Introgression

The breeding system of crossing *S. officinarum* with *S. spontaneum*, and repeatedly backcrossing the hybrids, as males, to *S. officinarum* was developed by Dutch breeders in Java at the start of the twentieth century, and is known as nobilisation. This process is commonly referred to as introgression breeding by sugarcane improvement scientists. Simmonds[22], however, defined introgression as the transfer of a relatively small number of specific genes from ill-adapted germplasm into current commercial germplasm. Most often, this was accomplished by backcross breeding, or a modification. Incorporation, or base broadening, on the other hand, was considered a more powerful approach, but required a long-term commitment so that less-adapted germplasm could be incorporated into commercially adapted germplasm. Sugarcane improvement, by this definition, has been by incorporation. Progress from nobilisation breeding from 1960 to 1987 is summarised by Berding & Roach[9].

History

As indicated above, modern sugarcane clones are the products of complex interspecific hybridisation among several *Saccharum* species. The concept of interspecific hybridisation in sugarcane originated with Soltwedel in Java in 1885. Since then, sugarcane improvement has been dominated by the use of *S. spontaneum* in introgression breeding; however, interest beyond *S. spontaneum* is increasing. Research in Taiwan focused on the incorporation of *Miscanthus* Andersson germplasm into commercial germplasm[9]. Chen & Lo[23] summarised the progress for sugar content and disease resistance. Unfortunately, this programme seems not to have delivered commercial material despite promising early results.

One of the disadvantages of using *S. robustum* for base-broadening breeding is that the advantages of $2n+n$ chromosome transmission are not available. Although *S. robustum* germplasm has been little exploited in *Saccharum* spp. hybrids, some research has focused on use of this resource. Hemaprabha & Ram[24] studied the variability of seven quantitative characters in a range of *S. robustum* derived progeny. They concluded that additional backcrossing to *Saccharum* spp. hybrids would be necessary to improve sucrose content, and that intercrossing diverse I_1 (first introgression) progeny would improve cane yield.

Currently, considerable interest is focused upon use of *Erianthus arundinaceus* (Retz.) Jeswiet. This is reflected in the number of recent papers on the use of molecular cytogenetic techniques to diagnose definitively the products of introgression breeding [25–28].

Recent progress

The use of basic germplasm in broadening the genetic base of the Louisianan industry from 1972 to 1991 was reviewed by Burner & Legendre[29]. Two commercial cultivars, 101 elite clones, and 255 parental clones were selected from 14 crosses. These primarily involved *S. spontaneum*, with one clone featuring in half of these crosses.

Researchers in Taiwan studied the performance of a BC_1 hybrid derived from an indigenous *S. spontaneum* clone in over 50 combinations with *Saccharum* spp. hybrid clones[30]. Over 33 000 seedlings were produced in a 5-year period. Their performance was compared with crosses in the core programme. The indigenous germplasm generated promising interspecific hybrids after three generations of backcrossing.

In the Queensland industry, considerable success has been enjoyed with use of a potent I_3 clone that derived from a cross between POJ2878 and the *S. spontaneum* clone Mandalay, collected in Burma. The I_3 clone represents a combination of Indonesian ($2n = 112$) and Burmese ($2n = 96$) *S. spontaneum*, but whether the potency of the clone as a parent can be attributed to this is not known. This parent has appeared in 14 'Q' cultivars of the 50 released from Q138 to Q187.

It could be concluded that recent use of *S. spontaneum* germplasm, or, for that matter, any other taxa within the *Saccharum* complex, has been relatively unsuccessful. The best contemporary efforts of base broadening in sugarcane improvement have occurred in a unique programme conducted in Barbados[9,31], and at the programme of CSR Ltd at Macknade, Queensland[8]. However, these programmes have been relatively unsuccessful. The limited sampling of gametes undertaken in the early programmes in Indonesia and India yielded very successful results that modern breeding programmes have not emulated.

The USDA group at Canal Point has accumulated knowledge on aspects of base-broadening breeding. Tai *et al.*[32] studied the genetic behaviour of morphological and juice quality traits in progenies derived from *Saccharum* spp. hybrid × *Miscanthus* spp., *Saccharum* spp. hybrid × *Miscanthidium* Stapf. spp., and *Saccharum* spp. hybrid × *Erianthus* spp crosses. Juice quality in $I_1 × I_1$ and BC_1 progenies was greatly improved, and selection should be effective. Improvement for stalk diameter required additional backcrosses.

Crossing of *S. spontaneum* germplasm with *Saccharum* spp. hybrid material is difficult because *S. spontaneum* flowers much earlier. Inheritance of flowering in I_1 progenies was studied by Tai *et al.*[33] to improve planning and management of this programme. On average, I_1 progenies flowered 43 days later than the *S. spontaneum* parents, and approximately 67 days earlier than the *Saccharum* spp. hybrid parents. The authors concluded that photoperiodic manipulation of flowering, or pollen storage, would be necessary to effect the required crosses.

Exploitation of *S. spontaneum* has been important in development of *Saccharum* spp. hybrid clones, but an improved understanding of the breeding behaviour of quality characteristics is still required. Tai *et al.*[34] studied this in I_1, $I_1 \times I_1$, and BC_1 generations from crosses between four *Saccharum* spp. hybrid clones and eight *S. spontaneum* clones. Results indicated that selection to improve juice quality traits, while reducing fibre levels, should be effective.

Resistance to disease is one of the desired characteristics for transfer from basic *Saccharum* complex germplasm. Few studies have identified the distribution of resistance to disease over the different taxa within this complex. Burner *et al.*[35] studied the distribution of resistance to sugarcane smut (*Ustilago scitaminea* Syd.) using 102 clones drawn from six taxa. In one experiment, clones of *Erianthus* sect. *Ripidium* were most resistant, while those from *S. officinarum* and *S. robustum* were most susceptible. Clones in the second experiment were predominantly *S. spontaneum* drawn from three geographic regions. Those from India were most resistant (1% infection), followed by those from Indonesia (37%), and Philippines (51%). Obviously, if smut resistance is an important criterion for a base-broadening programme, a study of this nature gives clear focus for a preferred source for basic germplasm, and perhaps for additional germplasm acquisition and collection.

In 1994, Jackson & Roach[36] evaluated sugar yield in 32 progenies derived from crosses among I_1 clones from crosses among *S. officinarum* × *S. spontaneum* and *Saccharum* spp. hybrids × *S. spontaneum*. The *S. spontaneum* clones were drawn from diverse geographic origins. Parents and progenies exhibited high variation for all characters. There were no significant differences between parent and progeny populations for either mean performance or variance. Little immediate gain in terms of favourable gene interactions seemed achievable by crossing I_1 clones from diverse introgression backgrounds. Use of *S. spontaneum* germplasm is directed to improve ratooning and stress tolerance. Jackson[37] studied the genetic relationships among yield attributes in a population of I_1 (*S. officinarum* × *S. spontaneum* and *Saccharum* spp. hybrid × *S. spontaneum*) and $I_1 \times I_1$ clones grown over a single crop cycle. The results indicated that sugar content was negatively associated with traits associated with ratooning ability. Intensive selection for sugar content in isolation of ratooning performance was considered possibly detrimental.

Cytogenetics

Cytogenetics of the *Saccharum* complex was reviewed most in 1987 by Sreenivasan *et al.*[38] and in 1991 by Heinz[39]. Cytogenetic studies have added considerably to an understanding of the taxonomy and evolution of the genera/species within the complex, and their interaction within the complex *Saccharum* spp. hybrids has been used commercially.

Burner[40] studied the meiotic stability of 31 clones of basic germplasm of the *Saccharum* complex used in sugarcane improvement. He used 14 clones of *S. spontaneum*, eight *Erianthus* clones, five *Miscanthus sinensis* Andersson clones, three *S. robustum* clones, and one clone of *Narenga porphyrocoma* (Hance *ex* Trimen) Bor. Previously, data on meiotic pairing were sparse. Burner[40] found bivalent pairing predominated in all clones. Meiotic irregularities tended to be associated with taxonomic grouping and ploidy level. Clones of *Erianthus*, *Miscanthus*, and *Narenga* appeared euploid, with fewer irregularities than *Saccharum*. In a further study of meiotic stability, Burner & Legendre[41] studied progeny from crosses between elite *Saccharum* spp. hybrids and four *S. spontaneum* clones. Cytological behaviour, pollen staining, and seed yield were measured in 23 clones in the parental and introgression generations. Chromosome counts were made for 13 clones. Chromosome transmission was primarily $n + n$, although meiotic irregularities resulted in aneuploid gametes. Chromosome number increased with generation, from $2n = 60.5$ in *S. spontaneum* to 108 in the BC_3. Chromosomes paired primarily as bivalents, although variable numbers of univalents and multivalents were observed. Interestingly, the authors suggested that *Saccharum* spp. hybrids and *S. spontaneum* clones share a common endosperm balance number, unlike *S. officinarum* and *S. spontaneum*. Because of the occurrence of $n + n$ rather than $2n + n$ transmission, the authors concluded that additional cycles of recombination may be necessary to recover desirable recombinants. In the light of the subsequent demonstrated recombination between *S. officinarum* and *S. spontaneum* chromosomes within a *Saccharum* spp. hybrid, this conclusion is doubtful[42].

One exciting application of new technology is the use of computer-aided imaging to quantitatively karyotype chromosomes of several *S. spontaneum* clones[43]. This research resolved that one clone, a haploid derivative ($2n = 32$) of SES 208, was a tetraploid. This agreed with independent findings from molecular mapping of SES 208. In conjunction with cytological developments using in situ hybridisation by D'Hont *et al.*[26], which is discussed in the following section, one can anticipate marked developments in the cytological understanding of the *Saccharum* complex.

Molecular cytogenetics

The potential power of molecular biology for genomic analysis of sugarcane is already evident in a number of applications[44,45]. Applications include surveying cytoplasmic and nuclear polymorphisms, genomic analysis through genetic mapping, comparative mapping with related species, identification of parental chromosomes in introgression breeding, as well as marking desired traits for selection.

Single dose restriction fragments have been shown to be necessary and sufficient for mapping of a cytogenetically complex group such as *Saccharum*[46,47]. Their application to a population derived from the *S. spontaneum* clone SES 208 was most revealing as to its cytological origin, and its relationship to other important Gramineae species[48]. The same population was subjected to further analyses using polymerase chain reaction based markers[49]. The data from these two marker analyses were combined to provide a single genetic linkage map. This revealed the clone displayed

polysomic inheritance, a genetic trait of an autopolyploid genome[50].

Use of DNA RFLP analysis on basic and commercial germplasm allowed the separation of the three basic species *S. spontaneum*, *S. robustum*, and *S. officinarum*[51]. Data for *S. spontaneum* were related to geographic origin. Data for *S. barberi* and *S. sinense* supported current hypotheses for these secondary species. Genomic mapping across genera within the tribe Andropogoneae using maize genomic probes has revealed differences between sugarcane and two other genera for gene spacing and recombination rates in syntenic genome regions[52].

Specific characterisation of the strong molecular differentiation between *S. officinarum* and *Erianthus arundinaceus* was undertaken using RFLP markers[26]. Numerous *Erianthus*-specific RFLP bands were identified in the intergeneric hybrids. Besse *et al.*[28] also demonstrated genus-specific markers in a study of 62 *Erianthus* clones drawn from 11 species. These markers provide a means for following *Saccharum* × *Erianthus* introgression. As well, this study provided new information on species relationships and evolution within the genus *Erianthus*, on relationships between *Erianthus* and *Saccharum*, and among these genera and sorghum *Sorghum* Moench. and maize.

A further exciting development in molecular-assisted cytology is the application of fluorescent in situ hybridisation that allows visualisation of chromosomes, intact or fragmented, in cytological preparations. D'Hont *et al.*[26] demonstrated the use of this in diagnosing *S. officinarum* × *E. arundinaceus* hybrids. Jenkin *et al.*[53] also applied this technology to visualise hybridisation sites of an rDNA probe coding for ribosomal genes on chromosomes of *Saccharum* and *Erianthus*. More recently, D'Hont *et al.*[42] demonstrated the power of fluorescent in situ hybridisation. They distinguished the chromosomal contribution of *S. officinarum* and *S. spontaneum* in an interspecific I_1 hybrid between these species, as well as in a commercial *Saccharum* spp. hybrid clone. About 10% of the chromosomes of the commercial clone ($2n = 107 - 115$) were from *S. spontaneum*. For the first time, this technique allowed verification of recombination between the two genomes. The rDNA sites also were located, verifying differences between the species for basic chromosome number and chromosome structure. This research provides a first bridge between physical and genetic mapping in sugarcane.

Concern has been expressed regarding the narrow cytoplasmic base on which *Saccharum* spp. hybrids are based, given the recurrent use of a small number of *S. officinarum* parental clones. A study of cytoplasmic diversity in a range of *Saccharum* complex germplasm, including *Saccharum* spp. hybrids revealed no variability in the chloroplast genome[54]. Some variation in the mitochondrial genome was detected among species, but there was no mitochondrial variation in *S. officinarum*. The large variability in the mitochondrial genome within *S. spontaneum*, and *Erianthus* and *Miscanthus* offers potential cytoplasmic germplasm sources for future crop improvement. Successful use of these sources may prove difficult because of the importance of $2n + n$ chromosome transmission.

Commercial (*Saccharum* spp. hybrid) germplasm

Nobilisation

Much of the genetic improvement embodied in *Saccharum* spp. hybrids used commercially since the production of POJ2878 in Java in 1921 can be ascribed to the use of unreduced gametes in the introgression of *S. spontaneum* into *S. officinarum*. This is known as $2n + n$ chromosome transmission. Sreenivasan *et al.*[38] summarised six hypotheses to explain the mechanism of $2n + n$ transmission. Heinz[39] considered the most logical mechanism was the formation of reduced *S. officinarum* gametes that are stimulated to double in chromosome number when fertilised by gametes of *S. spontaneum* or of hybrids involving *S. spontaneum*. Sugarcane improvement has been very dependent upon use of unreduced gametes, yet the controlling mechanism is relatively unstudied.

Genetic diversity

A major concern in an important clonal crop such as sugarcane is the fear of limited genetic diversity among cultivars. This results from a selection process that severely erodes the genetic variability created from recombination among selected parental clones. This vulnerability due to lack of genetic variability may not be realised until an industry is subjected to a new disease infection or pest infestation. Researchers have attempted to analyse, and interpret, genetic relationships among clones at the commercial level using pedigree relationships[55–57]. Such studies must be considered suspect given the poor pollination control for cross integrity used in many sugarcane improvement programmes, past and present. Until the tool of genomic analysis using molecular markers became available, a rigorous analysis of this problem was simply impossible. Harvey *et al.*[20] studied 20 *Saccharum* spp. hybrids and one *S. spontaneum* clone using polymerase chain reaction–random amplified polymorphic DNA techniques. The primers used generated 356 loci. Similarity between hybrids ranged from 71.7 to 89.7%. The authors concluded that very limited genetic diversity remained within the hybrid group. A comparison of one of the *Saccharum* spp. hybrids and the *S. spontaneum* clone revealed much greater diversity. Lu *et al.*[58] performed a similar analysis on 40 cultivars using 22 low-copy maize DNA clones that produced 411 polymorphic fragments from a total of 425. The average genetic similarity between clones was 0.61. The authors concluded that, although the commercial clones seemed to be closely related to *S. officinarum*, the genes of *S. spontaneum* seemed to contribute primarily to the genetic variability observed. A very weak global structuring among the 40 clones was observed. This was consistent with the active interchange of parental materials among major improvement programmes.

Germplasm conservation

The position regarding conservation of the *Saccharum* complex was reviewed by Berding & Roach[9]. The assessment was that the Miami World Collection is not optimally located in a stress-free environment, but that the Indian collection at Kannur is in a much better location. Erosion of entries from the Miami collection was excessive, and the integration of the World Collections with each other, and with secondary and tertiary collections, was less than desirable.

Sugarcane improvement scientists have been preoccupied with the preservation of clones rather than genes, and the World Collections are still field-based clonal collections. In an endeavour to conserve genes rather than genotypes, the USDA has produced selfed seed of *S. spontaneum* clones for long-term storage in Fort Collins[59]. The Hawaiian Agricultural Research Center (formerly HSPA) has embarked on a similar programme with *S. officinarum* under contract with the USDA, and selfed seed of 78 clones has been stored (P. Y. P. Tai, personal communication).

In vivo, or in-field, clonal maintenance is the norm in sugarcane. Such collections are in danger from environmental stresses and stresses imposed by possible pathological infections or entomological infestations. Excellent germplasm maintenance requires the use of a high level of in-field management, and is demanding of resources. *In vitro* conservation techniques are well developed[60], and have obvious appeal for the conservation of clonal germplasm, for which good examples exist[60]. *In vitro* storage removes germplasm from exposure to in-field stresses, particularly pathological ones. This offers the possibility of reducing maintenance costs significantly, particularly if reduced-growth storage at lowered temperatures is feasible. The technology has been developed for sugarcane[61], but has not been adopted as widely as in other tropical clonal crops. Technology to cryopreserve-dehydrated and -encapsulated apices has been developed[62].

If conservation of genes rather than genotypes is the most important criterion addressed by germplasm conservation, and *in vitro* preservation offers decided advantages, use of tissue culture rather than meristems, should be an acceptable alternative. Somaclonal variation may be a complicating, but perhaps unimportant, factor associated with tissue culture in *Saccharum*.

CROSS-POLLINATION

Introduction

Cross-pollination is the principal method for the creation of new genetic variability in most crops, including sugarcane. Most sugarcane clones grown commercially are hybrids of *Saccharum officinarum* and *Saccharum spontaneum*, although modern hybrids generally have fewer than 20% of their chromosomes derived from *S. spontaneum* (Piperidis, personal communication). Nevertheless, genes from both species are important for the commercial success of clones. *Saccharum officinarum* contributes most of the sucrose genes and *S. spontaneum* contributes disease resistance, vigour, ratooning ability, and the ability to withstand harsh environmental conditions.

Flower induction

Availability of flowers for production of sexual seed is essential for production of genetically variable populations through meiotic recombination. This requirement for flowers will not be replaced by emerging technologies such as somatic fusion[63] or genetic transformation[64] in the foreseeable future. *Saccharum* spp. hybrids and basic germplasm, such as *S. officinarum*, have poor or variable flowering, which is a major impediment for plant improvement programmes.

Temperate and sub-tropical sugar industries have found it difficult to conduct breeding programmes. Research in South Africa[65] showed that low temperature was detrimental to flower initiation and development of spikelet fertility, particularly maleness. From this result, South African research developed photoperiod facility technology that was adopted by sugar industries in Argentina, China, Florida, Louisiana, and Taiwan. Essentially, container-raised plants are induced in a nocturnal environment maintained at $\geq 21°C$, and subjected, in most instances, to a reducing photoperiod using either natural dusks and artificial dawns or artificial dusks and dawns. A major advantage of artificially induced flowering is that crossing activities can be planned, and flowering of clones manipulated selectively to ensure desired combinations are achievable[66]. Moore[67,68] and Moore and Nuss[69] published comprehensive reviews on flowering in sugarcane, which cover the physiology of the process and manipulative control, in both field and photoperiod facilities.

Manipulations of the flowering process have not been confined to controlled environments. Considerable effort has focused on manipulation of environmental variables to enhance field flowering. Recent research of this nature was presented by Berding[70]. An important result was that adequate soil moisture is not a sufficient condition for flowering when temperatures are extreme.

The problem of poor or variable flowering is not unique to temperate or subtropical industries. Poor flowering has been recognised as an impediment to genetic improvement of some tropical sugarcane industries. Photoperiod facilities have been commissioned in Australia, Colombia, and Cuba. In Australia, an average of 38.2% of clones in the breeding collection at BSES Meringa (17°04′S latitude) flowered from 1978 to 1996. Flowering is variable, ranging from 16% of clones in 1993 to 66% in 1984. This flowering pattern prevents planned cross-pollination. Many desirable parental clones cannot be used, and the most desired combinations are rarely possible. Clonal improvement in the Queensland industry has been most successful since the development in 1954[71] of cross-pollination techniques to minimise cross-contamination, and the subsequent development of scientifically based selection methods[72].

Eleven experiments have been conducted in the BSES photoperiod facility at Meringa in tropical north-eastern Australia. These examined variables such as quality of extension lighting, constant photoperiods, canopy structure, plant age, high day temperature, post-inductive photoperiods, and nutrient level – particularly nitrogen. Results for the first eight experiments were summarised by Berding[70]. The efficacy of this research was tested in a validation experiment using 192 clones, drawn from flowering and non-flowering groups[73]. The success of the controlled regime was ascribed to management of the cultured plants as well as the photoperiodic regime used. Most importantly, a working hypothesis emerged that linked the level of suppression of flowering with the number of days in the inductive window with maximum temperatures > 32°C. This is based on circumstantial and experimental evidence gathered from observations in the field (F. A. Martin & N. Berding, unpublished data) and the photoperiod facility.

Cross-pollination

Cross-pollination procedures were reviewed by Heinz & Tew[74]. Sugarcane breeders are divided between use of biparental (full-sib) or polycross (half-sib) progenies. Generally, programmes based on large original seedling populations (the extensive approach) use a polycross approach. The ease and cost effectiveness of seed production probably drive this strategy. The alternative, or intensive approach, makes use of biparental progenies. Intuitively, use of full-sib progeny should yield greater progress, as input to both sides of the cross is fully controlled, particularly if seed true to label is produced under isolation in quality cross-pollination lanterns[71]. However, many programmes continue to produce reputedly full-sib progeny in cross-pollination conditions that appear not to guarantee trueness to label.

Most cross-pollination activities are conducted in a centralised cross-pollination area, either outdoors in tropical areas or indoors in regions that are more temperate. Hawaiian acid crossing solution primarily is used to preserve severed flowering stalks in the former. Marcotted or canned stalks are preferred in the latter, although crossing solution is often used in conjunction. Extensive experimentation to improve the efficacy of the crossing solution by modifying the formula[75], or by exploring methods of prolonging solution longevity by enhancing SO_2 retention, have yielded little significant improvement.

Environmental conditions under which cross-pollination is conducted can have a significant effect on anther dehiscence, pollination, and fertilisation, but little research has been conducted. Berding & Skinner[11] found heating of the crossing lantern environment improved cross fertility. Subsequent research, in another season, to optimise this treatment failed to determine a combination among three temperature and two humidity treatments that yielded better cross fertility than ambient conditions (N. Berding & J. C. Skinner, unpublished data).

Pollen preservation has the potential to bridge gaps between taxa that are reproductively isolated because of temporal differences in flowering. If artificial flower induction is possible in the taxa concerned, manipulation of the inductive treatments can result in emergence of flowers together so crosses among taxa of interest can occur. This is not always possible, however, especially when natural flowering is used. If pollen could be collected and stored with minimal loss of viability, crosses could be made by using such pollen to fertilise panicles emerging at different times. Tai[76] successfully stored pollen from *S. spontaneum* clones at low temperature after drying, to produce seedlings in crosses. He later extended this successfully to preservation of pollen from I_1 progeny from crosses between *S. officinarum* and *S. spontaneum*, and *Saccharum* spp. hybrids and *S. spontaneum*[77].

Contamination of sugarcane crosses can arise from collection of pre-fertilised panicles, poor cross isolation techniques, or poor post-fertilisation handling of panicle material. This may be of little consequence given unsophisticated early generation clonal evaluation techniques and generally low heritabilities for traits of interest. However, extra variation generated in contaminated crosses certainly is not conducive to effective family selection or development of proven crosses, and should be avoided.

In introgression breeding involving either interspecific or intergeneric combinations, cross-contamination is unacceptable. Use of putative introgression progeny as parents in subsequent crosses is simply ineffective.

Zheng *et al.*[78] applied polyacrylamide gel electrophoresis to the analysis of three enzyme systems in progenies of a cross between *S. spontaneum* and *E. procerus* (Retz.) R.C. Bharadwaja and between a *Saccharum* spp. hybrid and a *Sorghum* spp. There were differences between progenies and parents for banding patterns. Progeny in the second cross presented bands of both parents and seemed to be true hybrids. Results showed isozyme markers could be used to verify hybrid progenies. Use of biochemical markers for diagnosing a wrongly substituted parent in an introgression study was demonstrated by Lee[25]. More recently, D'Hont *et al.*[26] demonstrated the application of isozymic, molecular, and in situ cytological hybridisation techniques for verifying the hybrid origin of progeny from a *S. officinarum* × *Erianthus arundinaceus*

cross. Variation at the 18S + 26S and 5S ribosomal DNA loci was assessed in 62 *Erianthus* Michx. clones representing 11 species[27]. Genus-specific markers were identified which will be useful in following *Saccharum* × *Erianthus* introgression. Applications of new technology such as that used by D'Hont *et al.*[26] and Besse *et al.*[27] will have a dramatic impact on improving the efficacy of *Saccharum* complex introgression breeding.

Parental selection

Accurate prediction of the performance of possible parental combinations would yield tremendous returns in any crop-improvement programme. Resources could be allocated to parents of greatest potential in terms of field or photoperiod facility space. Return from cross-pollination and selection activities could be maximised by concentrating only on the most desirable crosses.

Hsu *et al.*[79] and Wang *et al.*[80] described the development of computer databases for management of sugarcane improvement programmes. Wang *et al.* described a computer program that linked parental clone, census, and selection data to a database of desired crosses. However, these crosses had been compiled by consensus among a breeding team, and not by mathematical estimation of parental breeding value or cross value.

Hogarth & Skinner[81] developed a computerised algorithm for selection of sugarcane crosses using phenotypic data from parental clone evaluation as well as progeny evaluation data. This approach aimed to improve the efficiency of cross-combination selection by reducing the intensive inputs of a team of experienced crop-improvement scientists. This development used regression modelling techniques applied to data from a series of sequential, replicated evaluation trials of full-sib families.

Chang & Milligan[82] attempted to identify a reliable and easily obtained cross appraisal statistic, and in which selection stage and crop could best be determined. They examined four statistics, including best linear unbiased predictors (BLUPs). The potential of a cross to produce elite progeny was most accurately predicted by the cross mean of that trait. This was the most practical statistic because of ease of collection. Chang & Milligan[83] also examined three bivariate statistical methods for prediction of family potential to produce elite progeny. Estimation of the sum of ranks based on family mean values for two traits versus the calculation of BLUPs was comparatively easy. As there was no apparent loss of predictive value, the former was considered the most suitable statistic to use for bivariate predictions.

Stringer *et al.*[84] examined BLUPs for estimating the potential performance of parental combinations in sugarcane. The BLUP method was more effective than the empirical algorithm used by BSES[81,85]. Superior combinations were identified with less information, making a reduction of generation interval a real possibility. Further refinement of the application of BLUP methodology to sugarcane improvement may see further increases in the efficacy of prediction of cross potential.

In sugarcane, scientists are concerned with two generation intervals. The first is the time from a seedling being planted in the first selection stage until that clone has seedlings planted in the same stage. At BSES Meringa, and in Hawaii, this is 7 years. At BSES Meringa, use of a clone as a parent is dependent upon assessment in the final trial stage on the experiment station. In an environment not conducive to free flowering, this time would be a minimum. Rapid, generation-wise selection of parental material is vital to sustained progress. This will be difficult to reduce unless a concerted effort is made to use clones from selection stages earlier than the final yield trial on the experiment station.

The second generation interval of concern in sugarcane improvement is the period from seedling to commercial release. This has been as short as 8 years in Queensland, but more typically varies from 12 to 15 years. There are real advantages in reducing this interval, and more research is necessary on this topic. Conceivably, *in vitro* propagation of meristem propagules early in the selection process for potentially elite clones could reduce the interval by accelerating the bulking up of clean propagation material.

Seed storage

The storage of sexual seed produced by cross-pollination under optimum conditions is essential. Sugarcane cross-pollination is labour intensive, and therefore expensive. Many programmes are still plagued by variable flowering. Consequently, seed supply from crosses with certain parental clones may be erratic. Long-term preservation of seed under deep-freeze conditions, without loss of viability, is therefore essential. Breaux and Miller [86] considered all aspects of seed handling. Seed handling and drying technology have improved markedly since 1987. Sugarcane seed is small, and drying of such seed is now best accomplished by use of cold dehumidification technology [87]. In addition, such seed must be packed under reduced humidity conditions.

Panicles bearing seed are dried in open-weave Terylene bags in an insulated, tightly sealed chamber maintained at 11–13°C and 15–25% relative humidity using refrigeration and sorption-type dehumidification. Repeated experiments have confirmed fuzz moisture levels reach 4.5% moisture wet basis in 4 days. This is within the IBPGR recommendation of $5 \pm 1\%$ moisture as necessary for long-term storage of small seed. Use of multi-laminate aluminium foil and plastic film for seed packaging, and storage in deep-freeze conditions of $< -20°C$ optimise long-term storage of such low-moisture seed. This technology is a marked advance over forced-air drying at 36–38°C. Cold dehumidification seed drying technology was adopted at BSES Meringa when forced hot-air drying was found not to reduce seed moisture to within the IBPGR recommended range under all external ambient conditions.

SELECTION

Introduction

Selection systems were reviewed by Skinner *et al.*[88]. These have evolved as sugarcane breeders have understood more about the mode of inheritance of important characters, and the importance of using valid statistical designs with sound interpretation of results. Equipment for measuring cane yield and sugar content has also improved. Finally, methods have changed as breeders have gained a greater understanding of the importance of multi-environment selection at different stages of selection.

Sugarcane is a difficult plant to study with classical genetics[89], so most genetic studies have used quantitative genetic analyses. For example, studies such as those conducted by Brown *et al.*[90,91], Hogarth[92–94], and Hogarth *et al.*[95] have provided useful information that is used in modern selection programmes. In particular, these studies have shown that additive genetic variance is more important than non-additive genetic variance for most economically important characters, such as sugar content, resistance to diseases (i.e. smut, common rust, and Fiji disease), and fibre content. The major exception is cane yield, which seems to have similar amounts of additive and non-additive genetic variance. The implication of this finding is that parental selection based on the known performance of the parent will be successful for most characters apart from cane yield. For cane yield, prediction of progeny performance is less likely to be successful, and it is necessary to make a large number of crosses so that the highest yielding families can be identified. Quantitative genetic studies have also shown the relative importance of genotype, genotype × environment interaction, and experimental error in determining the value of the phenotype at different stages of selection.

Skinner *et al.*[88] presented data showing the degree of genetic determination (or broad sense heritability) for a range of characters on an individual and family basis for cane grown as original seedlings or in very small plots. For cane yield, the degree of genetic determination on an individual basis was 0.17 or less, but this increased to 0.75 on a family basis in Australia and to 0.48 in Fiji. For brix (total soluble solids, an estimate of sugar content), estimates on an individual basis ranged from 0.27 (Hawaii) to 0.65 (Australia) and on a family basis from 0.53 to 0.90. These figures suggest that selection for cane yield on an individual basis will not be very successful, but that progress can be made in selection for brix on an individual basis, and that more progress may be made using family selection.

At later stages of selection, clones are grown in larger plots and trials are often replicated. This reduces the importance of error variance, and there is a large increase in the degree of genetic determination on an individual basis. For example, Kang *et al.*[96] showed that degree of genetic determination for a range of agronomic characters in a well-replicated trial with large plots (four rows × 6 m) was high (0.77–0.91). Thus, individual selection at this stage of selection should be successful, as there is usually adequate genetic variation for the important characters.

It is important to test a large range of families at the original seedling stage of selection, rather than large populations of seedlings in individual experimental crosses. This is because of the importance of non-additive genetic variance for yield of cane, which makes it impossible to predict the cane yield of seedling families. By testing a large range of families in replicated and weighed trials, it is possible to identify the families with outstanding productivity. These families are then replanted in much larger populations in subsequent years as well as undergoing selection in the first ratoon crop after weighing families in the plant crop.

Rather than compare the sizes of original seedling populations, sugarcane breeders should compare the effective sizes of the programmes. It could be argued that the effective size of a programme is the number of clones weighed at the first stage in the programme where individual clones are weighed, because this is the first stage at which objective data are obtained. By this standard, the Australian programme has increased dramatically in the last 10 years from about 500 clones to about 12 000 clones. We believe that this has resulted in an improvement in the population of clones being selected, and there is evidence that many potentially commercial cultivars are being selected. This increase in the effective size of the programme has resulted without an increase in the size of the original seedling population.

The improvement in the effective size of the Australian programme has resulted from the development of mobile weighing equipment [97]. Without efficient methods for weighing plots, it would be prohibitively expensive to weigh plots at all stages of selection because of the cost of the labour involved. With good weighing systems, it is actually cheaper to run selection systems because the need for labour is greatly reduced. Weighing systems used in Australia are expensive to purchase, but we believe that this cost is justified by the saving in labour costs and the vast improvement in selection efficiency.

Selection in original seedlings

Many sugarcane breeding programmes have very large populations of original seedlings, and use individual selection based on the phenotype as the selection criterion. The programmes are successful, but must be inefficient because results from quantitative genetic studies would suggest that selection efficiency has to be low. In view of the importance of this stage of selection and the cost of conducting these trials in countries with high labour costs, breeders have been attempting to find more efficient methods of selection. Most attempts have involved the use of family selection.

Hogarth[93] suggested that family selection should be more efficient than individual selection at the original seedling stage of selection. However, at that time, family plots had to be cut and weighed manually, and the cost was prohibitive. With the development of mobile weighing machines in Australia[97], there was an opportunity to investigate the advantages of family selection in more detailed experiments, and a considerable amount of research has been completed. Interest in family selection is also evident in Indonesia[98], Cuba[99], South Africa[100], Hawaii[101], Florida[102], and Louisiana[82,83]. A review of family selection prepared by Jackson *et al.*[103] in 1996 described the implementation of family selection in Australian regional selection programmes. They concluded that family selection is well suited to the mechanical harvesting and weighing systems developed in Australia, it is efficient in terms of labour usage, and it is likely to be superior to individual selection in most situations. McRae *et al.*[104] and Cox *et al.*[105] showed that a combination of family and mass selection was likely to be more effective than family selection alone. Cox & Hogarth[106] showed that the most efficient method of family selection was likely to be based on the performance of families in replicated plant crop trials, followed by individual selection within the best families in the

first ratoon crop. In 1996, Simmonds[107] encouraged the use of family selection and pointed out that family selection is only used routinely in the Australian sugar industry and a Scottish potato breeding programme.

Hogarth *et al.*[108] showed that family selection was particularly useful in situations where the growth of crops has to be restricted to prevent lodging to enable visual selection to be conducted. Restriction of growth was shown to change the ranking of clones significantly, and it is probable that the most productive clones would not be selected. This work was done in the highly productive Burdekin region of Australia where crop lodging is regarded as normal. Progress from selection in this region had been slow compared to other regions of Australia, and it is likely that the selection programme was inefficient when crop growth was deliberately restricted. Since the introduction of family selection and weighing of plots at all stages of selection in the Burdekin region, impressive gains from selection have been recorded, and breeders are confident that commercial clones being released now have the potential to significantly improve productivity.

The importance of family × environment interaction on the assessment of families has been studied by Hogarth & Bull[109] and Bull *et al.*[110] in the Bundaberg region, Jackson *et al.*[111] in the Herbert River region, and McRae & Jackson[112] in the Burdekin region of Australia. In the Bundaberg and Herbert studies, family × location interactions seemed to be as important as family effects for yield of cane but not for sugar content. In the Burdekin study, the interaction was unimportant. In all studies, family × crop-year interactions were unimportant. Because of these studies, families were planted routinely on two or more sites in the Bundaberg and Herbert selection programmes. However, we have found that, when many families were planted in two sites in the Bundaberg region, there was no family × location interaction in two successive years. Similar results have been obtained in the Herbert region (P. A. Jackson, personal communication). Thus, family performance may be quite robust and less influenced by interaction with the environment than previously found. Jackson *et al.*[113] found that the levels of several soil nutrients, notably calcium and zinc, seemed to be associated with the family × location interactions. If nutrient levels are confirmed as a major reason for family × location interactions, correction of nutrient levels before trials are planted should minimise the interactions.

Selection of clones

Most countries have four to five stages of selection following selection of the original seedlings. In Australia, there were six selection stages until the mid 1980s, and it took about 15 years to release a variety. Since the adoption of mobile weighing equipment, the programme now has only three selection stages: original seedlings, an unreplicated clonal stage on the experiment station, and final assessment trials which are replicated and planted on a range of locations. All plots are weighed and assessed for sugar content. In addition, promising varieties are planted in a range of agronomy trials to assess response to treatments such as fertilisers, irrigation, and green cane harvesting. Similar trials are also conducted in South Africa where clones are also assessed for their response to ripeners.

In countries where visual selection is used, it is customary to have more stages of selection, as it is more difficult to discriminate between clones. However, visual selection has some advantages over weighing, as selectors are better able to assess non-yield characteristics of clones, such as ideotype, lodging propensity, suckering, and flowering. These characteristics are important when assessing a clone's performance, and may be overlooked when all plots are weighed.

Genotype × environment interaction

Genotype × environment (G×E) interactions have a major influence on selection strategies in most sugarcane selection programmes, but the nature of the interactions and how to avoid, minimise, or use these interactions have been poorly addressed. The problems of G×E interactions were discussed by Skinner *et al.*[88] who addressed the issue of resource allocation, i.e. the number of locations, years, and ratoons to assess at various stages of selection. They also discussed the confound-

ing of years and crop classes in sugarcane; such confounding is difficult to avoid, but may have a major effect on the interpretation of the results of G×E studies. Jones *et al.*[114] proposed a method for minimising the confounding of genotype × year and genotype × crop class interactions.

Until recent years, there have been relatively few G×E studies in sugarcane. Some examples of studies were those of Pollock[115], Espinosa & Galvez[116], Kang & Miller[117], Tai & Miller[118], and Milligan *et al.*[119]. These studies found significant first and second order interactions among genotypes, locations, and crop-years. In recent years, there has been considerable research on this topic in Australia and significant papers are those of Jackson *et al.*[120], Jackson & Hogarth[121], Jackson[122], Bull *et al.*[110], and Mirzawan *et al.*[123–125].

The Australian studies have shown that genotype × location interactions are generally far more important than genotype × crop-year interactions. In fact, Jackson & Hogarth[121] concluded that the similarity in information obtained across crop-years within most sites suggests that there may be limited gain in testing across multiple crops or years within a particular site. Jackson[122] suggested that for early stages of selection only plant crops need to be evaluated. In practice, in the Herbert region where this work was done, stage 3 trials (unreplicated 10 m plots) are now planted on two sites and assessed only in the plant crop. However, for the final stages of selection, Jackson[122] felt this strategy would not be appropriate as ratooning ability is an important characteristic of a commercial clone.

Mirzawan *et al.*[123,124] also found that emphasis should be placed on sampling a greater number of locations rather than testing clonal ratooning ability within locations. They concluded that this would improve the chances of obtaining both broadly and specifically adapted clones. Mirzawan *et al.*[125] studied the repeatability of interactions across years, and found that some aspects of the interactions were repeatable. The retrospective analysis used by Mirzawan *et al.*[125] is a powerful technique that could be used by most breeding stations, and may assist in determining the similarity between test environments. If several sites are similar, rationalisation of test sites should follow, which would have a beneficial effect on the allocation of resources.

The use of repeated standard (check) clones in clonal evaluation trials is standard practice in trials in Australia and elsewhere. Bull *et al.*[126,127] have used the repeated checks to determine appropriate group number in classification analyses that are used in G×E studies. Bull *et al.*[128] used a similar concept by treating the two blocks of a replicated trial as two different environments; when the two blocks were placed in the same group, it was found that this was a suitable stopping point for the classification. This technique is useful for determining the similarity between environments used in multi-environment trials.

Recently research has focused on determining the environmental factors giving rise to G×E interaction in multi-environment trials[113,128]. If the major environmental factors contributing to G×E interaction can be identified, they can be avoided when locating trials, ameliorated, or directly manipulated as managed environment selection trials [129,130]. The nature of the factor, the crop and the production system will determine whether the interaction should be avoided, minimised or exploited.

For example, Jackson *et al.*[113] reported that the availability of some micro-nutrients was correlated with principal component scores that represented G×E interaction. Since the particular micro-nutrients identified are relatively inexpensive, this source of variability can be avoided by simply monitoring the fertility of selection sites and applying additional nutrients as required.

Further, Bull *et al.*[129] found that both differential water and nitrogen availability gave rise to substantial G×E interaction in managed environment trials at one location in a plant crop. If a strong correlation can be established with core selection trials, managed environments would represent an efficient and repeatable method of screening clones[129].

Competition and plot technique

The importance of competition between neighbouring varieties in sugarcane trials is frequently underestimated. It is not unusual to see plots in advanced selection trials that are relatively unguarded with wide alleys between plots for convenient access. Such trials do not reflect commercial reality, and may give totally misleading estimates of the potential of varieties under test. Simmonds[131], in discussing selection efficiency, cautioned that 'any comparison of unguarded plots or rows leaves genotypic differences confounded with competitive effects, so any improved selection could be as much for competitive ability as for the desired performance in pure stand.' Later, in a discussion of trial format, he again emphasised that 'The need for guard rows springs from the fact that varieties differ widely in competitive ability'.

The theory of intergenotypic competitive effects in small-plot sugarcane evaluation trials was developed by Skinner[132]. He illustrated the importance of competition, particularly in ratoon crops. Subsequently, measurement of true yield in plot formats used for replicated trial evaluation was optimised using border rows[133]. Skinner *et al.*[88] summarised the implications of competition in sugarcane selection, and the implications of plot shape and weighing strategies upon trial design.

N. Berding & J. C. Skinner (unpublished data) compared four plot shapes, using four replicates of 60 clones. Plot shapes were four-rows long, four-rows short, two-rows long, and one-row long. Long rows were 9.2 m, with the short row being 4.6 m. The experiment was harvested over three crops, i.e. plant, first, and second ratoons. The coefficient of variation increased as plot size decreased, being lowest in the four-rows long, intermediate in the four-rows short and two-rows long, and highest in the one-row plots. The value for the latter was too high to be acceptable for normal replicated yield trials. At earlier selection stages, use of such plots is warranted, however, as the number of genotypes handled offsets the increased error of evaluation and sustains advance from selection. The ratio of end to lateral competition was treated as an unknown in the estimation model for the first time. End competition was 6.5 times that of lateral competition in four-rows long plots, and three times in four-rows short plots. Four-row plots were clearly superior to three-row plots for competition analyses, as even with a uniform trial using four replicates of four-row plots, the normal competition analysis gave an insensitive estimate of competition.

McRae & Jackson[134] also studied competition in a range of plot sizes and shapes. They concluded that estimation of cane yield in small plots was biased by competition between clones in neighbouring plots, and the effect was more serious in ratoon crops. In contrast to cane yield, sugar content was little affected by competition between neighbouring plots. Thus, greater emphasis should be placed on selection for sugar content at early stages of selection when small, unguarded plots are used. Matassa *et al.*[135] quantified the bias caused by competition in single-row plot trials with a statistical model. A moderate improvement in the predicted pure stand yield estimates over the unadjusted clonal mean yields resulted from the fit of this model to experimental data.

Experimental design

In all sugarcane selection programmes, there are stages of selection where clones are grown in small, unreplicated plots. These trials usually include check (or standard) clones at regular intervals so that test clones can be compared with a check clone. McDonald & Milligan[136] studied various adjustment models to control environmental heterogeneity in an unreplicated sugarcane trial in Louisiana in order to improve selection decisions. None of the models studied was better than unadjusted yield, but the authors believed their results were affected by severe weather and G×E interactions between the testing stages.

There has been interest in nearest neighbour models for some time. These models, which are based on adjusting plot values by covariance on neighbouring plot values, were discussed by Bartlett[137]. One possibility is to use moving averages as discussed by Weber & Stam[138]. However, the most promising method is a nearest-neighbour analysis described by Wilkinson *et al.*[139], which involves a continuous process of detrending the data. Authors such as Baird[140] have shown that improvements may be gained by using the analysis.

A method described by Cullis *et al.*[141], in which data in unreplicated trials are detrended, may be of benefit in early-stage trials. The analysis, termed spatial analysis of field experiments (SAFE), is widely used in grain crops in Australia. Matassa *et al.*[142] demonstrated that spatial analysis could be applied successfully to sugarcane trials. Spatial analysis resulted in a lower estimate of error variance than the more classical methods of analysis currently used. Consequently, there was a substantial improvement in the precision of the estimates of clonal effects.

Another possibility for increasing efficiency in early stage trials is to use replicates as suggested by Shaw & Hood[143]. If two replicates were used, only half the number of clones could be tested in the same area, but this would be worthwhile if there was an increased gain from selection.

Automation

Sugarcane improvement programmes have been relatively slow to adopt automation. However, data from juice laboratories can already be captured electronically, which eliminates transcription errors and errors of plot identification. Although mobile weighing machines with electronic readouts are used in Australia for weighing trials[97], these data are not yet captured automatically because of the time taken for the weight reading to stabilise. This system will improve with better technology, and all data will then be captured automatically.

The other obvious advance in data handling will come from improved database systems for storage and retrieval of all results collected from clones. Many countries already have excellent systems, but integration of all data sets stored has yet to be achieved in most.

Further automation is possible in the harvesting of clonal evaluation trials, particularly in countries that do not have mobile weighing equipment. A further advance envisioned for the Australian programme is to collect cane samples for juice analysis automatically. Currently, samples for juice analysis consist of random whole stalks cut by hand and stripped free of trash. This is atypical of the cane supply sent to the mill, which consists of short billets of cane (100–250 mm long), some cane tops, cane trash, suckers (water shoots), and dirt[144,145]. Thus, the sample analysed is a biased sample of the cane supply, and this may affect the validity of comparisons between clones.

It is proposed to construct a billet sampler, located on the mobile weighing device, to collect a random sample of all the material sent to the mill[144,145]. This will provide an unbiased sample for juice analysis, but will also reduce labour costs, and will provide a safer working environment because people will not be required to collect samples while harvesting equipment is operating nearby. In the future, it is hoped that the billet sample could be prepared for analysis in the field, and only prepared samples would be returned to the laboratory for analysis.

A further improvement in automation will be the use of new technology, such as near infra-red spectroscopy (NIR), to analyse cane samples for a range of constituents, including sugar content and fibre precentage. NIR has been shown to have great potential for analyses of cane samples[146–151]. These authors have shown that NIR can produce results that are highly correlated with routine laboratory analyses, and NIR has the great environmental and human safety advantage of not requiring the use of lead to clarify juices. In the foreseeable future, it is hoped that cane samples will be collected automatically by a billet sampler, prepared by a shredder in the field, and analysed for all constituents of interest by NIR spectroscopy.

BREEDING OBJECTIVES

Productivity

It is almost impossible to assess the improvement in productivity resulting from breeding, because positive and negative agronomic and environmental factors are confounded with gains from breeding. When a disease problem is solved with resistant clones, there may be an apparent fall in productivity, but breeding has prevented losses in productivity. It is often stated that gains from plant breeding are about 1% per annum[152], but this is difficult to prove.

Plucknett & Smith[153] in Hawaii estimated an average improvement of 1.1% per annum for the period from 1908 to 1984, but they did not attempt

to partition gains from breeding from other advances. In Barbados, Simmonds[131] estimated an annual gain from breeding of 0.64%, and most of this gain was from increased biomass rather than increases in sugar content. In Australia, Berding & Skinner[154] showed that the release of the new cultivar, Q90, over the existing cultivar, Pindar, resulted in an increase in productivity of 31%. However, when common rust entered Australia in 1978, Q90 suffered a loss of productivity of 25%, as measured against resistant clones in a range of trials. Q113, a rust resistant clone, produced only 95% of Q90 in the absence of rust, but was clearly more productive than Q90 in the presence of rust. In 1995, Cox & Hansen[155] reported on the substantial advances made by new commercial clones in southern and central Queensland, and showed that gains of 20% could be made with the release of superior clones.

Berding & Skinner[154] and Bull *et al.*[156] note that the apparent gains from plant breeding have not resulted in an increase in productivity in Australia. There have been suggestions that a productivity plateau has been reached, but there are strong indications that productivity, as measured by sucrose yield per hectare, is increasing again. However, it is worth examining the reasons for the apparent productivity plateau when there seem to have been substantial genetic gains. Berding & Skinner[154] stated that 'plant breeding had maintained productivity in a declining agro-environment, probably related to a deterioration in the soil environment'. Bull *et al.*[156] included other factors, such as: older ratoons, industry expansion of 47% since 1971 (mostly onto more marginal soil) and increased extraneous matter being sent to the mill. An additional factor is that the more powerful modern cane harvesters are likely to result in larger cane losses at harvest, and may damage the stool at harvest, which lowers ratoon performance. Cane losses of 10% are common, and there have been reports of cane losses up to 18%[157]. The latter problem is being addressed, and this should improve the apparent productivity per hectare.

Walker & Simmonds[158] compared the performance of sugarcane clones in trials and in commercial agriculture, and found that trials greatly overestimate the performance of new cultivars. This was attributed to the effect of G×E interactions, because it is not feasible to sample a sufficiently large range of environments in a testing programme. Bull *et al.*[156], however, showed that, under some circumstances, the correlation between gain predicted from trials and gain achieved in agriculture is not so poor.

Roach & Daniels[159] reported the results of trials in northern Queensland in which historical and modern clones were included to determine the progress made by breeding. It is difficult to interpret the results of this type of trial, as management practices have changed, and it is possible that the historical clones would be at a substantial disadvantage. The results suggested that the main effect was to increase cane yield, particularly in ratoon crops, although this is probably partly an artefact of mechanical harvesting. There was little evidence to suggest that much progress had been made in increasing sugar content. Chapman[160] presented evidence showing substantial increases in productivity from plant breeding in central Queensland from 1946 to 1994.

Most sugarcane breeding programmes emphasise the importance of selecting clones with higher sugar content, but the evidence suggests that progress has been generally disappointing. This is not surprising in view of the results of quantitative inheritance studies in Australia[92,94] and Hawaii [95]. These showed that there was relatively little genetic variation for sugar content compared with cane yield, although sugar content had a higher heritability.

The major exception for improved sugar content has been Louisiana, where substantial progress has been made. In that state, the crop-growing season extends for only 8–9 months, compared with 12–18 months in tropical areas and, consequently, the cane can be immature at harvest. Breaux [161,162] described the process of recurrent selection introduced into the Louisiana programme in an endeavour to improve sugar content. He showed that sugar content of the population of clones improved from a mean of 9.7% in 1936 to 10.9% in 1952, to 12.5% in 1963, and then to 14.1% in 1974. Breaux attributed this substantial improvement to a number of factors. The most significant of these may be that there is more variability in sucrose content among

clones during the incline phase of the maturity curve where the Louisiana breeding programme and industry operate. Legendre[163] has now completed a fifth cycle of recurrent selection, and the average sugar content marginally (and probably non-significantly) decreased to 13.6%. He believes that further progress in improving sugar content will be more difficult as sugar content of parent clones reaches an apparent plateau.

Cox & Hansen[155] showed that gains in sugar content have been made in south Queensland in recent years. In this region, sugar content had dropped following the demise of NCo310 owing to its susceptibility to Fiji disease.

Sugarcane breeders are interested in improving sugar content at the beginning of the harvest season. This would improve the economics of crushing cane early in the harvest season and could enable sugar factories to commence crushing earlier than is currently possible, which would improve the use of expensive milling equipment. Estimates of the degree of genetic determination by Mariotti[164], Cox *et al.*[165], and Cuenya & Mariotti[166] have shown that the potential to make genetic gain for sugar content is greater earlier in the crushing season than later. This is the incline phase of the maturity curve and is, therefore, consistent with the results of Breaux[162]. Cox *et al.*[167] described a recurrent selection programme in Australia designed to improve early season sugar content, and presented results that indicated that progress was being made. Cuenya & Mariotti[166] also reported improvements in sugar content (i.e. brix) of progenies, and found that new genotypic variation could be induced through recombination.

Disease resistance

Breeding for disease resistance was reviewed by Walker[31]. He pointed out that most diseases of sugarcane are controlled by resistant clones with the exception of pineapple disease *Ceratocystis paradoxa*, ratoon stunting disease *Leifsonia xyli* subsp. *xyli*, and grassy shoot disease. Historically, breeding for disease resistance is one of the most important reasons for sugarcane industries to engage in breeding programmes. For most diseases, there seems to be ample genetic variation for resistance or tolerance for a successful breeding programme to be conducted. Most sources of resistance come from wild canes, specifically *S. spontaneum* which has many clones with resistance to red rot, smut, leaf scald, and mosaic[31].

Since the beginning of the 1970s, there have been serious outbreaks of a number of diseases in many sugarcane industries. The most important of these have been common rust which appeared in the late 1970s in the Caribbean and Australia, and smut which spread to the USA, the Caribbean and South America in the 1970s. In Australia, there was also a devastating epidemic of Fiji disease and an increase of a root pathogen called *Pachymetra*. More recently, there has been considerable concern about a new disease called yellow leaf syndrome, particularly in Brazil. All of these diseases have been kept largely under control by resistant or tolerant clones.

Common rust *Puccinia melanocephala* H. & P. Sydow has been easily controlled by breeding resistant clones in most countries, despite the high degree of susceptibility of commercial clones in Australia and the Caribbean when the disease was first detected. Walker[31] pointed out that most clones of *S. spontaneum* and *S. officinarum* were resistant to common rust and that high susceptibility seems to be a *de novo* feature of hybrid sugarcanes. In Australia, Berding *et al.*[168] developed a rapid test using potted plants to screen for rust resistance, and Hogarth *et al.*[169] showed that there was a high heritability for rust resistance. Therefore, it would be an easy matter to breed for rust resistance. In practice, a planned programme has not been necessary, as seedling families showing a high level of susceptibility are easily identified and discarded, while susceptible clones are also easily identified and/or produce very poor crops of cane. Thus, the susceptibility of the population of clones under selection has rapidly decreased, and common rust has become a minor selection problem.

Smut *Ustilago scitaminea* Sydow has been more difficult to control. The disease was largely confined to Asia and southern Africa until the 1970s when there was a widespread distribution of smut to most sugarcane growing countries of the world. Simmonds[170] speculated on the probable airborne spread of the disease, which may also have been responsible for the spread of common rust. However, unlike common rust, smut has not been found in Fiji, or Papua New Guinea, and was only found in Western Australia in 1998. Yield losses of 20% have been reported in some clones, but Whittle[171] felt that this was only in highly susceptible clones and that modest levels of infection may not cause measurable losses. In Hawaii, Wu *et al.*[172,173] estimated heritability of smut resistance and found that, on a family basis, it was adequate for selection to be effective and that selection on an individual basis would also have some effect. There was evidence that selection of resistant parents would improve the level of resistance of progeny. The Hawaiian breeders discarded susceptible parents from their collection and reduced the proportion of susceptible progeny from 64% in 1972 to 15% in 1982[31]. Balance *et al.*[174] found that general combining ability for smut resistance was more important than specific combining ability, and concluded that a recurrent selection programme should be effective.

Fiji disease is caused by a reovirus and is spread by a planthopper, *Perkinsiella saccharicida*, or by planting infected cane. It was originally found in Fiji, but has caused devastating losses in both Australia and Fiji. It had been a serious disease in Australia up until the 1950s when it was thought to have been largely eradicated. In 1969, it was rediscovered in southern Queensland, and it spread throughout south Queensland at an alarming rate, threatening the existence of the industry in that area. At the time, the industry was largely dependent on the clone NCo310, which was susceptible to the disease and highly favourable to the vector which bred in massive numbers. The control strategy was to use more resistant clones as parents in the breeding programme, and to discard clones from selection if they showed more than a little disease. Inheritance studies conducted by Hogarth *et al.*[175] showed that there was a high degree of genetic determination on a family basis, and indeed breeding was successful. Resistant progeny in the selection population increased from about 20% in the early 1970s to 80% by the early 1980s. Resistant clones with lower productivity were released to bring the disease under control, but these have now been replaced by higher yielding, resistant clones[155]. In the late 1970s, there were millions of infected stools in the Bundaberg region, but no infected stools have been found for several years, which has been an outstanding achievement for sugarcane breeding.

Croft & Magarey[176] identified a soil-borne pathogen *Pachymetra chaunorhiza*, which attacks the root system of sugarcane and causes losses estimated to be as high as 30%. Research by N. Berding & B. J. Croft (unpublished data) has shown that it is possible to breed and select resistant clones. However, the screening procedure requires the testing of clones in pots in temperature-controlled benches, which is time consuming and expensive, so it is difficult to mount a major breeding and selection programme. Nevertheless, all clones in advanced stages of selection are tested for resistance to the disease, and more resistant clones are being released for commercial production.

A disease called yellow leaf syndrome may cause yield losses of up to 50% in susceptible clones in Brazil[177]. The disease has been detected in many cane growing countries. At present, the exact nature of the pathogen has not been determined, although it is suspected to be a luteovirus[178] or a phytoplasma[179]. Symptoms are more apparent after a period of stress[177]. However, there is evidence from Hawaii and Brazil that there is genetic variability for resistance to the disease, and it may be possible to find a breeding solution[177].

The other major disease, for which breeding and selection is conducted, is sugarcane mosaic virus, which is mainly a disease of the subtropics. Breeding for resistance is complicated by the presence of a large number of strains, but there are good levels of resistance to mosaic in *S. spontaneum* clones[31]. Several of these clones have been incorporated into the Louisiana breeding programme with promising results.

Pest resistance

Compared with sugarcane diseases, relatively little work has been done on breeding clones with resistance to pests, particularly insect pests. It has been more difficult to determine which characters to assess and how to relate these characters to genetic resistance. For example, root-feeding insects may cause damage to roots but may have relatively little effect on final yield. As for diseases, the simplest strategy is to select in the presence of the pest.

In South Africa, the eldana borer *Eldana saccharicida* Walker is a significant pest of sugarcane, and a programme for selecting for resistance has commenced. Bond[180] reported that there were differences in susceptibility between clones, and a high degree of genetic determination when clones were assessed in replicated trials. Some progress had been made in selecting resistant clones. Nuss[181] found that a large proportion of clones showed a tolerant or resistant reaction to eldana, indicating that resistance is common in the local germplasm. However, clonal reaction to eldana is highly variable across trials and within trials[181].

In Louisiana, the sugarcane borer *Diatraea saccharalis* is the only insect pest requiring routine spraying to prevent economic damage[182]. Because of the cost of insecticides and environmental concerns, plant resistance has become more attractive as a method for controlling the pest. Resistance does exist, but it is complex because economic damage results from an accumulation of damage over a relatively long growing period. White[182] developed a method of cluster analysis for identifying borer resistance in an unselected population of sugarcane clones.

Stem borers *Chilo* spp. are a major pest of sugarcane in Asia and Africa. Pathak[183] reported that there is no information on the genetics of resistance to this pest in sugarcane, but he speculated that host plant resistance offered the best method for control. Ashraf & Fatima[184] agreed that breeding may be the best method for control in view of the difficulty of controlling the insects with insecticides. They reported that genetic resistance does exist, but stated that there was no information on the mode of resistance.

In Australia, the most important insect pests of sugarcane are canegrubs that attack the roots of the plant. Traditionally, these pests have been controlled with insecticides, but there is now more emphasis on developing integrated pest management programmes for canegrubs[185]. They showed that there is variation in the tolerance of clones to canegrub feeding and antibiosis. They have studied two grub species and found a surprising repeatability of results across grub species, but these results need to be confirmed with more grub species. They have found no evidence of total resistance to canegrubs, but host plant tolerance would be a useful factor in an integrated pest management programme.

Weevil borer *Rhabdoscelus obscurus* Boisduval has become a significant pest of sugarcane in northern Australia since the widespread adoption of green cane harvesting. This borer is very difficult to control with insecticides, and the most promising method for control is to breed resistant cultivars. Berding[186] has conducted extensive tests in advanced selection trials and in parent collections, and has found a good range of reactions. Thus, it should be possible to make significant progress with selection but, as the genetics of resistance is not yet understood, it is not known how successful breeding is likely to be.

Milling characteristics

There is relatively little published information on breeding for milling charactersitics. The major character of interest is cane fibre, and mills are interested in both quantity and quality of fibre. Brown *et al.*[90] found that fibre content had a heritability of 0.34 on an individual seedling basis, while Hogarth & Cross[187] obtained a value of 0.45. Hogarth & Cross[187] also found that 80% of the genetic variance for fibre content was additive, indicating that progress could be made by breeding for the character. Most breeders try to minimise fibre content, because mills can have difficulties in disposing of excess bagasse. However, if fibre is found to have a value (e.g. for cogeneration of electricity), breeders may be required to breed for high fibre; most sugarcane breeders would regard this as a relatively easy selection criterion, and progress should be rapid.

Hogarth & Cross[187] also studied two fibre quality parameters. The first was a measure of the energy consumed when a cane sample was macerated in a Jeffco cutter grinder. The second parameter was an impact test that measured the energy absorbed in shearing a 10-mm diameter core of cane internode, as described by Brotherton *et al.*[188]. Both characters had high heritability and were predominantly additive in inheritance, so that breeding for or against the characters is feasible.

Sugar quality

Breeding for sugar quality has not been a high priority for cane breeders, but this may change as more emphasis is placed on product quality in the future. Brown *et al.*[90] showed that starch had a heritability of 0.53 on an individual basis, so that good progress should be made from breeding for low starch. However, starch can now be controlled in the mill, so breeders do not need to be concerned about this character.

Hogarth & Kingston[189] studied the inheritance of ash in juice and found that 95% of the genetic variance was additive. The character also has a heritability of 0.46 on an individual basis, so progress from selection should be assured. Few, if any, breeding programmes are actively selecting for low ash in juice, but prospects for success are good.

Crop ideotype

Clones can have a marked impact on the quantity and quality of material delivered by harvesters, particularly those delivering stalks cut into billets, as operating in Queensland. Ridge & Dick[190] clearly demonstrated the variation among clones for percentage cane loss and extraneous matter. Variation for the latter occurred for levels before and after harvesting. This research was conducted in a test rig using a static harvester and simulated harvest of a lodged crop. They were able to predict cane loss with a regression equation that used percentage extraneous matter, stalk diameter, and stalk density as independent variables. In a broader survey, Linedale & Ridge[157] demonstrated the importance of optimising harvester performance. More importantly, they showed the relationship that existed between cane losses from harvesting and the extraneous matter content of the harvested product.

There has been almost no research on plant ideotype or architecture and its effect on the suitability of sugarcane for harvesting. There has been considerable research on trying to improve the performance of harvesters, to reduce cane loss and extraneous matter. Some progress has been made, but the increasing power of modern harvesters makes it very difficult to clean the cane supply effectively.

As engineering has failed to solve the problem, it is likely that clones are required that improve the presentation of the crop to the harvester.

An ideal clone for mechanical harvesting is one that remains erect, does not sucker, and has trash that does not cling to the stalk. Such clones are also most suitable for hand cutting. With mechanical harvesting, it had been assumed that such restrictions are not necessary but, in retrospect, this may have been optimistic. Selection for such ideal ideotypes will minimise progress from selection for productivity, as most high yielding clones are likely to lodge. There is also a suggestion that clones with strong ratooning characteristics are also more likely to sucker.

Row spacing

In Australia, sugarcane row spacings have increased over time to accommodate mechanisation. The predominant row spacing in Australia is about 1.5 m[191]. Such wide row spacings reduce crop yield potential because they limit interception of incident light energy during the early stages of growth[192]. Cultivars have been selected to tiller rapidly to form a closed canopy at wide row spacings. However, tiller initiation requires a diversion of photosynthate away from the primary stalks. This leads to increased competition for light, water and nutrients amongst stalks of the same stool, and a marked loss of young tillers at canopy closure[193].

One way to avoid this potential yield loss is to grow crops at high planting density so that the crop consists of mostly primary stalks that grow rapidly. These compete actively with weeds, avoid death of tillers near canopy closure, and exploit soil water and nutrient reserves more efficiently than conventionally planted crops. Yadav[194] suggested tiller mortality could be checked by adjusting row spacing to provide uniform light distribution in the canopy and by providing optimum soil moisture during the tillering phase. A 'ring planting system' yielded 184 t/ha, while parallel and triangular sett placement geometries yielded 143 and 137 t/ha, respectively. This research suggested that crop improvement would have to develop clones with synchronised early tillering, rapid initiation of tiller rooting, and correct canopy architecture for better light transmission.

Bull[193] and Bull & Bull[195] found that substantial yield increases could be achieved by using a range of clonal material, particularly unselected seedling clones. Bull & Bull[195] demonstrated in field trials that there was potential to improve crop yields by up to 100% at higher planting densities. Irrigation and fertilisation were supplied on a row basis to limit resource-induced yield constraints. They found that unselected, or even rejected, clones grown at close row spacing outperformed cultivars grown at 1.5 m by 80% 200 days after planting, and by 64% 300 days after planting. These unselected clones grown at close row spacing also outperformed cultivars at 0.5 m by 18% at 200 days after planting and by 21% 300 days after planting. These results indicated there is genetic variability for the capacity to respond to high planting densities.

These results illustrate the point that use of cultivars alone would not have identified this potential. This was revealed by use of clones that had failed to perform at the 1.5 m row spacing. Sugarcane cultivars are the result of 10–15 years of selection. Traits or processes genetically correlated with fitness to perform under the management regime imposed on the selection programme are favoured. Obviously, the results and subsequent interpretation of physiological or agronomic research may be unduly influenced if only such cultivars are used.

Achievements

Heinz[196], Moore[197] and Muchow *et al.*[198] have used record yields from different sources in considering achievements of sugarcane production, physiological bases for sugarcane improvement, and yield potential, respectively. The sugar record of 24.2 t/ha per year from 87 ha by Oahu Sugar Co Ltd, Waipahu[196] seems to have the soundest commercial basis. This, rather than a theoretical potential yield, seems to be the best figure for comparative use in crop-improvement discussions. Heinz[196] and Moore[197] compared the average Hawaiian industry yield relative to this record yield. They contrasted this with the position of other major crops in the USA, and suggested that sugarcane was placed relatively well. At just over 50%, Hawaiian sugarcane was further advanced than the other crops. The highest average sucrose yield for the Queensland industry of 12.0 t/ha for the 5-year period ending in 1995 compares favourably with this figure. The maximum sucrose yield for the Burdekin region of the Queensland industry of 17.4 t/ha for the period ending 1995, is almost 72% of this record value in Hawaii.

The above figure suggests that sugarcane is well placed in terms of the level of improvement. Sustained genetic improvement in a crop is exemplified by the US maize industry. In the 25 years to 1990, the improvement trend has been 164.5 L/ha/year (= 1.89 bushels/acre/year)[199]. In the 34 years from 1961 to 1994, this improvement has been 153.2 L/ha/year (1.76 bushels/acre/year; A. F. Troyer, personal communication). In isolation, the example for the Hawaiian industry also seems to be significant[153]. Here, sucrose production increased from 12.5 t/ha in 1908 to 30.12 t/ha in 1984, although these data obviously are not on an annual basis. Moore[197] highlighted the reality that the rate of improvement of sugarcane is low relative to other crops. This supported a similar contention by Berding & Skinner[154] who contrasted the rate of improvement of sugarcane with that found in maize and wheat.

The ultimate goal of all crop-improvement programmes is to produce continual genetic gains that are reflected in improvement in yields, or other characters, of a commercial product. However, demonstration that genetic gains are being achieved at the commercial level, and that any given improvement programme is, or is not, economically justifiable, is often difficult.

Genetic gain is usually defined as the change in mean performance of a population that is realised after each round of selection[200]. In commercial or industrial terms, the concept of genetic gain can have a much wider meaning. This may be defined as the change in the mean industry performance which occurs when a new cultivar is adopted. Hence, the estimation of commercial genetic gain is vital for assessing the economic value of a crop-improvement programme.

The precise estimation of genetic gain on an industry basis is often difficult because of the lack of suitable records[158,196,201]. Furthermore, a reliable comparison of the yields of old and new cultivars grown at the same time over large areas under similar commercial conditions (i.e. similar environments or crop class structure) can be difficult. Crop-improvement scientists are often faced with the daunting task of assessing the effectiveness of their improvement programmes in highly variable environments, the presence of strong G×E interactions, and few suitable genotypic records. The G×E interaction structure experienced commercially must be adequately and appropriately sampled in regional variety trials for improvement programmes to be effective, and efficient.

Australia

Productivity trends for the Queensland industry from 1961 to 1996 for cane yield, sucrose content and sucrose yield show interesting trends (Fig. 2.1). Cane yield increased until the end of the 1978 period, and then drifted to a low in the period ending 1991. This period was marked by expanding production onto less fertile land, and an increasing number of ratoons in the crop cycle. Since then, there has been a rapid increase in productivity. Sucrose content does not present a brilliant picture. Performance before 1972 was static, but declined to a low in 1990. This was a period commencing with mechanical harvesting and greatly increased use of nitrogenous fertilisers. Sucrose content has increased steeply since 1990. Sucrose yield shows an increase and then a decrease from 1980 to 1991. This was followed by a marked increase. The increase in all measures in the 1990s can largely be ascribed to the results of increased breeding activities over the whole industry. An example of progress in the central and southern regions of the Queensland industry is given by Cox & Hansen[155].

Bull *et al.*[156] found, for Bundaberg, that the correlation between predicted and realised gains for tonnes of sugar per hectare (TSH) was 0.66 ($P > 0.05$) and for commercial cane sugar (CCS, which is related to percentage sucrose) was 0.94 ($P < 0.01$). For TSH, G×E interaction is largely relative to the main effect of genotype, whereas for CCS the reverse is usually true[156]. However,

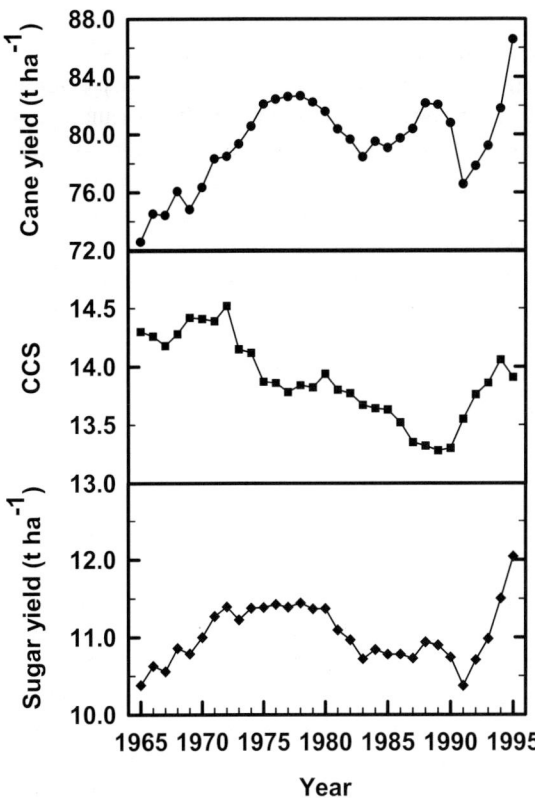

Fig. 2.1 Trends for cane yield (t/ha), commercial cane sugar (CSS) (%F.W.) and sucrose yield (t/ha), expressed as 5-year moving means for the Queensland industry 1965–95.

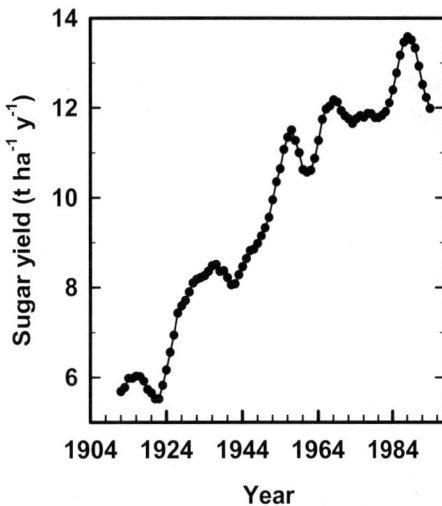

Fig. 2.2 Sucrose yield for the Hawaiian industry 1908–94 expressed as a 5-year moving mean.

for Tully the correlation between predicted and realised gains for TSH was 0.130 ($P > 0.05$) and for CCS was 0.604 ($P > 0.05$). Bull *et al.* [156] concluded that this difference in the correlations between trial and agricultural performance for Bundaberg (good correlation) and Tully (poor correlation) may be the result of differences in crop condition. In Bundaberg, cane rarely lodges, suckers (or water shoots) are few and small, and stool tipping is uncommon. These problems are common in Tully, and adversely affect the accuracy and repeatability of yield and sugar determinations.

Overall, these results suggest that some proportion of the genetic gains is being realised, and that the static (or negative) trend in productivity from 1971 to 1989 was probably the result of negative environmental factors.

Hawaii

Hawaiian production statistics were considered by Plucknett & Smith[153]. Productivity trends for sucrose yield for the Hawaiian industry to 1994 are presented as moving 5-year means (Fig. 2.2). Production per hectare per year in 1994 (12.0 t/ha) equalled that recorded in 1975 (11.9 t/ha), but had declined from the peak attained in 1988 (13.6 t/ha). Possible reasons suggested for this are the presence of yellow leaf syndrome in important commercial clones, suboptimal fertiliser usage, or a malaise besetting a declining industry. A significant reduction in the size and intensity of the crop-improvement programme must also be a possible reason.

FUTURE OPTIONS

Molecular-assisted selection

Lee[202] questioned whether crop improvement is at the dawn of the 'biology' era. He suggested that biotechnologies allowed more genotypes to be assessed in more environments, and that the new biotechnologies 'will require plant breeders and their colleagues to look within the plant and understand its architecture before routine, beneficial, and predictable advancements.' He considered that DNA markers should be regarded as the fundamental link between crop improvement and plant biology.

Moore[44] summarised the current status of application of molecular markers to sugarcane, and explained its apparently slow application. In a complex polyploid such as sugarcane, identification of traits coded by major genes is difficult, and most economically important traits are of a multigenic nature. Additionally, a valid assessment of phenotype is difficult and expensive because of the recognised large G×E interactions. Law[203] also cautioned that assessment cost and time will be a major limitation to development and use of market technology. Costs may be reduced by improvements in assessment technology, but the cost constraints imposed by phenotypic evaluation will limit marker applications to large improvement programmes. One possibility will be that application of marker technology itself may be automated, thus reducing costs and improving efficiency.

Moore[44] conceded that progress in identifying molecular markers for sugarcane applications was slow. The most promising application was an alternative approach for marker analysis called bulk segregant analysis. Msomi & Botha[204] assessed this technique for detecting polymorphisms in bulked DNA samples from individuals of extreme fibre values within a family. Eight of over 700 fragments amplified from the bulk samples were polymorphic. When DNA from 80 individuals in the population was amplified with the primers that had generated the polymorphisms in the bulk DNA, six were single dose fragments. Good progress has been made in converting these into sequence characterised amplified regions.

An excellent use of molecular markers was given in 1996 by Albert & Schenck[205]. The objective was detection of a foreign genome, sugarcane smut dikaryon DNA, within *Saccharum* spp. hybrid tissue, and not detection of portions of the host DNA. Sugarcane smut is an important disease, but symptom expression in clonal material inoculated in screening trials may take 2 years or more. This molecular screen is considered sensitive enough to detect the presence of the fungal genome in inoculated material in a much shorter time. Depending

on cost and speed of use of this molecular screen, this technology opens up the possibility of detecting fungal infection ahead of symptom expression. In effect, this is an example of the heritability of a screening test being maximised. If infection has occurred successfully, the result of the test is independent of pathogen × environment or pathogen × clone × environment interaction. The test also has possibilities for phytosanitary applications, e.g. for testing planting sources for pathogens.

Genetic transformation

The possibility of inserting a foreign gene into a plant, and obtaining successful expression of the gene, has tremendous appeal to crop-improvement scientists. The possibility of using such technology to rapidly correct defects in advanced selections, with minimal alteration to the host phenotype, is part of that appeal. An alternative use is to introduce *de novo* variation from other taxa, and exploit this by conventional hybridisation.

There already has been significant use of genetic transformation. Law[203] summarised annual worldwide approvals for genetically engineered plants. In the period 1986–1992, there was a total of 1257 approvals. He also listed current and future targets for this technology. These were classified under five categories: hybrid production, plant growth and development, altering inputs, altering products, and environment. Genetic transformation activities in sugarcane have been summarised by Moore[44].

There have been practical problems in development of gene-transfer systems for the family Poaceae (Gramineae) to which sugarcane, and the major cereal crops, belong. Three possible transformation systems can be used. The use of *Agrobacterium tumefaciens* Smith and Townsend to introduce and incorporate DNA into a plant genome, a method widely successful in dicotyledons, has been relatively unsuccessful in monocotyledons. The problems in this field, and the steps involved in transformation of monocotyledons, were considered by Smith & Hood[206]. Introduction of genetic material by electroporation of protoplasts has been successful, but there has been relatively little success in raising plants from protoplasts[207]. Until this can be overcome, this is not a viable path. Electroporation of intact cells also has been reported, with plants being raised[208].

The third option is to introduce DNA directly into the cell by force, using microprojectiles as carriers of DNA. Several tissue types, including apical meristem, embryogenic callus, and green callus, have been used as the target. Use of embryogenic callus seems to be the most successful. Bower & Birch[64] reported the successful production of transgenic sugarcane plants from bombardment of embryogenic callus. Birch and his co-workers have reported on further refinement of this technique[209]. The technology has been used to introduce a gene which detoxifies a phytotoxin produced by the xylem invading bacterium *Xanthomonas albilineans* (Ashby) Dowson causing leaf scald disease of sugarcane[210].

Details of transformation and screening processes were summarised by Moore[44]. While the techniques seem to be very successful in other crops, and the technology has been established in sugarcane, considerable fine tuning appears necessary. In the first option above, correction of a single defect requires that the host genome remains essentially unaltered except for the inserted gene. Problems with somoclonal variation arising from the culture phase to produce embryogenic callus are significant. If this variation occurs, it is a significant problem, as selection for agronomic desirability in transformed individuals cancels much of the perceived advantage. Minimisation of somoclonal variation from tissue culture is essential for any procedure used to produce target tissue for transformation[211]. In some instances, molecular analysis of the genomic stability of regenerants from a range of cell cultures revealed stability for sugarcane DNA after plant regeneration[212].

Two further aspects of this technology require consideration for full realisation of its potential. Appropriate genes must be identified, isolated, and placed in effective gene constructs. One concern is that the solution to certain problems, such as susceptibility to disease or insect infestation will, in effect, be single-gene solutions susceptible to breakdown under intense selection pressure. Widespread use of the technology based on a single gene approach will only enhance genetic vulnerability in a crop already inherently weak in this regard.

The application of genetic transformation to sugarcane improvement should not vary from the broad categories detailed by Law [203]. These are summarised below.

- Modification of resistance to disease or insect infestation.
- Modification of biochemical pathways that produce final products such as sucrose. Transformation of one or a small number of enzymes, resulting in overexpression to increase yield, is the objective of this strategy.
- Modification of biochemical pathways that produce metabolites detrimental to sucrose quality. Flavonoids, and components produced by polyphenol oxidase, are among such products that affect sugar colour. However, production of such metabolites may compromise essential plant functions, such as defence mechanisms.
- Modification of partitioning of the photosynthetic sink into sucrose or fibre. If the final proportions could be modified, or the temporal partitioning modified to affect ripening, final economic yield could be altered beneficially.
- The biomass bulk of sugarcane has appeal for use for the production of high value biochemicals. Novel genes that can divert a portion of the photosynthetic sink into alternative carbohydrates, for example, need only produce the final metabolite in micro-concentration. This assumes that extraction from the large volume of biomass is feasible and economically achievable from a high tonnage process stream. Examples are the transformation of high tonnage crops, such as potato and sugar beet, for biodegradable plastic production[213] and transformation of current oil seed crops for production of specialised industrial oils[214].

Functional genomics

Functional genomics is a relatively new branch of biotechnology, but it promises to have great significance in the future. It involves the sequencing of the genome of organisms. It requires high speed sequencing machines, automated by robotics, powerful computers with large data storage capability, and access to other large databases on the

World Wide Web. It culminates with laser powered scanners and new fluorescent dyes that can be read at different wavelengths. These technologies have been brought together to give scientists the capacity to do tens of thousands of experiments simultaneously, creating an explosion in information about how biological systems operate.

There are two different strategies for sequencing the genome of an organism:

(1) determine the sequence of all the DNA of the organism; and
(2) only sequence the messenger ribonucleic acid (mRNA).

For organisms with small genomes such as viruses, bacteria, fungi and plants such as *Arabidopsis*, whole genome sequencing is the preferred way. However, for higher organisms with large complex genomes such as sugarcane, mRNA sequencing is more productive and cost effective. This is because in higher organisms, much of the genome does not encode any proteins. Once sequence data are generated, they are checked against databases around the world to determine what the genes sequenced have homology to. Currently < 50% of the genes being sequenced have homology to genes with known function.

From the sequence data, the research can go in several directions. One is to study the expression patterns of the genes under different conditions. This is accomplished by arraying the DNA on to glass microscope slides or on to silicon chips, using robotics. On a single microscope slide, over 40 000 genes can be arrayed. Up to four different probes can then be used simultaneously to analyse the patterns of gene expression. For example, to study water stress in sugarcane, total mRNA would be isolated from a variety that is known to be tolerant to drought under stressed and non-stressed conditions. These two pools of mRNAs would be individually labelled with two different fluorescent labels. The same would be repeated for a variety known to be highly sensitive to water stress using two entirely different florescent labels. These four sets of labelled mRNA probes would simultaneously be hybridised to DNA arrays of the total genome of sugarcane. The different patterns of fluorescence are read using a confocal laser scanner, and the genes differentially expressed during water stress identified. By reference to the database, it is possible to determine what the identified genes encode for. Either the genes themselves or the patterns of expression in the arrays are now available for breeding programmes to rapidly screen potential new varieties for water stress tolerance.

Functional genomics is also being applied to evolution. Much of the diversity of life is believed to be brought about by combinations of single nucleotide polymorphisms (SNPs). A change in a single base pair in the DNA sequence results in a slightly different protein product from the encoded gene. In humans, it is estimated that one million SNPs are responsible for all the diversity, including susceptibility to many diseases. A similar phenomenon is probably true for plants. Using functional genomics to analyse these polymorphisms between varieties or closely related species will give biologists a much greater understanding of how plants adapt to different environmental conditions.

CONCLUSIONS

We stand on the threshold of a potential revolution in sugarcane improvement. This is available through the application of biotechnology, encompassing the fields of genetic transformation, and molecular markers. The basics for genetic transformation are in place, but further development is required for practical benefits to be delivered. Application of molecular marker technology gives us an exciting glimpse of the future. Analysis of genomic structure, identification of fragments of, or whole, chromosomes, measurement of genetic variability, and description of synteny over distant Gramineae taxa are now possible.

Most importantly, we must not forget that biotechnology is a tool and not a replacement for traditional plant improvement. There are many examples of plant improvement attaining major objectives through application of the empirical art and experience of crop-improvement traditionalists. Often these gains have been made with minimal science or technology. The benefits to be gained from a multidisciplinary and collaborative approach to sugarcane improvement are immense.

ACKNOWLEDGEMENTS

The authors thank Dr Grant Smith and Dr Stevens Brumbley for advice on the use of biotechnology in plant improvement programmes.

REFERENCES

1 Berding, N., Moore, P.H. & Smith, G.R. (1997) Advances in breeding technology for sugarcane. In: *Intensive Sugarcane Production: Meeting the Challenges Beyond 2000* (eds B.A. Keating & J.R. Wilson), pp. 189–220. CAB International, Wallingford, UK.

2 Hogarth, D.M., Cox, M.C. & Bull, J.K. (1997) Sugarcane improvement: past achievements and future prospects. In: *Crop Improvement for the 21st century* (ed. M.S. Kang), pp. 29–56. Research Signpost, Trivandrum, India.

3 Mukherjee, S.K. (1957) Origin and distribution of *Saccharum*. *Botanical Gazette*, **119**, 55–61.

4 Daniels, J., Smith, P., Paton, N. & Williams, C.A. (1975) The origin of the genus *Saccharum*. *International Society Sugar Cane Technologists' Sugarcane Breeders' Newsletter*, **36**, 24–39.

5 Daniels, J. & Roach, B.T. (1987) Taxonomy and evolution. In: *Sugarcane Improvement through Breeding* (ed. D. J. Heinz), pp. 7–84. Elsevier, Amsterdam.

6 Harlan, J.R. & de Wet, J.M.J. (1971) Towards a rational classification of cultivated plants. *Taxon*, **20**, 509–517.

7 Burner, D.M. & Webster, R.D. (1994) Cytological studies on North American species of *Saccharum* (Poaceae: Andropogoneae). *Sida*, **16**, 233–244.

8 Webster, R.D. & Shaw, R.B. (1995) Taxonomy of the native North American species of *Saccharum* (Poaceae: Andropogoneae). *Sida*, **16**, 551–580.

9 Berding, N. & Roach, B.T. (1987) Germplasm collection, maintenance, and use. In: *Sugarcane Improvement through Breeding* (ed. D. J. Heinz), pp. 143–210. Elsevier, Amsterdam.

10 Berding, N. & Koike, H. (1980) Germplasm conservation of the *Saccharum* complex: a collection from the Indonesian Archipelago. *Hawaiian Planters' Record*, **57**(7), 87–176.

11 Berding, N. & Skinner, J.C. (1980) Improvement in sugarcane fertility by modification of cross-pollination environment. *Crop Science*, **20**, 463–467.

12 Kandisami, P.A., Sreenivasan, T.V., Ramana Rao, T.C. et al. (1983) *Catalogue on Sugarcane Genetic Resources I. Saccharum spontaneum L.* Sugarcane Breeding Institute, Coimbatore.

13 Ramana Rao, T.C., Sreenivasan, T.V. & Palanichami, K. (1985) *Catalogue on Sugarcane Genetic Resources II. Saccharum barberi Jeswiet, Saccharum sinense Roxb. amend Jeswiet, Saccharum robustum Brandes et Jeswiet ex Grassl, Saccharum edule Hassk.* Sugarcane Breeding Institute, Coimbatore.

14 Sreenivasan, T.V. & Nair, N.V. (1991) *Catalogue on Sugarcane Genetic Resources – III Saccharum officinarum L.* Sugarcane Breeding Institute, Coimbatore.

15 Tai, P.Y.P., Miller, J.D. & Legendre, B.L. (1996) Evaluation of the World Collection of *Saccharum spontaneum* L. *Proceedings of the International Society of Sugar Cane Technologists* 22, Vol. 2, 250–260.

16 Gallacher, D.J. (1994) *Development of a minimum descriptor set for individuals of Saccharum spp. hybrid germplasm.* PhD thesis, James Cook University, Townsville.

17 Eksomtramage, T., Paulet, F., Noyer, J.L., Feldmann, P. & Glaszmann, J.C. (1992) Utility of isozymes in sugarcane breeding. *Sugar Cane*, 14–21.

18 Gallacher, D.J., Lee, D.J. & Berding, N. (1995) Use of isozyme phenotypes for rapid discrimination among sugarcane clones. *Australian Journal of Agricultural Research*, **46**, 601–609.

19 Burnquist, W.L. (1991) *Development and application of restriction fragment length polymorphism technology in sugarcane (Saccharum spp.) breeding.* PhD thesis, Cornell University, New York.

20 Harvey, M., Huckett, B.I. & Botha, F.C. (1994) Use of polymerase chain reaction (PCR) and random amplification of polymorphic DNAs (RAPDs) for the determination of genetic distances between 21 sugarcane varieties. *Proceedings of the South African Sugar Technologists' Association*, **68**, 36–40.

21 Smith, J.S.C. & Smith, O.S. (1991) Restriction fragment length polymorphisms can differentiate among U.S. maize hybrids. *Crop Science*, **31**, 893–899.

22 Simmonds, N.W. (1993) Introgression and incorporation: strategies for the use of crop genetic resources. *Biological Reviews*, **68**, 539–562.

23 Chen, Y.-C. & Lo, C.C. (1988) Disease resistance and sugar content in *Saccharum–Miscanthus* hybrids. *Report Taiwan Sugar Research Institute*, **132**, 1–7.

24 Hemaprabha, G. & Ram, B. (1993) Genetic variability in nobilization stages of *Saccharum* robustum Brandes et Jeswiet ex Grassl. *Sugar Cane*, **6**, 6–9.

25 Lee, D.J. (1995) *Enhancement of Saccharum spp. hybrid material by introgression with Erianthus arundinaceus germplasm.* PhD thesis, James Cook University, Townsville.

26 D'Hont, A., Rao, P.S., Feldmann, P., Grivet, L., Islam-Faridi, N., Taylor, P. & Glaszmann, J.C. (1995) Identification and characterisation of sugarcane intergeneric hybrids, *Saccharum officinarum* × *Erianthus arundinaceus*, with molecular markers and DNA in situ hybridisation. *Theoretical and Applied Genetics*, **91**, 320–326.

27 Besse, P., McIntyre, C.L. & Berding, N. (1996) Ribosomal DNA variations in *Erianthus*, a wild sugarcane relative (Andropogoneae–Saccharinae). *Theoretical and Applied Genetics*, **92**, 733–743.

28 Besse, P., McIntyre, C.L. & Berding, N. (1997) Characterisation of *Erianthus* sect. *Ripidium* and *Saccharum* germplasm (Andropogoneae – Saccharinae) using RFLP markers. *Euphytica*, **93**(3), 283–292.

29 Burner, D.M. & Legendre, B.L. (1993) Sugarcane genome amplification for the subtropics: a twenty year effort. *Sugar Cane*, 5–10.

30 Hsu, S.-Y., Lo, C.C. & Shih, S.C. (1988) Use of a *Saccharum spontaneum* derivative in sugarcane crossing. *Report*

Taiwan Sugar Research Institute, 1–8.

31 Walker, D.I.T. (1987) Breeding for disease resistance. In: *Sugarcane Improvement through Breeding* (ed. D. J. Heinz), pp. 455–502. Elsevier, Amsterdam.

32 Tai, P.Y.P., Gan, H., He, H. & Miller, J.D. (1991) Phenotypic characteristics of F_2 and BC_1 progenies from sugarcane intergeneric crosses. *Journal of the American Society Sugar Cane Technologists*, **11**, 38–47.

33 Tai, P.Y.P., He, H., Gan, H. & Miller, J.D. (1991) Flowering of hybrids from commercial sugarcane × *Saccharum spontaneum* crosses. *Journal of the American Society Sugar Cane Technologists*, **11**, 69–74.

34 Tai, P.Y.P., He, H., Gan, H. & Miller, J.D. (1992) Variation for juice quality and fibre content in crosses between commercial sugarcane and *Saccharum spontaneum*. *Journal of the American Society Sugar Cane Technologists*, **12**, 47–57.

35 Burner, D.M., Grisham, M.P. & Legendre, B.L. (1993) Resistance of sugarcane relatives injected with *Ustilago scitaminea*. *Plant Disease*, **77**, 1221–1223.

36 Jackson, P.A. & Roach, B.T. (1994) Performance of sugar-cane progeny from crosses between clones derived from diverse *S. spontaneum* sources. *Tropical Agriculture (Trinidad)*, **71**, 57–61.

37 Jackson, P. (1994) Genetic relationships between attributes in sugarcane clones closely related to *Saccharum spontaneum*. *Euphytica*, **79**, 101–108.

38 Sreenivasan, T.V., Ahloowalia, B.S. & Heinz, D.J. (1987) Cytogenetics. In: *Sugarcane Improvement through Breeding* (ed. D.J. Heinz), pp. 211–253. Elsevier, Amsterdam.

39 Heinz, D.J. (1991) Sugarcane cytogenetics. In: *Chromosome Engineering in Plants: Genetics, Breeding, Evolution* (eds T. Tsuchiya & P.K. Gupta), pp. 279–293. Elsevier, Amsterdam.

40 Burner, D.M. (1991) Cytogenetic analysis of sugarcane relatives (Andropogoneae: Saccharinae). *Euphytica*, **54**, 125–133.

41 Burner, D.M. & Legendre, B.L. (1993) Chromosome transmission and meiotic stability of sugarcane (*Saccharum* spp.) hybrid derivatives. *Crop Science*, **33**, 600–606.

42 D'Hont, A., Grivet, L., Feldmann, P., Rao, S., Berding, N. & Glaszmann, J.C. (1996) Characterisation of the double genomic structure of modern sugarcane cultivars, *Saccharum* spp., by molecular cytogenetics. *Molecular General Genetics*, **250**, 405–413.

43 Fukui, K., Ohimido, N., Ha, S. & Moore, P.H. (1994) Analysis and utility of chromosome information. 67 Complete identification of wild sugarcane chromosomes. *Japanese Journal Breeding*, **44** (Suppl. 2), 29.

44 Moore, P.H. (1996) Progress in sugarcane molecular biology. *Proceedings of the International Society of Sugar Cane Technologists* 22, Vol. 2, 353–362.

45 D'Hont, A., Grivet, L., Lu, Y.H. *et al.* (1996) The genome of modern sugarcane varieties. *Proceedings of the International Society of Sugar Cane Technologists* 22, Vol. 2, 363–367.

46 Sorrells, M.E. (1992) Development and application of RFLPs in polyploids. *Crop Science* 32, 1086–1091.

47 Wu, K.K., Burnquist, W., Sorrells, M.E., Tew, T.L., Moore, T.L. & Tanksley, S.D. (1992) The detection and estimation of linkage in polyploids using single-dose restriction fragments. *Theoretical and Applied Genetics*, **83**, 294–300.

48 Da Silva, J., Sorrells, M.E., Burnquist, W.L. & Tanksley, S.D. (1993) RFLP linkage map and genome analysis of *Saccharum spontaneum*. *Genome*, **36**, 782–791.

49 Al-Janabi, S.M., Honeycutt, R.J., McClelland, M. & Sobral, B.W.S. (1993) A genetic linkage map of *Saccharum spontaneum* L. 'SES208'. *Genetics*, **134**, 1249–1260.

50 Da Silva, J., Honeycutt, R.J., Burnquist, W. *et al.* (1995) *Saccharum spontaneum* L. 'SES208' genetic linkage map combining RFLP- and PCR-based markers. *Molecular Breeding*, **1**, 165–179.

51 Lu, Y.-H., D'Hont, A., Walker, D.I.T., Rao, P.S., Feldmann, P. & Glaszmann, J.C. (1994) Relationships among ancestral species of sugarcane revealed with RFLP using single copy maize nuclear probes. *Euphytica*, **78**, 7–18.

52 Grivet, L., D'Hont, A., Dufour, P., Hamon, P., Roques, D. & Glaszmann, J.C. (1994) Comparative genome mapping of sugar cane with other species within the Andropogoneae tribe. *Heredity*, **73**, 500–508.

53 Jenkin, M.J., Reader, S.M., Purdie, K.A. & Miller, T.E. (1995) Detection of rDNA sites in sugarcane by FISH. *Chromosome Research*, **3**, 444–445.

54 D'Hont, A., Lu, Y.-H., Feldmann, P. & Glaszmann, J.C. (1993) Cytoplasmic diversity in sugarcane revealed by heterologous probes. *Sugar Cane*, 12–15.

55 Chang, Y.-S. & Lo, C.C. (1993) Genetic relationships among major sugarcane cultivars in Taiwan. *Report Taiwan Sugar Research Institute*, **140**, 1–10.

56 Chang, Y.-S. (1994) Pedigree analysis of genetic relationships among sugarcane cultivars ROC1, ROC10, and ROC16 and comparisons of their sugar content. *Report Taiwan Sugar Research Institute*, **144**, 1–9.

57 Chang, Y.-S. (1996) Implications of inbreeding coefficient and coancestry in a sugarcane breeding program. *Proceedings International Society Sugarcane Technologists* 22, Vol. 2, 307–311.

58 Lu, Y.-H., D'Hont, A., Paulet, F., Grivet, L., Arnaud, M. & Glaszmann, J.C. (1994) Molecular diversity and genome structure in modern sugarcane varieties. *Euphytica*, **78**, 217–226.

59 Tai, P.Y.P., Miller, J.D. & Legendre, B.L. (1994) Preservation of *Saccharum spontaneum* germplasm through storage of true seed. *Sugar Cane*, **6**, 3–8.

60 Dodds, J.H. (1991) *In Vitro Methods for Conservation of Plant Genetic Resources*. Chapman & Hall, London.

61 Taylor, P.W.J. & Dukic, S. (1993) Development of an *in vitro* culture technique for conservation of *Saccharum* spp. hybrid germplasm. *Plant Cell Tissue Organ Culture*, **34**, 217–222.

62 Paulet, F., Engelmann, F. & Glaszmann, J.C. (1993) Cryopreservation of apices of *in vitro* plantlets of sugarcane (*Saccharum* spp. hybrids) using encapsulation/deyhydration. *Plant Cell Reports*, **12**, 525–529.

63 Glimelius, K. (1993) Evolution and crop improvement. In: *Focused Plant Improvement: Towards Responsible and Sustainable Agriculture* (eds B. C. Imrie & J. B. Hacker), *Proceedings Tenth Australian Plant Breeding Conference*, April 1993, Vol. 1, pp. 32–36.

64 Bower, R. & Birch, R.G. (1992) Transgenic sugarcane plants via microprojectile bombardment. *Plant Journal,* **2,** 409–416.

65 Brett, P.G.C. (1951) Flowering and pollen fertility in relation to sugarcane breeding in Natal. *Proceedings International Society Sugarcane Technologists,* **7,** 43–46.

66 Bond, R.S. (1987) Controlled environments produce new varieties in Natal. *Sugar Cane,* 28–30.

67 Moore, P.H. (1985) *Saccharum.* In: *CRC Handbook on Flowering,* Vol. 4 (ed. A. H. Halevey), pp. 243–262. CRC Press, Boca Raton, FL.

68 Moore, P.H. (1987) Physiology and control of flowering. In: *Copersucar International Sugarcane Breeding Workshop,* pp. 102–127. Copersucar, Brazil.

69 Moore, P.H. & Nuss, K.J. (1987) Flowering and flower synchronization. In: *Sugarcane Improvement through Breeding* (ed. D. J. Heinz), pp. 273–311. Elsevier, Amsterdam.

70 Berding, N. (1995) Improving flowering of sugarcane for breeding: progress and prospects. *Proceedings of the Australian Society of Sugar Cane Technologists,* **17,** 162–171.

71 Skinner, J.C. (1959) Controlled pollination of sugarcane. *Bureau of Sugar Experiment Stations (Queensland) Technical Communication,* **1,** 7–20.

72 Skinner, J.C. (1972) Selection in sugarcane: a review. *Proceedings of the International Society of Sugar Cane Technologists,* **14,** 149–162.

73 Berding, N. & Moore, P.H. (1996) Towards optimized induction of flowering in sugarcane. In: *Sugarcane: Research towards Efficient and Sustainable Production* (eds J. R. Wilson, D. M. Hogarth, J. A. Campbell & A. L. Garside), pp. 44–46. CSIRO Division of Tropical Crops and Pastures, Brisbane.

74 Heinz, D.J. & Tew, T.L. (1987) Hybridization procedures. In: *Sugarcane Improvement through Breeding* (ed. D. J. Heinz), pp. 313–342. Elsevier, Amsterdam.

75 Ramdoyal, K., Badaloo, G. & Mangar, M. (1995) Effect on seed setting of potassium metabisulphite in sugarcane crossing solution. *Sugar Cane,* 3–7.

76 Tai, P.Y.P. (1988) Long-term storage of *Saccharum spontaneum* L. pollen at low temperature. *Sugar Cane,* 12–16.

77 Tai, P.Y.P. (1993) Low temperature preservation of F_1 pollen in crosses between noble or commercial sugarcane and *Saccharum spontaneum* L. *Sugar Cane,* **5,** 8–11.

78 Zheng, D., Lin, Y. & Tai, P.Y.P. (1993) Electrophoretic analysis of intergeneric hybrids of *Saccharum* × related genera. *Journal of the American Society Sugar Cane Technologists,* **13,** 46–53.

79 Hsu, S.-Y., Lin, C.-J. & Lo, C.C. (1991) Personal computer as an aid in sugarcane crossing program. *Report Taiwan Sugar Research Institute,* **132,** 1–11.

80 Wang, T.-H., Hour, A.-L., Hsu, S.-Y. & Lo, C.C. (1993) Data base management system for sugarcane breeding program. *Taiwan Sugar,* **40,** 8–15, 20–21.

81 Hogarth, D.M. & Skinner, J.C. (1987) Computerization of parental selection. In: *Copersucar International Sugarcane Breeding Workshop,* pp. 87–101. Copersucar, Brazil.

82 Chang, Y.-S. & Milligan S.B. (1992) Estimating the potential of sugarcane families to produce elite genotypes using univariate cross prediction methods. *Theoretical and Applied Genetics,* **84,** 662–671.

83 Chang, Y.-S. & Milligan S.B. (1992) Estimating the potential of sugarcane families to produce elite genotypes using bivariate prediction methods. *Theoretical and Applied Genetics,* **84,** 633–639.

84 Stringer, J.K., McRae, T.A. & Cox, M.C. (1996) Best linear unbiased prediction as a method of estimating breeding value in sugarcane. In: *Sugarcane: Research towards Efficient and Sustainable Production* (eds J. R. Wilson, D. M. Hogarth, J. A. Campbell & A. L. Garside), pp. 39–41. CSIRO Division of Tropical Crops and Pastures, Brisbane.

85 Hogarth, D.M. & Skinner, J.C. (1986) Computerization of cane breeding records. *Proceedings of the International Society of Sugar Cane Technologists,* **19,** 478–491.

86 Breaux, R.D. & Miller, J.D. (1987) Seed handling, germination and seedling propagation. In: *Sugarcane Improvement through Breeding* (ed. D. J. Heinz), pp. 385–407. Elsevier, Amsterdam.

87 Cromarty, A.S., Ellis, R.H. & Roberts, E.H. (1982) *The design of seed storage facilities for genetic conservation.* IBPGR Secretariat, Rome.

88 Skinner, J.C., Hogarth, D.M. & Wu, K.K. (1987) Selection methods, criteria, and indices. In: *Sugarcane Improvement through Breeding* (ed. D. J. Heinz), pp. 409–453. Elsevier, Amsterdam.

89 Hogarth, D.M. (1987) Genetics of sugarcane. In: *Sugarcane Improvement through Breeding* (ed. D. J. Heinz), pp. 255–271. Elsevier, Amsterdam.

90 Brown, A.H.D., Daniels, J. & Latter, B.D.H. (1968) Quantitative genetics of sugarcane. I. Analysis of variation in a commercial hybrid sugarcane population. *Theoretical and Applied Genetics,* **38,** 361–369.

91 Brown, A.H.D., Daniels, J. & Latter, B.D.H. (1969) Quantitative genetics of sugarcane. II. Correlation analysis of continuous characters in relation to hybrid sugarcane breeding. *Theoretical and Applied Genetics,* **39,** 1–10.

92 Hogarth, D.M. (1971) Quantitative inheritance studies in sugarcane. I. Estimation of variance components. *Australian Journal of Agricultural Research,* **22,** 93–102.

93 Hogarth, D.M. (1971) Quantitative inheritance studies in sugarcane. II. Correlations and predicted responses to selection. *Australian Journal of Agricultural Research,* **22,** 103–109.

94 Hogarth, D.M. (1977) Quantitative inheritance studies in sugarcane. III. The effect of competition and violation of genetic assumptions on estimation of genetic variance components. *Australian Journal of Agricultural Research,* **28,** 257–268.

95 Hogarth, D.M., Wu, K.K. & Heinz, D.J. (1981) Estimating genetic variance in sugarcane using a factorial cross design. *Crop Science,* **21,** 21–25.

96 Kang, M.S., Miller, J.D. & Tai, P.Y.P. (1983) Genetic and phenotypic path analyses and heritability in sugarcane. *Crop Science,* **23,** 643–647.

97 Hogarth, D.M. & Mullins, R.T. (1989) Changes in the BSES plant improvement program. *Proceedings of the International Society of Sugar Cane Technologists,* **20,**

956–961.

98 Sukarso, G. (1986) Assessment of family selection on original seedlings of sugarcane at Pasuruan. *Proceedings of the International Society of Sugar Cane Technologists*, **19**, 440–446.

99 Ortiz, R. & Caballero, A. (1989) Effectiveness of early sugarcane selection procedures in Cuba. *Proceedings of the International Society of Sugar Cane Technologists*, **20**, 932–937.

100 Bond, R.S. (1989) Observations on family selection in the Mount Edgecombe, Durban sugarcane breeding programme. *Proceedings of the South African Sugar Cane Technologists' Association*, **63**, 132–135.

101 Wu, K.K. & Tew, T.L. (1989) Evaluation of sugarcane crosses by family yields. *Proceedings of the International Society of Sugar Cane Technologists*, **20**, 925–931.

102 Tai, P.Y.P. & Miller, J.D. (1989) Family performance at early stages of selection and frequency of superior clones from crosses among Canal Point cultivars of sugarcane. *Journal of the American Society Sugar Cane Technologists*, **9**, 62–70.

103 Jackson, P.A., Bull, J.K. & McRae, T.A. (1996) The role of family selection in sugarcane breeding programs and the effect of genotype × environment interactions. *Proceedings of the International Society of Sugar Cane Technologists*, 22, Vol. 2, 261–270.

104 McRae, T.A., Hogarth, D.M., Foreman, J.W. & Braithwaite, M.J. (1993) Selection of sugarcane seedling families in the Burdekin district. In: *Focused Plant Improvement: Towards Responsible and Sustainable Agriculture* (eds B. C. Imrie & J. B. Hacker), *Proceedings Tenth Australian Plant Breeding Conference*, April 1993, Vol. 1, pp. 77–82, Gold Coast, Australia.

105 Cox, M.C., McRae, T.A., Bull, J.K. & Hogarth, D.M. (1996) Family selection improves the efficiency and effectiveness of as sugarcane improvement program. In: *Sugarcane: Research towards Efficient and Sustainable Production* (eds J. R. Wilson, D. M. Hogarth, J. A. Campbell & A. L. Garside), pp. 42–43. CSIRO Division of Tropical Crops and Pastures, Brisbane.

106 Cox. M.C. & Hogarth, D.M. (1993) The effectiveness of family selection in early stages of a sugarcane improvement program. In: *Focused Plant Improvement: Towards Responsible and Sustainable Agriculture* (eds B. C. Imrie & J. B. Hacker), *Proceedings Tenth Australian Plant Breeding Conference*, April 1993, Vol. 2, pp. 53–54, Gold Coast, Australia.

107 Simmonds, N.W. (1996) Family selection in plant breeding. *Euphytica*, **90**, 201–208.

108 Hogarth, D.M., Braithwaite, M.J. & Skinner, J.C. (1990) Selection of sugarcane families in the Burdekin district. *Proceedings of the Australian Society of Sugar Cane Technologists*, **12**, 99–104.

109 Hogarth, D.M. & Bull, J.K. (1990) The implications of genotype × environment interactions for evaluation of sugarcane families. I. Effect on selection. In: *Genotype-by-environment interaction and plant breeding* (ed. M. S. Kang), pp. 335–344. Louisiana State University, Baton Rouge, LA.

110 Bull, J.K., Hogarth, D.M. & Basford, K.E. (1992) Impact of genotype × environment interaction on response to selection in sugarcane. *Australian Journal Experimental Agriculture*, **32**, 731–737.

111 Jackson, P.A., McRae, T.A. & Hogarth, D.M. (1995) Selection of sugarcane families across variable environments I. Sources of variation and an optimal selection index. *Field Crops Research*, **43**, 109–118.

112 McRae, T.A. & Jackson, P.A. (1995) Selection of sugarcane families for the Burdekin River Irrigation area. *Proceedings of the Australian Society of Sugar Cane Technologists*, **17**, 134–141.

113 Jackson, P.A., McRae, T.A. & Hogarth, D.M. (1995) Selection of sugarcane families across variable environments. II. Patterns of response and association with environmental factors. *Field Crops Research*, **43**, 119–130.

114 Jones, P.N., Ferraris, R. & Chapman, L.S. (1993) A technique for minimizing confounding of genotype × year and genotype × crop type effects in sugarcane. *Euphytica*, **67**, 199–204.

115 Pollock, J.S. (1979) *Variety–environment interaction and selection of sugarcane*. MAgrSc thesis, University of Queensland, Australia.

116 Espinosa, R. & Galvez, G. (1980) Study of genotype × environment interaction in sugarcane. 1: The interaction of genotypes with plant dates and harvest cycles. *Proceedings of the International Society of Sugar Cane Technologists*, **17**, 1161–1167.

117 Kang, M.S. & Miller, J.D. (1984) Genotype × environment interactions for cane and sugar yield and their implications in sugar cane breeding. *Crop Science*, **24**, 435–440.

118 Tai, P.Y.P. & Miller, J.D. (1986) Genotype by environment interaction for cold tolerance in sugar cane. *Proceedings of the International Society of Sugar Cane Technologists*, **19**, 454–462.

119 Milligan, S.B., Gravois, K.A., Bischoff, K.P. & Martin, F.A. (1990) Crop effects on broad-sense heritabilities and genetic variances of sugarcane yield components. *Crop Science*, **30**, 344–349.

120 Jackson, P.A., Horsley, D., Foreman, J., Hogarth, D.M. & Wood, A.W. (1991) Genotype × environment (GE) interactions in sugarcane variety trials in the Herbert. *Proceedings of the Australian Society of Sugar Cane Technologists*, **13**, 103–109.

121 Jackson, P.A. & Hogarth, D.M. (1992) Genotype × environment interactions in sugarcane. I. Patterns of response across sites and crop-years in North Queensland. *Australian Journal of Agricultural Research*, **43**, 1447–1459.

122 Jackson, P.A. (1992) Genotype × environment interactions in sugarcane. II. Use of performance in plant cane as an indirect selection criterion for performance in ratoon crops. *Australian Journal of Agricultural Research*, **43**, 1461–1470.

123 Mirzawan, P.D.N., Cooper, M. & Hogarth, D.M. (1993) The magnitude of genotype by environment interactions for cane yield, sugar yield and ccs in southern Queensland and their impact on selection. In: *Focused Plant Improvement: Towards Responsible and Sustainable Agriculture* (eds B. C. Imrie & J. B. Hacker), *Proceedings*

Tenth Australian Plant Breeding Conference, April 1993, Vol. 1, pp. 57–61, Gold Coast, Australia.

124 Mirzawan, P.D.N, Cooper, M. & Hogarth, D.M. (1993) The impact of genotype × environment interactions for sugar yield on the use of indirect selection in southern Queensland. *Australian Journal Experimental Agriculture*, **33**, 629–638.

125 Mirzawan, P.D.N., Cooper, M., DeLacy, I.H. & Hogarth, D.M. (1994) Retrospective analysis of the relationships among test environments of the Southern Queensland sugarcane breeding programme. *Theoretical and Applied Genetics*, **88**, 707–716.

126 Bull, J.K., Basford, K.E., DeLacy, I.H. & Cooper, M. (1992) Classifying genotypic data from plant breeding trials: a preliminary investigation using repeated checks. *Theoretical and Applied Genetics*, **85**, 461–469.

127 Bull, J.K., Basford, K.E., DeLacy, I.H. & Cooper, M. (1993) Determining appropriate group numbers and composition for data sets containing repeated check cultivars. *Field Crops Research*, **31**, 369–383.

128 Bull, J.K., Basford, K.E., Cooper, M. & DeLacy, I.H. (1994) Enhanced interpretation of pattern analyses of environments: the use of blocks. *Field Crops Research*, **37**, 25–32.

129 Bull, J.K., Bull, T.A. & Cooper, M. (1996) The importance of water and nitrogen in generating clone by environment interaction in sugarcane: a preliminary investigation based on plant crop results. In: *Sugarcane: Research towards Efficient and Sustainable Production* (eds J. R. Wilson, D. M. Hogarth, J. A. Campbell & A. L. Garside), pp. 49–51. CSIRO Division of Tropical Crops and Pastures, Brisbane.

130 Cooper, M., Woodruff, D.R., Eisemann, R.L., Brennan, P.S. & DeLacy I.H. (1995) A selection strategy to accommodate genotype-by-environment interaction for grain yield of wheat: managed-environments for selection among genotypes. *Theoretical and Applied Genetics*, **90**, 492–502.

131 Simmonds, N.W. (1979) *Principles of Crop Improvement.* Longman, London.

132 Skinner, J.C. (1961) Sugarcane selection experiments. 2 Competition between varieties. *Bureau of Sugar Experiment Stations (Queensland) Technical Communication*, **1**, 1–26.

133 Skinner, J.C. & Hogarth, D.M. (1978) Efficiency of border rows in replicated sugar cane variety trials. *Euphytica*, **27**, 629–643.

134 McRae, T.A. & Jackson, P.A. (1998) Competition effects in selection trials. *Proceedings Australian Sugar Cane Technologists*, **20**, 154–161.

135 Matassa, V.J., Basford, K.E. & Jackson, P.A. (1999) Intergenotypic competition in single-row plots of sugarcane variety trials. *Proceedings Australian Sugar Cane Technologists*, **21**, 234–240.

136 McDonald, L.M. & Milligan, S.B. (1994) Field evaluation of check plot adjustments to control environmental heterogeneity in an unreplicated sugarcane trial. *Journal of the American Society Sugar Cane Technologists*, **14**, 40–52.

137 Bartlett, M.S. (1978) Nearest neighbour models in the analysis of field experiments. *Journal Royal Statistical Society B*, **40** (2), 147–174.

138 Weber, W.E. & Stam, P. (1988) On the optimum grid size in field experiments without replications. *Euphytica*, **39**, 237–247.

139 Wilkinson, G.N., Eckert, S.R., Hancock, T.W., Mayo, O., Rathgen, A.J. & Sparrow, D.H.B. (1983) A new statistical methodology for design and analysis of plant breeding and varietal field trials. *Proceedings 8th Australian Plant Breeding Conference*, pp. 59–65.

140 Baird, D. (1984) The use of nearest neighbour (NN) models in field trials. *Proceedings Agronomy Society of New Zealand*, **14**, 47–50.

141 Cullis, B.R., Gleeson, A.C. & Thomson, F.M. (1992) The response to selection of different procedures for the analysis of early generation variety trials. *Journal of Agricultural Science*, **118**, 141–148.

142 Matassa, V.J., Basford, K.E., Stringer, J.K. & Hogarth, D.M. (1998) The application of spatial analysis to sugarcane variety trials. *Proceedings of the Australian Society of Sugar Cane Technologists*, **20**, 162–168.

143 Shaw, D.V. & Hood, J.V. (1985) Maximising gain per effort by using clonal replicates in genetic tests. *Theoretical and Applied Genetics*, **71**, 392–399.

144 McRae, T.A., Bull, J.K., Robotham, B.G. & Sweetnam, R.C. (1996) Measuring sugar content in variety trials. In: *Sugarcane: Research towards Efficient and Sustainable Production* (eds J. R. Wilson, D. M. Hogarth, J. A. Campbell & A. L. Garside), pp. 55–56. CSIRO Division of Tropical Crops and Pastures, Brisbane.

145 McRae, T.A., Bull, J.K. & Robotham, B.G. (1998) Billet samples for measuring sugar content in variety trials. *Proceedings of the Australian Society of Sugar Cane Technologists*, **20**, 211–217.

146 Berding, N., Brotherton, G.A., le Brocq, D.G. & Skinner, J.C. (1989) Application of near infrared reflectance (NIR) spectroscopy to the analysis of sugarcane in clonal evaluation trials. *Proceedings of the Australian Society of Sugar Cane Technologists*, **11**, 8–15.

147 Berding, N., Brotherton, G.A., le Brocq, D.G. & Skinner, J.C. (1991) Near infrared reflectance spectroscopy for analysis of sugarcane from clonal evaluation trials. I. Fibrated cane. *Crop Science*, **31**, 1017–1023.

148 Berding, N., Brotherton, G.A., le Brocq, D.G. & Skinner, J.C. (1991) Near infrared reflectance spectroscopy for analysis of sugarcane from clonal evaluation trials. II. Expressed juice. *Crop Science*, **31**, 1024–1028.

149 Berding, N. & Brotherton, G.A. (1995) Analysis of fibrated sugarcane by near infrared reflectance spectroscopy. In: *Leaping Ahead with Near Infrared Spectroscopy* (eds G. D. Batten, P. C. Flinn, L. A. Welsh & A. B. Blakeney), pp. 199–203. Royal Australian Chemical Institute, Near Infrared Spectroscopy Group, Melbourne.

150 Berding, N. & Brotherton, G.A. (1996) Analysis of high-moisture material – fibrated sugarcane. In: *Near infrared spectroscopy: The Future* (eds A. M. C. Davies & P. Williams), *Proceedings 7th International Conference on Near Infrared Spectroscopy*, pp.648–654, Montreal.

151 Berding, N. & Brotherton, G.A. (1996) Analysis of samples from sugarcane evaluation trials by near infra-

red spectroscopy using a new at-line, large cassette presentation module. In: *Sugarcane: Research towards Efficient and Sustainable Production* (eds J. R. Wilson, D. M. Hogarth, J. A. Campbell & A. L. Garside), pp. 57–58. CSIRO Division of Tropical Crops and Pastures, Brisbane.

152 Hogarth, D.M. (1976) New varieties lift sugar production. *Producers Review*, **66** (10), 21–22.

153 Plucknett, D.L. & Smith, N.J.H. (1986) Sustaining agricultural yields. *BioScience*, **36**, 40–45.

154 Berding, N. & Skinner, J.C. (1987) Traditional breeding methods. In: *Copersucar International Sugarcane Breeding Workshop*, pp. 269–320. Copersucar, Brazil.

155 Cox, M.C. & Hansen, P.B. (1995) Productivity trends in southern and central regions and the impact of new varieties. *Proceedings of the Australian Society of Sugar Cane Technologists*, **17**, 1–7.

156 Bull, J.K., Mungomery, V.E. & Hogarth, D.M. (1993b) Realisation of genetic gain for yield in commercial production. In: *Focused Plant Improvement: Towards Responsible and Sustainable Agriculture* (eds B. C. Imrie & J. B. Hacker), *Proceedings Tenth Australian Plant Breeding Conference*, April 1993, Vol. 1, pp. 92–103.

157 Linedale, A.I. & Ridge, D.R. (1996) A successful campaign to minimise harvesting losses within the Queensland sugar industry. *Proceedings Australian Society of Sugar Cane Technologists*, **18**, 1–5.

158 Walker, D.I.T. & Simmonds, N.W. (1981) Comparisons of the performance of sugar-cane varieties in trials and in agriculture. *Experimental Agriculture*, **17**, 137–144.

159 Roach, B.T. & Daniels, J. (1987) A review of the origin and improvement of sugarcane. In: *Copersucar International Sugarcane Breeding Workshop*, pp. 1–31. Copersucar Technology Center, Piracicaba, Brasil.

160 Chapman, L.S. (1996) Increase in sugar yield potential from plant breeding from 1946 to 1994. In: *Sugarcane: Research Towards Efficient and Sustainable Production* (eds J. R. Wilson, D. M. Hogarth, J. A. Campbell & A. L. Garside), pp. 37–38. CSIRO Division of Tropical Crops and Pastures, Brisbane, Australia.

161 Breaux, R.D. (1984) Breeding to enhance sucrose content of sugarcane varieties in Louisiana. *Field Crops Research*, **9**, 59–67.

162 Breaux, R.D. (1987) Some breeding strategies with biparental and polycrosses. In: *Copersucar International Sugarcane Breeding Workshop*, pp. 71–85. Copersucar Technology Center, Piracicaba, Brasil.

163 Legendre, B.L. (1995) Potential for increasing sucrose content of sugarcane: an assessment of recurrent selection in Louisiana. *Sugar Cane*, 4–8.

164 Mariotti, J.A. (1977) Sugarcane clonal selection research in Argentina. A review of experimental results. *Proceedings of the International Society of Sugar Cane Technologists*, **16**, 121–136.

165 Cox, M.C., Hogarth, D.M. & Mullins, R.T. (1990) Clonal evaluation of early sugar content. *Proceedings of the Australian Society of Sugar Cane Technologists*, 251–255.

166 Cuenya, M.I. & Mariotti, J.A. (1996) Breeding sugarcane for early high-sugar contents in subtropical climates. *Proceedings of the International Society of Sugar Cane*

Technologists, **22**, 316–320.

167 Cox, M.C., Hogarth, D.M. & Hansen, P.B. (1994) Breeding for high early season sugar content in a sugarcane (*Saccharum* spp. hybrids) improvement program. *Australian Journal of Agricultural Research*, **45**, 1569–1575.

168 Berding, N., Skinner, J.C. & Ledger, P.E. (1984) A naturally-infected bench test for screening sugarcane clones against common rust (*Puccinia melanocephala* H. & P. Syd.). *Protection Ecology*, **6**, 101–114.

169 Hogarth, D.M., Ryan, C.C. & Taylor, P.W.J. (1993) Quantitative inheritance of rust resistance in sugarcane. *Field Crops Research*, **34**, 187–193.

170 Simmonds, N.W. (1994) Some speculative calculations on the dispersal of sugarcane smut disease. *Sugar Cane*, No. 1, 2–5.

171 Whittle, A.M. (1982) Yield loss in sugarcane due to culmicolous smut. *Tropical Agriculture (Trinidad)*, **59**, 239–242.

172 Wu, K.K., Heinz, D.J. & Meyer, H.K. (1983) Heritability of sugarcane smut resistance and correlation between smut grade and yield components. *Crop Science*, **23**, 54–56.

173 Wu, K.K., Heinz, D.J. & Hogarth, D.M. (1988) Association and heritability of sugarcane smut resistance to races A and B. *Theoretical and Applied Genetics*, **75**, 754–760.

174 Balance, M.C., Milanés Ramos, N. & Mesa López, J.M. (1996) Diallelic analysis of resistance of five sugarcane parents to *Ustilago scitaminea* Sydow. *Proceedings of the International Society of Sugar Cane Technologists*, **22**, 340–345.

175 Hogarth, D.M., Reimers, J.F., Ryan, C.C. & Taylor, P.W.J. (1993) Quantitative inheritance of Fiji disease resistance in sugarcane. *Field Crops Research*, **34**, 175–186.

176 Croft, B.J. & Magarey, R.C. (1989) A review of research into *Pachymetra* root rot, an important new fungal disease of sugarcane. *Proceedings of the International Society of Sugar Cane Technologists*, **20**, 686–694.

177 Comstock, J.C., Irvine, J.E. & Miller, J.D. (1994) Yellow leaf syndrome appears on the United States mainland. *Sugar Journal*, March 1994, 33–35.

178 Smith, G.R., Fraser, T.A., Braithwaite, K.S. & Harding, R.M. (1995) RT-PCR amplification of RNA from sugarcane with yellow leaf syndrome using luteovirus group-specific primers. In: *Proceedings 10th Biennial Australasian Plant Pathology Society Conference*, Lincoln, NZ, p. 84 [abstract].

179 Cronje, C.P.K., Tymon, A.M., Jones, P. & Bailey, R.A. (1998) Association of a phytoplasma with a yellow leaf syndrome of sugarcane in Africa. *Annals of Applied Biology*, **133**, 177–186.

180 Bond, R.S. (1988) Progress in selecting for eldana resistance. *Proceedings of the South African Sugar Cane Technologists' Association*, **62**, 129–133.

181 Nuss, K.J. (1991) Screening sugarcane varieties for resistance to eldana borer. *Proceedings of the South African Sugar Cane Technologists' Association*, **65**, 92–95.

182 White, W.H. (1993) Cluster analysis for assessing sugarcane borer resistance in sugarcane line trials. *Field Crops Research*, **33**, 159–168.

183 Pathak, R.S. (1990) Genetics of sorghum, maize, rice and

sugar-cane resistance to the cereal stem borer, *Chilo* spp. *Insect Science Applications*, **11**, 689–699.

184 Ashraf, M. & Fatima, B. (1990) Breeding for resistance to *Chilo* spp. in sugarcane. *Insect Science Applications*, **11**, 683–687.

185 Allsopp, P.G., McGhie, T.K., Cox, M.C. & Smith, G.R. (1996) Redesigning sugarcane for resistance to Australian canegrubs: a potential IPM component. *Integrated Pest Management Reviews*, **1**, 79–90.

186 Berding, N. (1996) Sugarcane weevil borer resistance: breeding strategy development using survey data. *Proceedings of the Australian Society of Sugar Cane Technologists*, **18**, 90–99.

187 Hogarth, D.M. & Cross, K.W.V. (1987) The inheritance of fibre characteristics in sugar cane. *Proceedings of the Australian Society of Sugar Cane Technologists*, **9**, 93–98.

188 Brotherton, G.A., Cross, K.W.V. & Stewart, P.N. (1986) Development of test methods to predict the handling characteristics of cane fibre during the milling process. *Proceedings of the Australian Society of Sugar Cane Technologists*, **8**, 17–24.

189 Hogarth, D.M. & Kingston, G. (1983) The inheritance of ash in juice from sugar cane. *Proceedings of the Australian Society of Sugar Cane Technologists*, **5**, 21–27.

190 Ridge, D.R. & Dick, R.G. (1988) Current research on green cane harvesting and dirt rejection by harvesters. *Proceedings of the Australian Society of Sugar Cane Technologists*, **10**, 19–25.

191 Ridge, D.R. & Hurney, A.P. (1994) A review of row spacing research in the Australian sugar industry. *Proceedings of the Australian Society of Sugar Cane Technologists*, **16**, 63–69.

192 Irvine, J.E., Richard, C.A., Garrison, D.D. *et al.* (1980) Sugarcane spacing III. Development of production techniques for narrow rows. *Proceedings of the International Society of Sugar Cane Technologists*, **17**, 368–375.

193 Bull, T.A. (1975) Row spacing and potential productivity in sugarcane. *Agronomy Journal*, **67**, 421–423.

194 Yadav, R.L. (1991) High population density management in sugarcane. *Proceedings Indian National Science Academy*, **3 & 4**, 175–182.

195 Bull, T.A. & Bull, J.K. (1996) Increasing sugarcane yields through higher planting density – preliminary results. In: *Sugarcane: Research towards Efficient and Sustainable Production* (eds J. R. Wilson, D. M. Hogarth, J. A. Campbell & A. L. Garside), pp. 166–168. CSIRO Division of Tropical Crops and Pastures, Brisbane.

196 Heinz, D.J. (1987) Sugarcane improvement: current productivity and future opportunities. In: *Copersucar International Sugarcane Breeding Workshop*, pp. 55–70. Copersucar, Brazil.

197 Moore, P.H. (1989) Physiological basis for varietal improvement in sugarcane. In: *Sugarcane Varietal Improvement* (eds K. Mohan Naidu, T. V. Sreenivasan & M. N. Premachandran), pp. 19–55. Sugarcane Breeding Institute, Coimbatore.

198 Muchow, R.C., Hammer, G.L. & Kingston, G. (1991) Assessing the potential yield of sugarcane. *Proceeding Australian Society Sugar Cane Technologists*, **13**, 146–151.

199 Troyer, A.F. (1990) A retrospective view of corn genetic resources. *Journal of Heredity*, **81**, 17–24.

200 Fehr, W.R. (1987) Heritability. In: *Principles of Cultivar Development*, Vol. 1, *Theory and Technique*, pp. 95–105. Macmillan, New York.

201 Specht, J.E. & Williams, J.H. (1984) Contribution of genetic technology to soybean productivity – retrospect and prospect. In: *Genetic Contributions to Yield Gains of Five Major Crop Plants* (ed. W. R. Fehr), pp. 49–74. Crop Science Society of America, Special Publication No. 7, Madison, Wisconsin, .

202 Lee, M. (1995) DNA markers and plant breeding programs. In: *Advances in Agronomy* (ed. D. L. Sparks), pp. 265–344. Academic Press, San Diego.

203 Law, C.N. (1995) Genetic manipulation in plant breeding – prospects and limitations. *Euphytica*, **85**, 1–12.

204 Msomi, N. & Botha, F.C. (1994) Identification of molecular markers linked to fibre using bulk segregant analysis. *Proceedings of the South African Sugar Technologists' Association*, **68**, 41–45.

205 Albert, H.H. & Schenck, S. (1996) PCR amplification from the bE mating-type gene as a sensitive assay for the presence of sugarcane smut (*Ustilago scitaminea*) DNA. *Plant Disease*, **80**, 1189–1192.

206 Smith, R.H. & Hood, E.E. (1995) *Agrobacterium tumefaciens* transformation of monocotyledons. *Crop Science*, **35**, 301–309.

207 Liu, M.C. (1994) A novel method of plant regeneration from suspension culture protoplasts of sugarcane. *Journal of Plant Physiology*, **143**, 753–755.

208 Arencibia, A., Molina, P.R., de la Riva, G. & Selman-Housein, G. (1995) Production of transgenic sugarcane (*Saccharum officinarum* L.) plants by intact cell electroporation. *Plant Cell Reports*, **14**, 305–309.

209 Birch, R.G., Bower, R., Elliott, A., Potier, B., Franks, T. & Cordeiro, G. (1996) Expression of foreign genes in sugarcane. *Proceedings of the International Society of Sugar Cane Technologists*, **22**, 368–373.

210 Zhang, L. & Birch, R.G. (1995) Genetic engineering of sugar cane for leaf scald phytotoxin and disease resistance. *Proceedings of the International Society of Sugar Cane Technologists*, **22** (2), 397–402.

211 Sweby, D.L., Huckett, B.I. & Botha, F.C. (1994) Minimising somaclonal variation in tissue cultures of sugarcane. *Proceedings of the South African Sugar Technologists' Association*, **68**, 46–50.

212 Chowdhury, M.K.U. & Vasil, I.K. (1993) Molecular analysis of plants regenerated from embryogenic cultures of hybrid sugarcane cultivars (*Saccharum* spp.). *Theoretical and Applied Genetics*, **86**, 181–188.

213 Nawrath, C., Poirier, Y. & Somerville, C. (1994) Targeting of the polyhydroxybutyrate biosynthetic pathway to the plastids of *Arabidopsis thaliana* results in high levels of polymer accumulation. *Proceedings of the National Academy Science of the United States of America*, **91**, 12760–12764.

214 Murphy, D.J. (1994) Transgenic plants – a future source of novel edible and industrial oils. *Lipid Technology*, 84–92.

Chapter 3
Diseases

R. A. Bailey

INTRODUCTION

Many serious outbreaks of diseases have occurred in sugarcane since the early days of commercial production. For example, Bourbon or Otaheite cane succumbed to a complex of diseases in Mauritius in the 1840s and to gumming disease in Brazil in 1869. The first outbreak of smut disease occurred in 'China' cane in South Africa in 1877. Since these early days, the control of diseases in order to minimise their economic impact has remained essential to successful sugarcane production. In 2000, there was an outbreak of orange rust in the widely grown variety Q124 in Queensland which devastated large parts of that sugar industry.

There are several reasons why sugarcane is relatively susceptible to outbreaks of damaging diseases. The crop is often grown over large, contiguous areas, which favours disease build-up and spread. Sugarcane is propagated vegetatively (i.e. through seedcane) and serious diseases are often systemic (occur within plant tissues) and can spread in seedcane. The crop is also effectively grown as a perennial in that crops are 'ratooned', often for many years, before fields are replanted. Thus systemic diseases can build up and re-occur when fields are ratooned.

Therefore in all major sugarcane industries, pre-emptive measures are taken to prevent or minimise disease outbreaks. This is achieved through variety resistance and by routine field control measures, such as attention to seedcane health and effective crop eradication before fields are replanted.

During the variety selection process, seedling varieties grown at breeding stations are often inoculated or exposed to high levels of serious local disease pathogens against which protection is specially required so that only new varieties with adequate resistance are released. Strict quarantine measures are often enforced, both to minimise the risk of diseases being introduced on imported varieties and to minimise spread among regions within an industry. The danger inherent in planting a disproportionately large area with one variety is often recognised, although short-term economic dictates often result in one variety dominating variety dispositions.

Because serious diseases are often systemic, when symptoms of a serious disease do appear, usually the first action is to rogue and destroy infected plants and to adopt procedures to produce 'clean' seedcane. Depending on the disease, this may involve treating seedcane with hot water and/ or fungicide immediately before planting. Meanwhile resistant varieties are sought.

Stevenson & Rands[1] prepared the first comprehensive, annotated list of fungi and bacteria associated with sugarcane and its products. In 1951, Martin[2] compiled the first checklist of sugarcane diseases of the world, and assigned the causal organism to each pathological disease and included a number of physiological diseases. Since then, the Pathology Section of the International Society of Sugar Cane Technologists (ISSCT) has assumed the responsibility of updating the international list of sugarcane pathogens and their distribution, and publishes this in the proceedings of the triennial congress. The authoritative work *Sugar-cane Diseases of the World*[3,4] was published in two volumes under the auspices of the ISSCT in the 1960s and is still a useful reference. Descriptions of major diseases were revised in *Sugarcane Diseases of the World – Major Diseases*[5], again co-ordinated by ISSCT.

A general publication, published in 2000 by CIRAD in association with ISSCT, is *A Guide to Sugarcane Diseases*[6]. This volume is a valuable source of up-to-date information on the causes, symptoms, and control measures for almost all sugarcane diseases, together with the most recent list of diseases occurring in different countries. It is profusely illustrated and is strongly recommended to all researchers and field staff interested in sugarcane diseases and their control. Also available from CIRAD (see References) is a companion interactive CD-ROM with the same title, which contains many further illustrations and permits the identification of diseases from the symptoms observed. The safe movement of sugarcane through quarantine, including methods for diagnosing pathogens and eliminating them from propagation material, is described by Frison & Putter[7].

Many countries and regions have published descriptions and lists of local cane diseases. References to these can be found in *Sugarcane Diseases of the World*[3,4] and *A Guide to Sugarcane Diseases*[6]. Only a few of the more recent ones are given in Table 3.1. For most regions producing sugarcane the list of local diseases that have been recorded is long; however, most are of little economic consequence. Invariably in most regions, only a few diseases are regarded as current hazards that require attention by those involved in sugarcane breeding and selection. There are even fewer diseases that require routine attention by cane managers and farmers. Major diseases of international distribution and importance include ratoon stunting disease (RSD), leaf scald, smut, red rot, common rust, and mosaic.

Yellow leaf syndrome (YLS) is widely distributed but its economic importance is uncertain and it is currently the subject of intensive research. Certain diseases of more limited distribution are locally important problems. These include white leaf disease in Thailand and, most recently, orange rust in Australia.

While most diseases are under satisfactory control in most cane industries, as a result of the on-going efforts by researchers to produce suitably resistance varieties, the possibility of a sudden outbreak of a damaging disease is ever-present. This can occur because of the spread of a disease to an area where it was previously absent. Examples of this include the outbreak of severe smut in the Caribbean in the 1970s[18] and in north-west Australia in 1998[19]. A further problem is the capacity for genetic change to more virulent strains by certain pathogens. This is the probable explanation for the outbreak of orange rust in Queensland in 2000[20], and it may explain the apparently sudden appearance of YLS in many countries in the 1990s. Constant vigilance and exchange of information is therefore necessary to minimise the risks from of sugarcane diseases. To this end, the regular meetings and co-operative activities of pathologists under the umbrella of the ISSCT serve a most valuable purpose.

PRINCIPLES OF DISEASE CONTROL IN SUGARCANE

The following section has been reprinted with amendments from *Sugarcane Diseases in South Africa*[17].

Table 3.1 Some regional descriptions of sugarcane diseases, after 1954.

Region	Reference	Title
Caribbean	Baker *et al.* (1954)[8]	Sugarcane diseases in the Caribbean
India	Chona (1956)[9]	Chairman's address, Pathology section ISSCT
Louisiana	Abbott (1963)[10]	Problems in sugar cane disease control in Louisiana
Mauritius	Wiehe (1963)[11]	The control of sugar cane diseases in Mauritius
Réunion	Horau (1967)[12]	Sugar cane diseases in Réunion island
Puerto Rico	Liu *et al.* (1967)[13]	Diseases of sugarcane in Puerto Rico
Fiji	Daniels *et al.* (1972)[14]	The control of sugarcane disease in Fiji
Brazil	Planalsuçar (1977)[15]	A guide to identification of sugarcane diseases and nutritional deficiencies in Brazil
Australia	Bureau of Sugar Experiment Stations (1991)[16]	Diseases of sugarcane
South Africa	SASA Experiment Station (2003)[17]	Sugarcane diseases in South Africa

Variety resistance

Disease control in sugarcane is mainly achieved by using resistant varieties. Frequent inspection of new varieties at various selection stages in the breeding programme, together with screening trials against some of the most important pathogens, are intended to eliminate susceptible varieties. This ensures that the varieties eventually released to growers have a high measure of general resistance to disease. However, the resistance of new varieties may not be permanent. New problems or the reappearance of diseases that were previously important can occur, and varieties once adequately resistant may not remain so under changing circumstances.

The breeding and selection of new varieties to meet the industry's requirements is a lengthy process. It also takes a long time for growers to replace existing varieties with ones that are more resistant to a disease, particularly if the susceptible variety is widely planted. For both these reasons a variety cannot always be rapidly withdrawn from production. Disease problems, therefore, must often be contained by other means, pending the eventual planting of resistant varieties. In the case of RSD, few varieties possesses adequate resistant or tolerance, and control of this disease depends mainly on methods other than varietal resistance.

The incidence of many diseases is related to specific environmental conditions. For example in South Africa, smut is most prevalent in the warmer, northern areas where susceptible varieties, although suitable elsewhere, are not recommended. Mosaic is most likely to occur in cooler, southern areas.

Growers should try all new, resistant varieties that become available to see if they will be useful under the growing conditions on their farms. Dependence on one dominant variety should be avoided where this is economically possible.

Seedcane quality

Most important sugarcane diseases, including RSD, smut, mosaic, leaf scald, and to some degree red rot, are systemic, that is they are present within the cane stalk. These diseases, therefore, can be spread by planting infected seedcane, they can persist in the stubble to recur after cutting and they also survive in volunteer regrowth to contaminate newly planted fields. Similar control measures are used to combat all of these systemic diseases.

The planting of healthy seedcane is essential for general disease control. Growers should establish 'nurseries' with heat-treated stock to provide healthy, high quality seedcane to meet their annual planting requirements. Hot-water treatment, at 50°C for two hours, is essential for control of RSD and eliminates several other diseases, including smut and chlorotic streak. Seedcane requirements should be estimated well in advance, so that adequate stocks can be produced. Seedcane fields must be inspected regularly to ensure that they remain free of disease and only the plant and first ratoon crops should be used as seedcane.

Field control practices

Healthy seedcane must be planted into fields that are free of volunteer regrowth from old stubble. If any volunteers present are diseased, much of the benefit of planting good seedcane will be lost. It is essential to destroy the old crop effectively and to prepare the land thoroughly so that volunteers are eliminated before replanting.

The inspection and roguing of cane fields to remove diseased plants can do much to contain diseases at a low level and these are recommended practices, particularly for smut. The periodic inspection of fields also gives early warning of new problems as they develop and enables action to be taken at the most appropriate time. The ploughing out of severely diseased fields also contributes greatly to reducing the amount of infective material.

The control of some important diseases, notably RSD, smut and mosaic, is most effective when control measures are applied in an integrated manner. Attention should be routinely paid to variety resistance, a mixed disposition of varieties, seedcane health effective eradication of old crops, and roguing (where applicable). This integrated approach greatly minimises the risk of these and other diseases reaching damaging levels.

SELECTION OF VARIETIES FOR DISEASE RESISTANCE

Resistance to many diseases is significantly heritable. This is taken into consideration when breeders choose parents for crossing to increase the numbers of resistant progeny. Over time, accumulated data on progeny reactions guide the choice of parental combinations. Such decision making is often computer-aided. In South Africa, parental choice has been effective in increasing general resistance to leaf scald, smut and mosaic. Following the outbreak of smut in north-west Australia in 1998, efforts are being made to reduce the risk that this disease presents to the Queensland industry (where smut does not yet occur) by increasing the numbers of resistant progeny[21].

The reactions of new progeny to the more important diseases in a region are determined at various stages during selection. The scheme followed in South Africa is given as an example (Fig. 3.1). This uses specialised trials running concurrently with the routine selection programme and using inoculation techniques or exposure to high levels of pathogens by 'spreader' plants. A new development in Florida concerning RSD is resistance screening using inoculation and subsequent assessment of reactions using a serological technique. This is reported to have increased the general resistance of released varieties to this important disease[22]. Useful selection against susceptibility to common and highly infectious foliar diseases, such as rust, often occurs by natural exposure of progeny in routine selection trials.

In this chapter only the more important diseases causing significant damage or regarded as potential hazards, either world-wide or in specific regions, are described. They have been grouped according to their causal agents as fungal, bacterial and viral. Within each group the diseases are arranged in alphabetical order of common name. The names of pathogens are those currently approved by the Pathology section of ISSCT.

FUNGAL DISEASES

Pineapple disease

- Perfect state: *Ceratocystis paradoxa* (Dade) Moreau.
- Imperfect state: *Thielaviopsis paradoxa* (de Seynes) von Höhnel.

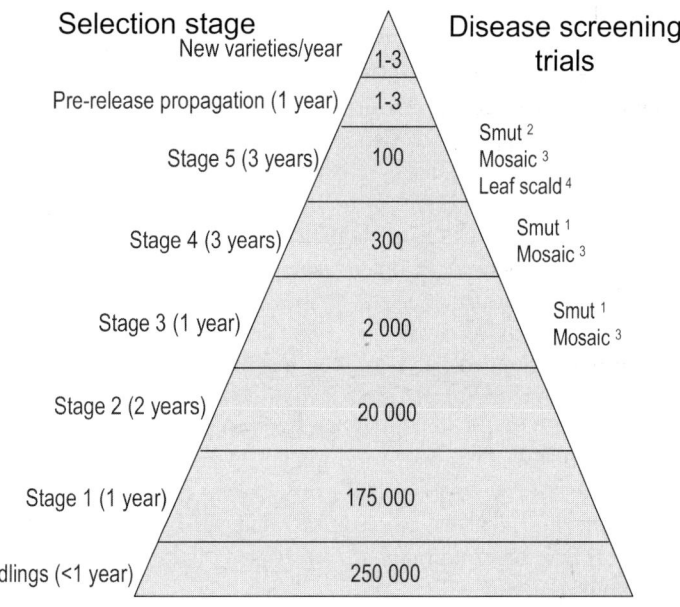

Fig. 3.1 Screening for resistance to sugarcane diseases during variety selection in South Africa. Number of varieties shown at each stage: 1, smut by inoculation; 2, smut by natural exposure; 3, mosaic by natural exposure with spreader plants; 4, leaf scald by inoculation.

A full account of pineapple disease has been given by Wismer & Bailey[23]. The causal agent, the soil-inhabiting fungus *Ceratocystis paradoxa*, infects the setts through the cut ends shortly after they have been planted and spreads rapidly through the parenchyma. The tissue first turns red and at this stage a distinctive smell similar to that of ripe pineapples may be noticed. Later, the interior of the infected setts becomes hollow and black because of the production of dark chlamydospores. Infected setts fail to germinate or the shoots die shortly after emergence, causing poor or patchy germination.

The presence of *C. paradoxa* has been reported in most sugarcane producing countries. Serious losses resulting from the failure of infected setts to germinate have been reported from many countries, but are most likely to occur when conditions for sett germination are poor, for example if soil temperature is low or the soil is excessively wet. Pineapple disease is rarely a problem when conditions for germination are good.

Control was first achieved by treating the setts with a fungicide at the time of planting. Organo-mercury compounds were found to be highly effective, but they are a serious environmental hazard because of their toxicity to mammals and they were eventually banned in most countries between the 1970s and the 1980s. The systemic fungicide benomyl (Benlate) was the first effective alternative to mercury-based compounds. The efficacy of benomyl was first reported by Wismer[24] and it soon became the seedcane treatment of choice at a rate of 150–500 µg/ml a.i. Benomyl is still widely used, both as a cold-water bath before seedcane is planted and as an addition to the hot water treatment for RSD control. Other fungicides, including guazatine (Panoctine), propiconazole (Tilt) and carbendazim + difenoconazole (Eria) have also been found to be effective.

Improving the prospects of good germination, for example by avoiding planting in cold soils or too deeply, should be the first approach to controlling pineapple disease. The use of a fungicide, either as a seedcane treatment or applied to the soil in the furrow during planting, is recommended whenever germination conditions are likely to be less than optimal.

Pokkah boeng

- Perfect state: *Gibberella fujikuroi* (Sawada) Wollenweber.
- Imperfect state: *Fusarium moniliforme* Sheldon.
- Perfect state: *Gibberella subglutinans* (Edwards) Nelson, Tousson & Marasas.
- Imperfect state *Fusarium subglutinans* (Wollenweber & Reinking) Nelson, Tousson & Marasas.

Pokkah boeng, sometimes referred to as Fusarium pokkah boeng, was first comprehensively described by Van Dillewijn[25]. It is induced by the fungi *Gibberella fujikuroi* and *G. subglutinans* and has been recorded in almost all countries in which cane is grown commercially. The name is derived from the Javanese for a malformed or twisted top.

Although symptoms of pokkah boeng are common, few cases of serious economic damage to commercial varieties have been reported. In Java, minor symptoms on 90% of stalks and losses of 38% were reported in POJ 2878[25]. In South Africa, serious damage to the newly released variety N34 caused its withdrawal from production in 2000.

Symptoms

As its name implies, the most obvious effect of pokkah boeng is to distort the cane tops. The earliest symptoms are seen on the young leaves, which become chlorotic towards the base, twisted, wrinkled, and narrower and shorter than normal leaves. Irregular reddish stripes may develop within the chlorotic areas. If the infection is limited to the leaves, the plants usually recover with little damage. However, infection may progress into the stalks, where internal and external ladder-like (knife-cut) lesions can develop. The most serious damage occurs when the fungus penetrates the growing-points, which can die, resulting in a top rot.

Infection is by airborne spores being washed between partially unfolded leaves to the base of the spindle during periods of rapid growth in hot conditions, followed by rain or irrigation. The spores then germinate and infect the young tissues of the spindle.

Control

Control is achieved only by planting resistant varieties. In Java, seedlings were tested for resistance to pokkah boeng by injecting a suspension of the conidia of *G. fujikuroi* into the leaf spindles 10 cm below the highest visible leaf joint[26]. At most breeding institutes, susceptible new varieties are discarded during the selection programme and this provides adequate control.

Red rot

- Perfect state: *Glomerella tucumanensis* (Speg.) Arx and Muller (formerly *Physalospora tucamanensis* Speg.).
- Imperfect state: *Colletotrichum falcatum* Went.

Red rot is one of the oldest and most serious diseases of sugarcane. It was first reported in Java by Went in 1893[27] and has been fully described by several authors, including Singh & Singh[28] and Singh & Lal[29]. Red rot occurs in most countries where sugarcane is grown and is often common. It may attack any part of the plant but is especially important as a stalk rot and a disease of setts.

Symptoms

The typical symptom of red rot is a distinctly red discoloration of the internal tissues of the stalk. Particularly in susceptible varieties, the reddening is often interspersed with white blotches to give a characteristically mottled appearance (Fig. 3.2). If the fungus has gained entry through the nodes, or via wounds or pest injury such as borer damage, rotting will begin at those points and extend through the stalk tissues. Infection can also spread from the base of the stalks. The rate of spread through the internodes and along the stalk depends on the susceptibility of the variety. External symptoms of ill-defined red-brown patches on the rind may also be present. In the late stages of the disease in susceptible varieties, the stalks dry out and shrink, appearing mummified.

The leaf canopy becomes yellow and eventually desiccates if stalk rotting is severe. Elongated lesions on the leaf midribs, caused by the red rot

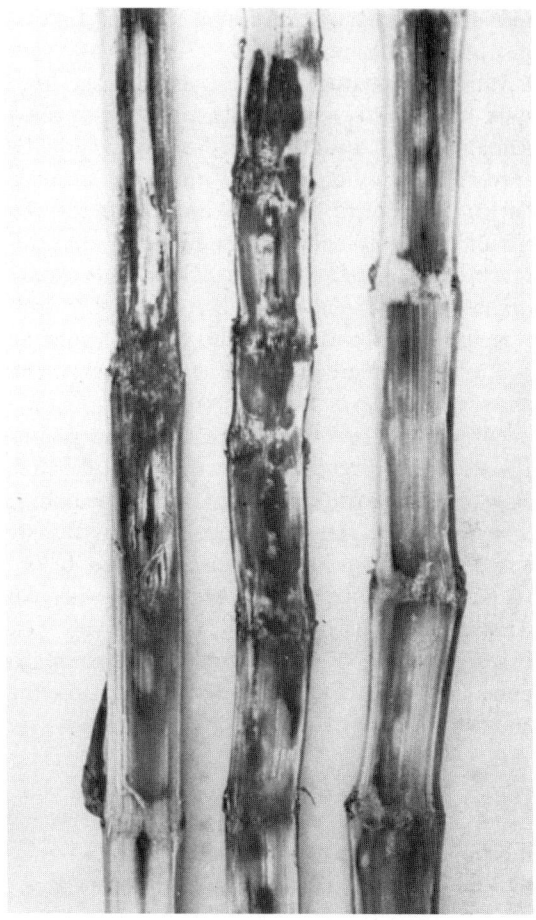

Fig. 3.2 Internal stalk symptoms of red rot. Source: South African Sugar Association Experiment Station.

pathogen following insect feeding, are commonly seen but are relatively unimportant.

Damage

Economic losses are caused by a reduction in stalk mass and a reduction in the sucrose content of infected stalks, and can be very severe. Severe losses have been reported from many countries. The epidemic of 1938–40 in India was disastrous when the most widely grown variety in the states of Uttar Pradesh and Bihar, Co213, was wiped out. Today, red rot is still regarded as the most important disease of sugarcane in India.

Wherever sugarcane is subject to damage from stalk boring insects there is a risk of associated damage from red rot because the wounds provide entry points for infection[30]. Surveys in South Africa have shown that approximately 50% of red rot infection was incited by the borers *Eldana saccharina* and *Sesamia calamistis*, which are common in the industry[31].

Red rot can also affect the establishment of young plant cane in subtropical countries where temperatures can be too low for quick germination but are suitable for the development of the fungus. If seedcane is infected at the time of planting, such as at the buds, rotting proceeds from these points. The buds on the setts are killed, or may germinate and then die, resulting in poor stands of cane. Red rot is an important disease of seedcane in Australia, Taiwan and Louisiana.

In relatively cool regions, red rot infection can penetrate the stubble and result in poor ratooning and 'gappy' stands[32].

Control

The most effective means of reducing or avoiding damage from red rot is by planting resistant varieties. Chemical control is not practical.

Two types of varietal resistance are recognised: morphological, associated with physical qualities of the stalk tissues that hamper the development of the fungus, such as a thick rind, and physiological. Physiological resistance is the more important. Most clones of *S. officinarum* are susceptible to the disease; as are those of *S. robustum* and *S. sinense*. Physiological resistance seems to be restricted to *S. spontaneum* and some clones of *S. barberi* and is governed by one or more genes. After inheritance studies, Azab & Chilton[33] suggested that in some commercial hybrids a dominant susceptible gene from *S. officinarum* masks the resistance genes from *S. spontaneum*.

There can be extensive variation in the virulence of different strains of the pathogen, more so than with most other sugarcane pathogens, hence the testing of new selections for resistance is an important aspect of control. Sugarcane varieties are screened for resistance by artificial inoculation with the pathogen. Virulent isolates from genetically different groups of the pathogen should be used. Plants that are 5–6 months old are inoculated at the nodes or in internodal tissues of standing stalks, which are evaluated later by examining the internal tissues. The methods used to test for resistant varieties in Louisiana were described by Abbott *et al.*[34] and those in India by Kar *et al.*[35].

Heat treatment eliminates infection from seedcane and is an important aspect of control in India.

In addition to the use of resistant varieties and healthy seedcane, the occurrence of red rot may be reduced by various agronomic practices. Appropriate planting times and soil conditions that favour germination will minimise rotting caused by seed-borne infection. Thorough eradication of old crops and even crop rotation may be necessary to manage the disease if stubble is severely infected.

Common rust *Puccinia melanocephala* Sydow and orange rust *P. kuehnii* Butler

There are two distinct rust diseases of sugarcane, common rust caused by *Puccinia melanocephala* and orange rust, caused by *P. kuehnii*. Serious outbreaks caused by these rust diseases since the 1980s have led to the recognition that both should now be included among the more important diseases of sugarcane. Common rust was fully described by Ryan & Egan[36] in 1989 and then by Raid & Comstock[37] in 2000. A full description of orange rust was given by Magarey *et al.*[20] in 2001.

Common rust is the more widely distributed of the two forms of rust disease. It occurs in almost all countries where sugarcane is grown and has caused serious economic damage to susceptible varieties in most regions. Common rust has occurred as a serious disease in the Old World for many decades, and its spread to the Caribbean and the Americas in the 1970s caused severe damage to crops in, for example, Cuba, which was then almost entirely reliant on the susceptible variety B4362, and Florida.

Orange rust is restricted to the Asian-Oceanic sugar industries and until recently was generally thought to be of minor significance. However, in 2000, a severe outbreak of orange rust devastated the predominant variety Q124 in Queensland and caused enormous losses in productivity over entire districts to the extent that a rapid programme of variety replacement was instituted. This epidemic was probably caused by the emergence of a new strain of the pathogen[20]. Strain variation is also a feature of common rust.

Symptoms

The two rust diseases seem to be superficially similar as both cause linear pustules that erupt through the lower surface of the leaf lamina to expose the urediniospores. The pustules of common rust are orange to a reddish-brown and later often turn dark-brown owing to the development of dark brown teliospores, whereas those of orange rust are initially orange and later orange-brown, but do not become dark in colour. The urediniospores can readily be distinguished by microscopic examination, those of orange rust being slightly larger and having an apical thickening.

When severe, the pustules of common rust may cover the entire lower surface of the leaf, whereas the pustules of orange rust tend to occur in clumps. In the later stages of disease development, both rusts can cause premature drying-off of infected leaves and can result in severe loss of photosynthetic leaf area. Common rust is most frequently seen in relatively young cane and as crops mature, symptoms become less obvious, so that it seems that the cane is 'growing out of the disease', although by this stage there can have been considerable suppression of growth. By contrast, symptoms of orange rust more commonly occur in semi-mature to mature cane.

Both species of rust pathogen are favoured by prolonged spells of humid, relatively cool weather. Because it is most common in more mature cane, orange rust tends to occur in autumn, whereas common rust can occur in young cane in both spring (after early season harvesting) and autumn (after late season harvesting).

Damage

Both rust diseases can cause severe reductions in yield because of the loss of effective leaf area. Common rust can cause yield losses of 40–50% in susceptible varieties[38]. Similar losses were reported for the outbreak of orange rust variety in Q124 in Queensland in 2000[20].

Control

Varietal resistance is the only economic method of control of both rust diseases in the long term. Because the pathogens are highly prevalent, once rust has occurred in a region, infection of new clones during selection tends to occur readily, and susceptible ones can be easily recognised and discarded.

Smut *Ustilago scitaminea* Sydow

Smut or culmicolous smut, is one of the major diseases of sugarcane worldwide. It was first reported from Natal, South Africa in 1877 and for many years was mainly restricted to the Old World, where it caused repeated outbreaks. It is still the most important disease problem in many African sugar industries and is often a serious problem in India. In the mid to late 1900s, smut spread throughout the Americas and today the only industries where smut has not been recorded are in Papua New Guinea and eastern Australia (an outbreak occurred in western Australia in 1998).

Symptoms

An authoritative description of the disease was made by Ferreira & Comstock[39] in 1989. The symptoms are so distinctive that smut is usually the most easily diagnosed disease of sugarcane. The characteristic symptom is the brown, whip-like fungal sorus that develops from the apex of an infected stem and can be > 1 m long (Fig. 3.3). Each of these structures comprises a core of parenchyma and fibrovascular tissue surrounded by a mass of brown chlamydospores, enclosed at first in a thin, silvery sheath. When the membranous covering splits, the exposed spores resemble a thick layer of soot. They are then dispersed, mainly by

Fig. 3.3 A smut whip, the characteristic symptom of sugarcane smut. Source: South African Sugar Association Experiment Station.

The formation of whips is determined by the age and physiological condition of the crop and the season. The main flush appears in mid-summer in relatively young cane and results from primary infection (i.e. infection of the stubble, or the seed-cane). Whips can continue to form later in the season as a result of secondary infection of the standing cane. The number of whips is greatest when the crop suffers from stress. Following primary infection, the number of whips usually increases from the plant crop to the early ratoons.

Damage

Determining the effect of smut on yield in the field is not easy because of the likelihood of latent infection and the effect of growing conditions on the pathogen–host plant interaction. The extent of damage from smut depends markedly on environmental and seasonal weather conditions as well as on varietal susceptibility. Within one variety, disease expression can vary dramatically from year to year and at different times of the year. Generally, smut is favoured by hot, dry conditions. Chona[9] reported a reduction in cane yield of 29% in plant cane, rising to 70% in ratoons of variety Co312. In a field trial in southern Africa with variety NCo376 in which different levels of smut were attained on a shallow soil under irrigation, reductions in plant cane were 48% and in the first ratoon approximately 90%[40]. Total crop failure resulting from smut infection is therefore possible in susceptible varieties under conditions that favour the disease.

wind. Stems or shoots with latent infection (before the whips form) are often thin and pencil-like, and infected stools may develop a grass-like appearance. Recognition of this symptom is important when roguing is used as a control measure.

Smut is transmitted mainly by wind-borne spores infecting the standing cane and by spores in the soil infecting planted setts. Infection takes place through the buds. Infected buds may develop quickly into whips, or the mycelia may remain dormant within the buds as a latent infection. The use of such infected stalks as seedcane is an important means of spread.

Control

The risk of severe damage usually precludes susceptible varieties from being grown in situations where smut is common. Variety resistance is therefore usually the preferred method of control. However, in industries where the economics of labour permit it, and unified field management is possible, smut can be contained by rigorous application of field control measures in all but highly susceptible varieties. For example, productive but susceptible varieties are successfully grown in several sugar industries in southern Africa, including Swaziland and Zimbabwe. Although conditions are favourable for smut development in these industries, intensive roguing (four to eight operations per season performed in young cane) is found to be effective and economically worthwhile[41]. This practice forms part of an integrated control programme, which also involves seedcane health to minimise primary infection. Hot water treatment, as used to control ratoon stunting disease, eliminates any smut infection from seedcane but may increase susceptibility to subsequent infection by smut. Hence the use of fungicides in hot water treatment is often an important aspect of such integrated control[42].

The use of resistant varieties is the only practicable method of control in most countries. Where smut is a problem, most variety development programmes include the screening of new clones during selection to identify those with satisfactory resistance. Screening usually involves special field trials in which new clones are exposed to high levels of the pathogen from infected 'spreader' plants, or the clones are inoculated with the pathogen, or a combination of both. Inoculation with the pathogen usually involves immersing the setts in a suspension of spores before planting. It is important that such screening trials are conducted in an environment that is favourable for disease development and that they are planted at the optimum time of year to maximise symptoms to avoid 'escapes'. However, inoculation conditions should not be so harsh that varieties with useful resistance are discarded unnecessarily. In South Africa, for example, the results of immersing setts for 5 min at 5×10^6 spores/ml correlated well with those from natural infection[43].

Resistance to smut is sufficiently heritable that careful choice of parent varieties can increase the numbers of resistant progeny. Following the outbreak of smut in Western Australia in 1998, it was found that the majority of new progeny in the Queensland sugarcane breeding programme was susceptible and efforts are underway to improve the resistance of parent varieties.

Reports of distinct strains of *U. scitaminea* have been made from several countries, including Hawaii and Taiwan[39]. However, the significance of this for smut epidemiology is unclear. In an international project conducted under the auspices of ISSCT, a set of 11 differential varieties was tested against local strains of the pathogen in 14 locations in 10 countries. The results of this project showed broad uniformity in the ranking of the resistance of the varieties in most countries and evidence for differences in the reactions of the set of varieties was observed only in Taiwan[44]. Although it is possible that genetic variation in *U. scitaminea* can arise, screening new clones against the current prevalent strain or strains will minimise its impact.

Miscellaneous fungal diseases

There are many other fungal diseases of sugarcane, a complete list of which is given in *A Guide to Sugarcane Diseases*[6]. Most of those not described in this chapter are either of usually minor economic importance today or of only local occurrence. Widely distributed but usually minor diseases include some causing leaf spots and leaf blotches, e.g. brown spot *Cercospora longipes*, ring spot *Leptosphaeria sacchari* and eye spot *Bipolaris sacchari*. Common but usually minor stalk diseases include Fusarium sett and stem rot (*Gibberella* and *Fusarium* spp.), and there are several root diseases of uncertain importance.

Among those leaf diseases of local economic importance that require control intervention is downy mildew *Peronosclerospora sacchari*, which currently important in Fiji and Papua New Guinea and where resistance is achieved by the screening of new clones in the field. Pachymetra root rot *Pachymetra chaunorhiza* is known to occur only in the Queensland sugar industry, where it can cause severe damage in tropical areas and where control is achieved by screening new clones by inoculation.

BACTERIAL DISEASES

Gumming disease *Xanthomonas axonopodis* pv *vasculorum* (Cobb) Vauterin

Gumming disease, or gumming, is one of the oldest recorded diseases of sugarcane and at various times has been a serious problem in several countries. The first outbreaks occurred in Brazil, Mauritius, Réunion, Australia and Fiji in the nineteenth century. Orian[45] postulated that the collapse of Otaheite cane in Mauritius and Réunion in the 1840s was probably caused by gumming, and that this disease was primarily one of palms native to the Mascarene islands, which formed a reservoir of infection responsible for the continuing widespread occurrence of gumming disease of sugarcane in Mauritius. Saumtally & Dookun[46], however, considered that the role of these other host plants in the epidemiology of gumming is still uncertain.

Gumming still occurs in many sugarcane growing countries as a common but relatively minor disease. It has been eradicated from Australia, where it was once very serious, and has not been reported for many years from several of the Caribbean islands. Gumming still has the potential to be serious in Mauritius, if cyclonic conditions favour the build-up and spread of the pathogen.

In 1989 Ricaud & Autrey[47] gave a full description of gumming, and in 2000 Saumtally & Dookun[46] summarised gumming.

Symptoms

Gumming has two distinct stages: a foliar phase following initial infection, and a systemic infection phase. There is, therefore, a range of symptoms. In its early stages, yellowish, somewhat irregular longitudinal stripes develop on the leaves, usually from the margins, and later become straw-coloured. Narrower, more regular reddish-brown stripes can also develop on the leaves. In humid weather, a silvery sheen of exuded bacteria may be seen on the leaf stripes. Following systemic infection of the vascular bundles, a distinctive chlorosis of young, newly emerged leaves can develop in susceptible varieties. Again in susceptible varieties, a characteristic symptom is the development of pockets of a gum-like bacterial exudate in the internal tissues of the stalk, from which the disease gained its name. A copious exudate can also be seen when infected stalks of susceptible varieties are cut transversely (Fig. 3.4). Some of the vascular bundles of infected stalks may become red, but this symptom is not specific to gumming. Infected stools may be stunted and produce weak, chlorotic shoots.

Spread

Gumming is spread locally within and between fields by wind-blown rain and on contaminated implements, e.g. cane knives and harvester blades. Spread from one locality to another can occur if infected setts are used for planting. Strong winds and storms favour the spread of the disease, as leaf damage to the leaves that occurs under such conditions provides infection points for the pathogen. Epidemics in Mauritius usually follow cyclonic activity.

Severe yield losses caused by gumming have been reported in the past. In the early twentieth century reductions in cane yield of 40% occurred in New South Wales, Australia[48]. Losses of 45% have been reported from Mauritius[49]. Additionally, the xanthan gum that is present in infected stalks can interfere with processing in the factory.

Control

As with most sugarcane diseases, the most effective means of controlling gumming disease is by planting resistant varieties. Where the disease occurs, the pathogen is relatively common and susceptible clones are readily identified during selection.

There is considerable variation among strains of the pathogen. Three distinct races are known to occur in Mauritius[50]. Further variation was reported by Saumtally[51] who, using serological and polymerase chain reaction (PCR) techniques, distinguished five groups among isolates of *X. axonopodis* pv *vasculorum* collected from different countries. This variability within the pathogen necessitates careful implementation of quarantine procedures when varieties are exchanged between countries.

Leaf scald *Xanthomonas albilineans* (Ashby) Dowson

Although leaf scald is currently not causing significant losses in most sugarcane growing countries, it is still regarded as an important disease in many, requiring attention by pathologists and breeders for control to be maintained. There are accounts of leaf scald by Ricaud & Ryan[52] and Rott & Davis[53].

Fig. 3.4 Bacterial gum of *Xanthomonas axonopodis* pv *vasculorum* exuding from an infected stalk. Source: Mauritius Sugar Industry Research Institute.

Leaf scald occurs in most regions where sugarcane is grown. It was recognised as a disease of sugarcane in Indonesia in the 1920s[54] and is thought to have originated in the Old World. Serious outbreaks occurred in Australia in the 1920s and 1930s and it was reported from Fiji, Hawaii, Malagasy, Mauritius, the Philippines and Réunion in the same period. It was first reported in South America in the 1940s, and in southern Africa and the continental USA in the 1960s. As with ratoon stunting disease (RSD), disease infection can occur without symptoms being expressed, hence in the past leaf scald was probably often spread inadvertently during the exchange of varieties between countries. Without sophisticated diagnostic tests, effective quarantine is difficult because of the frequent lack of symptoms and the difficulty of eliminating the pathogen from infected setts.

Variation in the virulence of different strains of the pathogen is known to occur. Autrey *et al.*[55] de-scribed different strains of *X. albilineans* in Mauritius, and an outbreak of leaf scald in Florida in the mid 1990s was attributed to a new strain[56].

Symptoms

Leaf scald can be one of the more difficult diseases to identify, as the symptoms are often not conspicuous or specific. The disease can also remain as a latent (symptomless) infection for a considerable period after infection. Two phases of the disease are recognised: chronic and acute. The external symptoms of the chronic phase are narrow, white 'pencil-line' stripes on the leaves (Fig. 3.5), which may become reddened, and side-shoots, which develop from the bottom of the stalk. This latter symptom is almost diagnostic for leaf scald. As the leaves mature, the pencil-line stripes might broaden and become more diffuse, and the leaves tend to curl inwards and wither from the tips downwards,

Fig. 3.5 Characteristic white 'pencil lines' on the leaves are a symptom of leaf scald. Source: South African Sugar Association Experiment Station.

giving a 'scalded' appearance from which the name of the disease is derived. Infected shoots tend to be stunted and may die-back. Internal examination of the shoots with other symptoms reveals the vascular bundles as fine red streaks, which are most pronounced in the nodal areas and at the attachments of the side-shoots.

In the acute phase, large areas of cane rapidly wilt and die, often without showing any previous symptoms of disease, as if affected by drought. This is restricted to highly susceptible varieties. It is its latent nature and the possibility of large areas of cane suddenly dying which give leaf scald its reputation as a hazardous disease.

Accurate identification is dependent on the use of serological or PCR tests, and such tests are ob-

ligatory for effective quarantine of material that is exchanged as conventional setts.

Spread and control

Leaf scald is mainly transmitted by infected setts and in the field by cane knives and harvesting machines. Aerial spread of the pathogen, similar to that of gumming, was first demonstrated in Mauritius[55].

Effective control of leaf scald is dependent on the use of resistant varieties. The production of resistant progeny can be enhanced by careful choice of the parent varieties used for crossing. Many selection programmes include the routine screening of new clones for resistance. The methods used are usually based on the 'decapitation' method first described for field use by Antoine & Ricaud[57], in which young shoots of the clones under test are cut above the growing point and inoculated with a suspension of a pure culture of *X. albilineans*. The clones are then assessed according to the severity of symptoms in the regrown plants. It is important that ratings are not based on the initial symptoms shortly after inoculation but on those resulting from systemic infection by the pathogen. When conducted in the field, these trials give the best results if inoculation is followed by a period of cloudy, relatively cool weather, which favours infection by the pathogen.

Disinfecting harvesting implements reduces the spread of the disease in the field. The pathogen is not eliminated from seedcane by the standard hot water treatment (HWT) of 2 h at 50°C that is widely used for control of RSD. In quarantine, a serial HWT involving a long soak in cold water is recommended to minimise the risk of transmission in symptomless material[7].

Ratoon stunting disease *Leifsonia xyli* subsp. *xyli*

Ratoon stunting disease (RSD) is widely regarded as causing greater economic loss to sugarcane industries throughout the world than any other pathogenic disease; yet few other diseases of sugarcane are less conspicuous. RSD was first recognised in Queensland, Australia in 1944–45 and its presence was soon reported from other countries. Today, with the possible exception of Papua New Guinea, RSD probably occurs in all countries where sugarcane is grown.

For many years the cause of RSD remained unknown, although a virus was widely suspected. However, in 1974 Gillaspie *et al.*[58] found that a small, non-motile, rod-shaped bacterium was associated with RSD in Louisiana. Teakle[59] reported similar findings from Queensland and concluded that this bacterium was probably the causal agent. A strong association between the distinctive bacterium and RSD symptoms was soon reported from other countries and thereafter consensus was soon reached that the bacterium was the cause of this hitherto puzzling disease.

Proof of the bacterial aetiology was provided by Davis *et al.*[60] and this group went on to name the organism *Clavibacter xyli* subsp. *xyli*[61]. In a recent reclassification, the generic name *Leifsonia* was been proposed and this is now accepted[62]. No strains of the pathogen have been reported.

The early history of RSD and the first control measures have been described by Steindl[63]. A more up-to-date review and a description of the pathogen were provided in 1989 by Gillaspie & Teakle[64]. A summary was given in 2000 by Davis & Bailey[65].

Symptoms

RSD produces no external symptoms that can be easily recognised in the field, although a non-specific stunting is common and may be severe. Diseased stalks of some varieties may exhibit internal discoloration of the vascular bundles at the nodes. These nodal symptoms appear as yellow to reddish-brown to black dots, 'commas', or short streaks when the nodes are sliced longitudinally (Fig. 3.6). This can be conspicuous in some varieties but is inconspicuous or absent in others and is not a reliable diagnostic symptom. The marks do not extend into the internodal tissues as, for example, is the case with leaf scald. If available, healthy stalks should be examined for comparison when looking for nodal symptoms. An orange to pink-red discoloration of the internal tissues just below the apical meristem ('juvenile stalk symp-

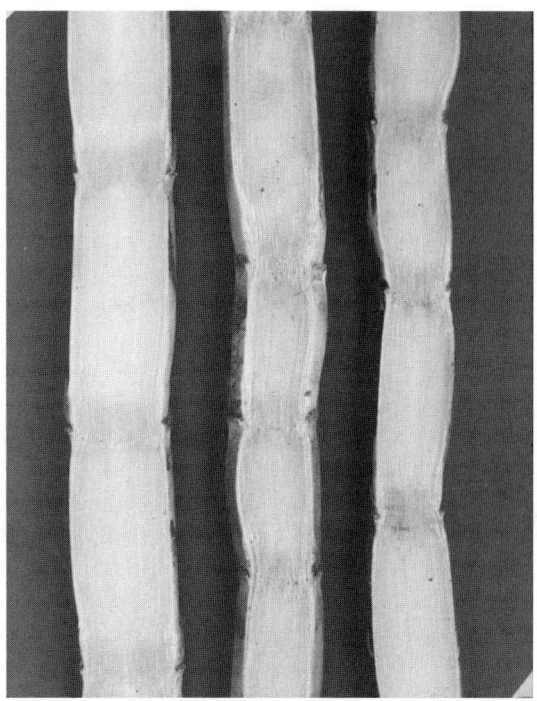

Fig. 3.6 Orange-reddish to dark brown dots or streaks in the nodal tissues of RSD-infected stalks can be seen in some varieties. Source: South African Sugar Association Experiment Station.

toms') may occur in some varieties under certain conditions, but this is not a consistent or useful symptom.

The lack of consistent, easily recognisable symptoms is the main reason that RSD spread so rapidly among cane producing countries as well as from farm to farm before accurate routine diagnosis was possible through microscopy, serology or DNA-based techniques.

Diagnosis

Because of the lack of reliable symptoms, RSD is usually detected using laboratory techniques. The pathogen is a small, xylem-inhabiting bacterium that can be observed directly in extracts of xylem sap using phase-contrast microscopy (PCM) at ×1000 magnification. The effectiveness of PCM largely depends on the skill of the microscopist, but given training, non-scientists can produce reliable results and the technique is suitable for large-scale use. PCM formed the basis of the successful industry-scale control scheme in South Africa from 1977[66], until it was replaced by newer technology.

The bacterium is now usually detected with various immunochemical tests. Two procedures that permit multiple samples to be examined are the tissue-blot enzyme immunoassay (TB-EIA)[67] and the evaporative binding enzyme immunoassay (EB-EIA)[68]. TB-EIA permits detection of colonised vascular bundles in cross sections of stalks or leaves. EB-EIA is a modified ELISA procedure for analysis of vascular sap extracts, is approximately as sensitive as PCM, and is easier to use for diagnosis on a large scale[69].

PCR assays based on detection of the 16S ribosomal RNA gene of the pathogen have been developed for detection and identification (e.g. Fegan *et al.*[70]). PCR tests are more sensitive than microscope and serological tests and are useful for highly accurate diagnosis and research on the pathogen, but have limitations for mass use.

Economic effects

Yield reduction due to RSD is mainly due to thinner and shorter stalks (Fig. 3.7) and sometimes a reduction in the number of stalks when the disease is severe[71]. In ratoon crops, infected plants are slower to regrow and reductions in yield are usu-ally greater. In highly susceptible varieties, plants may die if crops are severely stressed. The disease has no consistent effects on juice quality.

Yield losses have frequently been estimated at 5–10% on a field, farm and district basis. Hughes[72] estimated that where control measures were not being carried out in Queensland, yield reductions of 10% were probable under normal conditions. In Florida, an average yield reduction of 5% of the entire crop was estimated in 1988–89[73]. In South Africa, industrial losses are now currently estimated at 1% of production, but losses in some other African sugarcane industries, where infection is very common, are estimated to be 10–20% of annual production[74].

The effects of RSD on growth are more severe when crops are stressed; thus farm and industrial losses tend to be greatest under rainfed conditions when rainfall is erratic, or when irrigation management is poor. In field experiments in South Africa, losses of up to 45% were recorded in popular varieties grown under rainfed conditions and up to 32% under well-managed irrigation[71,75]. The death of plants can occur if infected, susceptible varieties suffer severe moisture stress.

In industries or regions where effective diagnosis is not available and control is lacking, RSD can reach very high levels, with all the stalks in a field infected, and the cause of the substantial yield reductions may not be appreciated.

Fig. 3.7 Effect of RSD on growth of three varieties: left to right, NCo376 healthy and infected, N53/216 healthy and infected, N55/805 healthy and infected. Source: South African Sugar Association Experiment Station.

Spread

Leifsonia xyli subsp. *xyli* has only been detected in sugarcane in nature. RSD is mainly transmitted in setts taken from diseased plants and by cane knives and other implements at harvest. Systemic infection of the xylem takes place through wounds. The rate of spread during harvesting can be rapid[76] and, once introduced into a field, most plants can become infected after only a few harvests. In southern, central and east Africa, fields are often replanted after only a short, if any, break between crop cycles. In these circumstances, the persistence of the disease in fields in the form of infected regrowth from old crops (volunteers) has been identified as an important factor in maintaining high levels of infection[74]. The pathogen can remain infectious for up to several months in either moribund plant debris or the soil itself, contributing to the persistence of infection in areas where the disease is common[77].

Control

Planting healthy seedcane is a key factor, without which effective control is not possible. The large-scale testing of seedcane to determine its health using PCM or EB-EIA is therefore widely practised. Because of the rapid spread of RSD in the field, difficulties in diagnosis and serious effects on production, many estates, regions or industries operate seedcane health schemes. Sanitation is important in keeping healthy cane from becoming infected, since the pathogen is easily transmitted mechanically. Cane knives and harvesting machines are often disinfected to minimise spread in the field, particularly for seedcane sources.

Seedcane can be heat-treated to eliminate the pathogen[63,64]. Hot-water treatment (HWT) at 50°C for 2–3 h is the most commonly used method. The temperature and time of treatment are a compromise between the need to eliminate the pathogen without severe effects on germination. However, a single treatment, even at 3 h, does not provide complete control. Consequently, heat-treatment is often used to establish pathogen-free 'nurseries' that are then used to supply planting material for commercial fields. Continued vigilance in the production of seedcane using heat treatment over several years, such as in seedcane production schemes, together with effective destruction of old crops is needed to achieve meaningful and sustainable reductions in RSD incidence.

Although there are large differences among varieties in susceptibility to and tolerance of infection, highly resistant varieties are relatively rare in germplasm collections. Thus breeding and selection for resistance is not yet common and it is debatable whether variety resistance will entirely displace traditional control methods, which are known to be effective in well-managed situations, in most sugarcane industries. The use of the serological technique TB-EIA to aid in selecting resistant varieties is most advanced in Florida, reportedly with some success for reduced RSD incidence in commercial fields[67,78].

The use of tissue culture-generated disease-free plants, produced by proprietary technology, forms the basis of successful control of RSD in the United States[79].

Other bacterial diseases

There are several widely distributed diseases caused by other bacterial pathogens that cause foliar symptoms, which can be seen in the field, but these are usually of minor importance. This group includes red stripe *Acidovorax avenae* subsp. *avenae*, which has been reported from most cane growing countries and causes numerous fine, red stripes on the leaves. These stripes are difficult to distinguish in the field from those that can be caused by gumming. Occasionally varieties that are highly susceptible to red stripe occur in selection programmes and in these, the disease can progress to cause a 'top rot', in which the growing point of the stem is killed. Mottled stripe *Herbaspirillum rubrisubalbicans* causes narrow, cream to reddish mottled stripes on the leaves and occurs in many countries.

VIRAL DISEASES

Until recently a number of important sugarcane diseases were ascribed to viruses. Some of these are no longer common or no longer cause serious damage, while others are now ascribed to a fourth category of pathogens, the phytoplasmas (see section that follows). However, mosaic is still a widespread and serious, or potentially serious, viral disease in many sugarcane industries. In the last decade, yellow leaf syndrome has been found to occur in many industries and one form of this disease is now known to be caused by a virus.

Mosaic, sugarcane mosaic virus and sorghum mosaic virus

Mosaic has been identified in almost all sugarcane-producing countries. It is one of the more distinctive diseases of sugarcane and has caused serious losses from time to time in many industries. Descriptions of mosaic are given by Koike & Gillaspie[80] and Grisham[81], the latter giving an update to 2000 on current understanding on the identities and relationships of the different strains of the viral pathogens that cause this important disease.

The term 'mosaic' refers to the symptoms caused by sugarcane mosaic virus (SCMV) and sorghum mosaic virus (SrMV), both of which occur in several distinct strains. These viruses belong to the Potyviruses. This viral family includes relatively long (*c.* 750 nm), thread-like viruses that are spread by various species of aphid. Mosaic strains were previously identified by the use of differential host plants. SCMV and SrMV can be differentiated by serological tests, but the identification of specific strains and establishing their relationships is now mainly based on PCR analysis[82].

Symptoms and effects

The leaves of infected stalks develop a typical 'mosaic' pattern of pale green to yellow elongated steaks or patches, interspersed with similar streaks of normal colour. The conspicuousness of the symptoms varies widely according to the strain of the virus and the sugarcane variety that is infected. Sometimes the symptoms are strikingly evident (Fig. 3.8a) but they can be more subtle (Fig. 3.8b). The symptoms are usually most readily seen towards the base of young leaves, but in some highly susceptible varieties symptoms may occur on the leaf sheaths and even on the stalks.

The main effect of mosaic infection on crop growth is to cause a stunting of infected stalks and this is the main component of reduced sugar yield. The extent of damage varies widely according to the variety being grown and the strain of virus involved. Infection has little effect on cane quality.

In the past, mosaic caused the near collapse of the sugar industry in Louisiana in the 1920s and severe losses occurred in Argentina and elsewhere in the Americas[80]. In 2000, Grisham[81] reported losses in sugar yield of 7–21% in different varieties over a 3-year crop cycle in Louisiana. In South Africa, reductions in sugar yield of a susceptible variety amounted to 42%, and to 30% in a less susceptible variety[83].

Spread

As with other systemic diseases, primary spread of mosaic occurs through the planting of infected seedcane. Secondary transmission occurs through a various species of aphids, among which *Dactynotus ambrosiae*, *Hysteroneura setarie* and *Rhopalosiphum maydis* are the most important. The main vector species differ from region to region, for example, *D. ambrosiae* was found to be the most efficient vector in Louisiana[84], whereas *H. setarie* was by far the most common vector in South Africa[85]. Where vector populations are high, secondary transmission within and between fields can occur very rapidly. Vector aphids acquire and transmit the virus very rapidly on feeding, which involves the probing of the phloem elements of host plants. Many grass species are hosts of the mosaic viruses and such grasses are epidemiologically significant as sources of virus for the infection of cane fields and in terms of the build-up vector populations. In this regard, mosaic differs from most other sugarcane pathogens, which are specific pathogens of sugarcane.

Control

Planting resistant varieties is the only satisfactory method of controlling mosaic in regions where the risk of aphid transmission is high. In general,

(a)

(b)

Fig. 3.8 Symptoms of mosaic may be conspicuous (a) or more subtle (b).

clones of *S. officinarum* are highly susceptible to mosaic, hence the crop failures of the 1920s when noble canes were widely grown. *Saccharum spontaneum* has been the main source of the genes that confer resistance in commercial hybrids. The suitable choice of parent varieties is important wherever mosaic is a hazard.

Where mosaic is an important disease, the screening of new clones to identify resistance is usually practised. This can be done by artificial inoculation of young plants in the glasshouse using an air-brush or sprayer, as in the United States[86], or exposing new clones to natural infection in the field, as in South Africa, or both. Artificial inoculation requires carefully controlled experimental conditions and the use of viral isolates of current field importance. Natural exposure trials are dependent on a high rate of aphid transmission at trial sites. However, such trials automatically cater for possible changes in virus strains in the field and the results are directly applicable to the commercial farming situation.

Progress has been made recently in developing genetically modified (GM) clones with resistance to mosaic. This is based on transfer of the genes coding for the virus coat protein, which conveys resistance to subsequent infection. Although not yet used in commercial practice, this is likely to be the first use of GM technology for the purpose of disease control in sugarcane.

A range of field control practices can be useful in minimising the spread of mosaic. As with other systemic diseases, the planting of healthy seedcane is essential for effective control, but it may be difficult to achieve this in outbreak areas. Apical meristem culture followed by rapid micro-propagation to produce virus-free planting material is used in the USA. In South Africa, choosing planting dates to avoid young, susceptible growth stages coinciding with peak aphid populations is used successfully[87]. Because infective aphids transmit the virus rapidly on feeding and because their populations build up outside cane fields, applying insecticides to cane crops is not a useful option.

Yellow leaf syndrome, sugarcane yellow leaf virus; sugarcane yellows phytoplasma

Yellow leaf syndrome (YLS) is of interest because identical symptoms are caused by pathogens from two very distinct groups, viruses and phytoplasmas, because the symptoms appeared in many countries in the late 1980s and 1990s, and because there are now experimental indications of effects on yield. Both forms of the disease are described here under virus diseases. YLS is probably the same as the condition previously known as 'yellow wilt' in Africa and 'autumn decline' in Brazil in the 1960s. Both forms are described by Lockhart & Cronjé[88].

The viral pathogen is now termed sugarcane yellow leaf virus (SCYLV)[89] and the phytoplasmal form as sugarcane yellows phytoplasma (SCYP)[90]. SCYLV has been reported from many sugarcane producing countries. SCYP has been reported from many countries in Africa, Cuba and Réunion. In a large-scale international survey of numerous samples collected from a total of 20 countries, SCYP was detected by PCR in samples of numerous varieties that had YLS symptoms and from all the countries[91]. Mixed infections of the two pathogens were also reported.

Symptoms and effects

The characteristic symptom of YLS is a distinct yellowing of the lower surface of the leaf midribs (Fig. 3.9), which may extend laterally into the leaf lamina. The midrib yellowing may be intense or in some varieties may have a reddish tinge, and is associated with sucrose accumulation in the midribs. Symptoms are best expressed when the crop is subject to stress, for example from low temperatures,

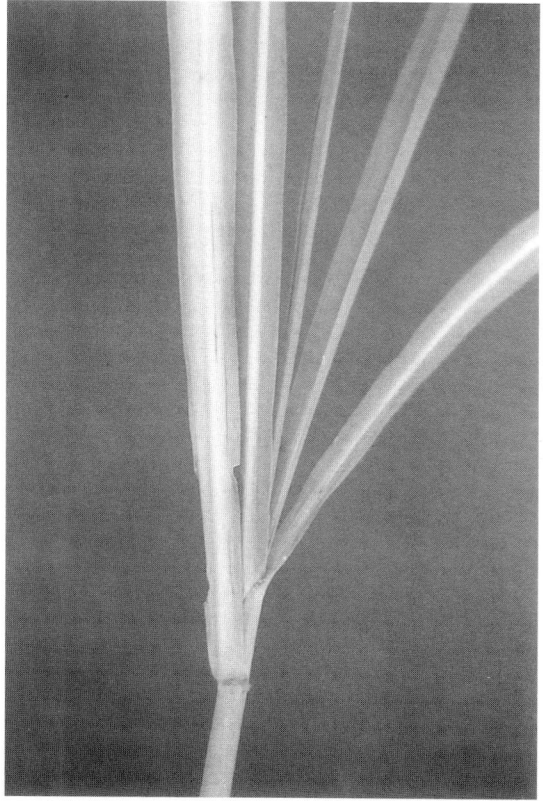

Fig. 3.9 A yellow colour of the lower surface of the leaf midribs and a yellowing of the leaf lamina is the characteristic symptom of yellow leaf syndrome. Source: South African Sugar Association Experiment Station.

and tend to fade in all but highly susceptible varieties with the onset of good growing conditions.

Both pathogens are very common and can be present without the expression of symptoms. SCYLV can be readily identified serologically, whereas identification of SCYP relies on a nested PCR protocol that requires exactitude in application.

It is a common observation that plants with conspicuous symptoms of YLS can be severely stunted. The first report of substantial yield loss associated with YLS was in variety SP71–6163 in Brazil. However, it is difficult to determine cause and effect from these field observations because symptoms are most common in poorly grown or stressed cane. Controlled experiments have demonstrated yield losses resulting from SCYLV of

2–20%, depending on variety[92], and there is little doubt that at least the viral form of YLS can affect the productivity of susceptible varieties.

Spread

Both forms of YLS are spread by planting infected seedcane, and this distinguishes YLS from the physiological conditions that cause similar symptoms. SCYLV is spread by the common aphids *Melanaphis sacchari* and *Rhopalosiphum maidis*. There is evidence that SCYP is spread by leafhoppers, including the sugarcane hopper *Perkinsiella saccharacida*. Both forms of YLS can be transmitted in seedcane.

Control

As yet there is no treatment to eliminate the phytoplasma SCYP for propagation material, but the virus SCYLV can be eliminated by apical meristem culture. The appropriate choice of parent varieties and discarding progeny with conspicuous symptoms during selection can improve the general resistance of new varieties to YLS.

Further studies on the transmission of both forms of YLS and critical studies on their effects on yield are necessary to better determine the hazard that YLS presents to sugarcane productivity.

Miscellaneous viral diseases

As with the other groups of diseases, there are some sugarcane diseases caused by viruses that are now regarded to be of minor importance or, although once serious in one or more regions, are now firmly under control. One of the latter is Fiji disease (FDV), which is spread by leafhoppers of the genus *Perkinsiella* and is restricted in distribution to Australia, east Asia and the Pacific islands. The characteristic symptoms of FDV include the presence of elongated galls on the undersurfaces of the leaves and a gross stunting of the growth of ratoon crops of susceptible varieties. It was the most serious disease in the Australian sugar industry in the 1970s, sometimes causing up to 100% losses in ratoon crops of otherwise productive varieties in central Queensland. FDV was eventu-

ally brought under control in Australia by a strict policy of eradicating infested fields, withdrawing susceptible varieties from production, breeding for resistance and screening new clones for resistance, which is still practised[93].

Sugarcane bacilliform virus (SCBV) is thought to be distributed world-wide in clones of *Saccharum officinarum*[94]. In commercial sugarcane hybrids, the symptoms of SCBV are usually inconspicuous white freckles, but more conspicuous freckles may be seen in noble canes in variety collections. SCBV is generally regarded as a minor pathogen.

PHYTOPLASMAL DISEASES

A number of sugarcane diseases that were previously of unknown cause or were ascribed to viruses are now known to be caused by members of a fourth group of pathogens – the phytoplasmas. These micro-organisms are simple prokaryotes, related to but distinct from bacteria. This group includes the phytoplasma form of yellow leaf syndrome (YLS), caused by the sugarcane yellows phytoplasma (SCYP), which for convenience is described in the section on viral diseases above. The best known of the other phytoplasma diseases of sugarcane are grassy shoot disease and white leaf disease, which are widespread in some Asian countries. Grassy shoot disease has been described by Viswanathan[95] and white leaf disease by Chen & Kusalwong[96].

The symptoms of grassy shoot and white leaf disease are similar in that both cause stunting of infected plants, profuse tillering, and chlorotic stripes on the leaves. As its name implies, white leaf disease can cause a particularly conspicuous chlorosis of the shoots (Fig. 3.10). Damaging outbreaks of white leaf disease occurred in the past in Thailand, although it is now regarded as being under control[96]. Severe yield losses owing to grassy shoot disease have been reported from India[95]. Identification of diseases caused by phytoplasmas is usually based on field symptoms, with confirmation based on serological or PCR techniques.

Phytoplasmas colonise the phloem vessels of infected plants and thus these diseases are systemic and can be spread in seedcane. This method of

Fig. 3.10 Stunting, multiple tillering and leaf chlorosis caused by sugarcane white leaf disease. Source: R. A. Bailey.

spread has usually been implicated in severe outbreaks. Secondary spread occurs through transmission by leafhoppers. Several grass species are also hosts of the causative agents. Control of grassy shoot and white leaf diseases is based on integrated strategies, which include seedcane health as a principal component.

An interesting member of this group of diseases is Ramu stunt, which devastated the new sugarcane industry in Papua New Guinea in the 1980s[97], but is not known to occur elsewhere. Ramu stunt is spread by the leafhopper *Eumetopina flavipes* and by the planting of infected seedcane. Control has been achieved by rigorous screening of varieties in natural exposure trials to identify those with suitable resistance.

REFERENCES

1 Stevenson, R.A. & Rands, R.D. (1938) An annotated list of fungi and bacteria associated with sugarcane and its products. *Hawaii Planters Record*, **42**, 247–313.

2 Martin, J.P. (1951) Sugarcane diseases and their world distribution. *Proceedings of the International Society of Sugar Cane Technologists*, **7**, 435–452.

3 Martin, J.C., Abbott, E.V. & Hughes, C.G (eds) (1961) *Sugar-cane Diseases of the World*, Vol. 1. Elsevier, Amsterdam.

4 Hughes, C.G., Abbott, E.V. & Wismer, C.A. (eds) (1964) *Sugar-cane Diseases of the World*, Vol. 2. Elsevier, Amsterdam.

5 Ricaud, C., Egan, B.T., Gillaspie, A.G. & Hughes, C.G. (eds) (1989) *Diseases of Sugarcane – Major Diseases*. Elsevier, Amsterdam.

6 Rott, P., Bailey, R.A., Comstock, J.C., Croft, B.J. & Saumtally, A.S. (eds) (2000) *A Guide to Sugarcane Diseases*. CIRAD and International Society of Sugar Cane Technologists, Montpellier.

7 Frison, E.A. & Putter, C.A.J. (1993) Leaf scald. In: *FOA/ IBPGR Technical Guidelines for the Safe Movement of Sugarcane Germplasm*, pp. 32–34. Food and Agriculture Organisation/IBPGR, Rome.

8 Baker, R.E.D., Martyn, E.B. & Stevenson, G.C. (1954) Sugarcane diseases in the Caribbean. *Proceedings of the International Society of Sugar Cane Technologists*, **8**, 895–902.

9 Chona, B.L. (1956) Chairman's address, Pathology section. *Proceedings of the International Society of Sugar Cane Technologists*, **9**, 975–986.

10 Abbott, E.V. (1963) Problems in sugar cane disease control in Louisiana. *Proceedings of the International Society of Sugar Cane Technologists*, **11**, 739–742.

11 Wiehe, P.O. (1963) The control of sugar cane diseases in Mauritius. *Proceedings of the International Society of Sugar Cane Technologists*, **11**, 743–748.

12 Horau, M. (1967) Sugar cane diseases in Réunion island. *Proceedings of the International Society of Sugar Cane Technologists*, **12**, 1232–1235.

13 Liu, L.J., Rosario, T. & Roig, F.M. (1967) Diseases of sugarcane in Puerto Rico. *Proceedings of the International Society of Sugar Cane Technologists*, **12**, 1236–1240.

14 Daniels, J., Husain, A.A. & Hutchinson, P.B. (1972) The control of sugarcane diseases in Fiji. *Proceedings of the International Society of Sugar Cane Technologists*, **14**, 1007–1014.

15 Planalsuçar (1977) *A Guide to Identification of Sugarcane Diseases and Nutritional Deficiencies in Brazil*. Planalsuçar, Piracicaba, Brazil.

16 Bureau of Sugar Experiment Stations (1991) *Diseases of Sugarcane*. BSES Manual, Queensland, Australia.

17 SASA Experiment Station (2003) *Sugarcane Diseases in South Africa*. South African Sugar Association, Mount Edgecombe, Durban, South Africa.

18 Clarke, M., Reece, N.E. & Elcock, H.L. (1976) The yield of sugarcane in Barbados in 1976. *Ministry of Agricultural Science and Technology, Barbados Bulletin*, 61.

19 Riley, I.T., Jubb, T.F., Egan, B.T. & Croft, B.J. (1999) First outbreak of sugarcane smut in Australia. *Proceedings of the International Society of Sugar Cane Technologists*, **23**, 333–337.

20 Magarey, R.C., Croft, B.J. & Willcox, T.G. (2001) An

epidemic of orange rust in Australia. *Proceedings of the International Society of Sugar Cane Technologists*, **24**, 410–416.

21 Cox, M. & Croft, B.J. (2002) BSES Plant breeding. *Bureau of Sugar Experiment Stations Bulletin*, **80**, 4–5.

22 Comstock, J.C., Shine, J.M., Tai, P.Y.P. & Miller, J.D. (2001) Breeding for ratoon stunting disease resistance: is it both possible and effective? *Proceedings of the International Society of Sugar Cane Technologists*, **24**, 471–476.

23 Wismer, C.A. & Bailey, R.A. (1989) Pineapple disease. In: *Diseases of Sugarcane – Major Diseases* (eds C. Ricaud, B. T. Egan, A. G. Gillaspie & C. G. Hughes), pp. 145–155. Elsevier, Amsterdam.

24 Wismer, C.A. (1968) Benlate – a promising new fungicide. *Annual Report of the Experiment Station of the Hawaii Sugar Planters Association*, p. 67.

25 Van Dillewijn, J. (1951) Fusarium pokkahboeng. *Proceedings of the International Society of Sugar Cane Technologists*, **7**, 473–499.

26 Han Lioe Hong (1956) Fusarium-pokkah boeng resistance trial. *Proceedings of the International Society of Sugar Cane Technologists*, **9**, 1023–1029.

27 Went, F.A.F.C. (1893) Het rood snot. *Archiv Java Suiker-industrie*, **1**, 265–282.

28 Singh, K. & Singh, R.P. (1989) Red rot. In: *Diseases of Sugarcane – Major Diseases* (eds C. Ricaud, B. T. Egan, A. G. Gillaspie & C. G. Hughes), pp. 169–188. Elsevier, Amsterdam.

29 Singh, R.P. & Lal, S. (2000) Red rot. In: *A Guide to Sugarcane Diseases* (eds P. Rott, R. A. Bailey, J. C. Comstock, B. J. Croft & A. S. Saumtally), pp. 153–158. CIRAD and International Society of Sugar Cane Technologists, Montpellier.

30 Sandhu, S.S., Mehan, V.K. & Singh, K. (1969) Extent of losses in sugarcane caused by red rot (*Physalospora tucumanensis* Speg) and smut *Ustilago scitaminea* Sydow. *Journal of Research Ludhiana*, **6**, 341–344.

31 Trenor, K.L. & Bailey, R.A. (1989) A preliminary report on the incidence of red rot in the South African sugar industry. *Proceedings of the South African Sugar Technologists Association*, **63**, 111–116.

32 Edgerton, C.W. (1955) *Sugarcane and its Diseases*. Louisiana State University Press, Baton Rouge.

33 Azab, Y.E. & Chilton, S.J.P. (1952) Studies on inheritance of resistance to red rot disease of sugarcane. *Phytopathology*, **42**, 282.

34 Abbott, E.V., Zummo, N. & Tippett, R.L. (1967) Methods of testing sugarcane varieties for disease resistance at the US sugar cane field station, Houma, Louisiana. *Proceedings of the International Society of Sugar Cane Technologists*, **12**, 1138–1143.

35 Kar, K., Gupta, S.C. & Kureel, D.C. (1974) Screening of varieties for red rot resistance. *Proceedings of the International Society of Sugar Cane Technologists*, **15**, 189–193.

36 Ryan, C.C. & Egan, B.T. (1989) Rust. In: *Diseases of Sugarcane – Major Diseases* (eds C. Ricaud, B. T. Egan, A. G. Gillaspie & C. G. Hughes), pp. 189–210. Elsevier, Amsterdam.

37 Raid, R.N.& Comstock, J.C. (2000) Common rust. In: *A Guide to Sugarcane Diseases* (eds P. Rott, R. A. Bailey, J.

C. Comstock, B. J. Croft & A. S. Saumtally), pp. 85–89. CIRAD and International Society of Sugar Cane Technologists, Montpellier.

38 Comstock, J.C., Shine, J.M. & Raid, R.N. (1992) Effect of rust on sugarcane growth and biomass. *Plant Disease*, **76**, 175–177.

39 Ferreira, S.A. & Comstock, J.C. (1989) Smut. In: *Diseases of Sugarcane – Major Diseases* (eds C. Ricaud, B. T. Egan, A. G. Gillaspie & C. G. Hughes), pp. 211–229. Elsevier, Amsterdam.

40 Anon (1979) Smut and cane yield. *Annual Report of the South African Sugar Association Experimental Station 1978–79*, 71.

41 Pearse, T.L. (1989) Influence of intensive roguing on the incidence of smut in Swaziland. *Proceedings of the South African Sugar Technologists Association*, **63**, 117–121.

42 Bailey, R.A. (1983) The effect of soil and seedcane applications of triadimefon on the incidence of sugarcane smut (*Ustilago scitaminea* Sydow). *Proceedings of the South African Sugar Technologists Association*, **59**, 99–104.

43 Anon (1980) Screening for smut resistance. *Annual Report of the South African Sugar Association Experimental Station 1979–80*, 70.

44 Grisham, M.P. (2001) An international project on genetic variability within sugarcane smut. *Proceedings of the International Society of Sugar Cane Technologists*, **24**, 459–461.

45 Orian, G. (1954) The probable origin of the gumming disease of the sugarcane. *Proceedings of the International Society of Sugar Cane Technologists*, **8**, 862–876.

46 Saumtally, S. & Dookun, A. (2000) Gumming. In: *A Guide to Sugarcane Diseases* (eds P. Rott, R. A. Bailey, J. C. Comstock, B. J. Croft & A. S. Saumtally), pp. 32–37. CIRAD and International Society of Sugar Cane Technologists, Montpellier.

47 Ricaud, C. & Autrey, L.J.C. (1989) Gumming disease. In: *Diseases of Sugarcane – Major Diseases* (eds C. Ricaud, B. T. Egan, A. G. Gillaspie & C. G. Hughes), pp. 21–38. Elsevier, Amsterdam.

48 North, D.S. (1935) The gumming disease of the sugarcane, its dissemination and control. *Agriculture Report (Technology)*, **10**. CSR, Sydney.

49 Autrey, L.J.C., Dhayan, S. & Sullivan, S. (1986) Effect of race 3 of gumming disease pathogen on growth and yield in two sugarcane varieties. *Proceedings of the International Society of Sugar Cane Technologists*, **19**, 420–428.

50 Anon (1983) Diseases. *Annual Report of the Maur Sugar Industry Research Institute 1983*, 34–41.

51 Saumtally, S. (1996) *Genetic variability in Xanthomonas campestris pv vasculorum (Cobb) Dye, causal agent of gumming diseases of sugarcane*. PhD thesis. University of Reading, UK.

52 Ricaud, C. & Ryan, C.C. (1989) Leaf scald. In: *Diseases of Sugarcane – Major Diseases* (eds C. Ricaud, B. T. Egan, A. G. Gillaspie & C. G. Hughes), pp. 39–58. Elsevier, Amsterdam.

53 Rott, P. & Davis, M.J. (2000) Leaf scald. In: *A Guide to Sugarcane Diseases* (eds P. Rott, R. A. Bailey, J. C. Comstock, B. J. Croft & A. S. Saumtally), pp. 38–44. CIRAD and International Society of Sugar Cane Technologists,

Montpellier.

54 Wilbrink, G. (1920) De gomziekte van het suikeriet, harr oorzaak en hare bestrrijding. *Archief voor de Suikerindustrie Nederlandsch-Indie*, **28**, 1399–1525.

55 Autrey, L.J.C., Saumtally, S., Dookun, A., Sullivan, S. & Dhayan, S. (1995) Aerial transmission of the leaf scald pathogen, *Xanthomonas albilineans*. *Proceedings of the International Society of Sugar Cane Technologists*, **21**, 508–525.

56 Davis, M.J., Rott, P., Warmuth, C.J., Chatenet, M. & Baudin, P. (1997) Intraspecific genomic variation within *Xanthomonas albilineans*, the sugarcane leaf scald pathogen. *Phytopathology*, **87**, 316–324.

57 Antoine, R. & Ricaud, C. (1962) Cane diseases – a method for inoculating leaf scald in field trials. *Annual Report of the Maur Sugar Industry Research Institute 1961*, 55–56.

58 Gillaspie, A.G., Davis, R.E. & Worley, J.E. (1974) Nature of the ratoon stunting disease agent. *Proceedings of the International Society of Sugar Cane Technologists*, **15**, 218–224.

59 Teakle, D.S. (1974) The causal agent of ratoon stunting disease. *Proceedings of the International Society of Sugar Cane Technologists*, **14**, 813–826.

60 Davis, M.J., Gillaspie, A.G. Jr, Harris, R.W. & Lawson, R.H. (1980) Ratoon stunting disease of sugarcane: isolation of the causal bacterium. *Science*, **210**, 1365–1367.

61 Davis, M.J., Gillaspie, A.G. Jr, Vidaver, A.K. & Harris, R.H. (1984) *Clavibacter*: a new genus containing some phytopathogenic coryneform bacteria, including *Clavibacter xyli* subsp. *xyli* sp. nov., subsp. nov. and *Clavibacter xyli* subsp. *cynodontis* subsp. nov., pathogens that cause ratoon stunting disease of sugarcane and Bermuda grass stunting disease. *International Journal of Systematic Bacteriology*, **34**, 107–117.

62 Evtushenko, L.I., Dorofeeva, L.V., Subbotin, S.A., Cole, J.R. & Tiedje, J.M. (2000) *Leifsonia poae* gen. nov., sp. nov., isolated from nematode galls on *Poa annua*, and reclassification of '*Corynebacterium aquaticum*' Leifson 1962 as *Leifsonia aquatica* (ex Leifson 1962) gen. nov., nom. rev., comb. nov. and *Clavibacter xyli* Davis *et al.* 1984 with two subspecies as *Leifsonia xyli* (Davis *et al.* 1984) gen. nov., comb. nov. *International Journal of Systematic and Evolutionary Microbiology*, **50**, 371–380.

63 Steindl, D.R.L. (1961) Ratoon stunting disease. In: *Sugarcane Diseases of the World*, Vol. I (eds J. P. Martin, E. V. Abbott & C. G. Hughes), pp. 433–459. Elsevier, Amsterdam.

64 Gillaspie, A.G. & Teakle, D.S. (1989) Ratoon stunting disease. In: *Diseases of Sugarcane – Major Diseases* (eds C. Ricaud, B. T. Egan, A. G. Gillaspie & C. G. Hughes), pp. 59–80. Elsevier, Amsterdam.

65 Davis, M.J. & Bailey, R.A. (2000) Ratoon stunting. In: *A Guide to Sugarcane Diseases* (eds P. Rott, R. A. Bailey, J. C. Comstock, B. J. Croft & A. S. Saumtally), pp. 49–54. CIRAD & International Society of Sugar Cane Technologists, Montpellier.

66 Bailey, R.A. & Fox, P.H. (1984) A large-scale diagnostic service for ratoon stunting disease of sugarcane. *Proceedings of the South African Sugar Technologists Association*, **58**, 204–10.

67 Davis, M.J., Dean, J.L., Miller, J.D. & Shine, J.M. (1994) A method to screen for resistance to ratoon stunting disease of sugarcane. *Sugar Cane*, **6**, 9–16.

68 Croft, B.J., Greet, A.D., Leaman, T.M. & Teakle, D.S. (1994) RSD diagnosis and varietal resistance screening in sugarcane using the EB-EIA technique. *Proceedings of the Australian Society of Sugar Cane Technology*, **14**, 143–151.

69 McFarlane, S.A., Bailey, R.A. & Subramoney, D.S. (1999) The introduction of a serological method for large-scale diagnosis of ratoon stunting disease in the South African sugar industry. *Proceedings of the South African Sugar Technologists Association*, **73**, 123–127.

70 Fegan, M., Croft, B.J., Teakle, D.S., Hayward, A.C. & Smith, G.R. (1998) Sensitive and specific detection of *Clavibacter xyli* subsp. *xyli*, causal agent of ratoon stunting disease of sugarcane, with a polymerase chain reaction-based assay. *Plant Pathology*, **47**, 495–504.

71 Bailey, R.A. & Bechet, G.R. (1997) Further evidence of the effects of ratoon stunting disease on production under irrigated and rainfed conditions. *Proceedings of the South African Sugar Technologists Association*, **71**, 97–101.

72 Hughes, C.G. (1974) The economic importance of ratoon stunting disease. *Proceedings of the International Society of Sugar Cane Technologists*, **15**, 213–217.

73 Dean, J.L. & Davis, M.J. (1990) Losses caused by ratoon stunting disease of sugarcane in Florida. *Journal of the American Society of Sugar Technologists*, **10**, 66–72.

74 Bailey, R.A. & McFarlane, S.A. (1999) The incidence and effects of ratoon stunting disease of sugarcane in southern and central Africa. *Proceedings of the International Society of Sugar Cane Technologists*, **23**, 338–346.

75 Bailey, R.A. & Bechet, G.R. (1986) Effect of ratoon stunting disease on the yield and components of yield of sugarcane under rainfed conditions. *Proceedings of the South African Sugar Technologists Association*, **60**, 143–147.

76 Bailey, R.A. & Tough, S.A. (1992) Rapid spread of RSD during manual harvesting of sugarcane and the effect of knife cleaning on the rate of spread. *Proceedings of the South African Sugar Technologists Association*, **66**, 78–81.

77 Bailey, R.A. & Tough, S.A. (1992) Ratoon stunting disease: survival of *Clavibacter xyli* subsp. *xyli*, in field soil and its spread to newly planted sugarcane. *Proceedings of the South African Sugar Technologists Association*, **66**, 75–77.

78 Comstock, J.C., Miller, J.D., Shine, J.M. Jr & Tai, P.Y.P. (1995) Screening for resistance to ratoon stunting disease in Florida. *Proceedings of the International Society of Sugar Cane Technologists*, **22**, 520–526.

79 Hoy, J.W. & Flynn, J.L. (2001) Control of ratoon stunting disease of sugarcane in Louisiana with seedcane produced through micropropagation and resistant cultivars. *Proceedings of the International Society of Sugar Cane Technologists*, **24**, 417–421.

80 Koike, H. & Gillaspie, A.G. (1989) Mosaic. In: *Diseases of Sugarcane – Major Diseases* (eds C. Ricaud, B. T. Egan, A. G. Gillaspie & C. G. Hughes), pp. 301–322. Elsevier, Amsterdam.

81 Grisham, M.P. (2000) Mosaic. In: *A Guide to Sugarcane Diseases* (eds P. Rott, R. A. Bailey, J. C. Comstock, B. J. Croft & A. S. Saumtally), pp. 249–254. CIRAD and International Society of Sugar Cane Technologists,

Montpellier.

82 Yang, Z.N. & Mirkov, T.E. (1997) Sequence and relationships of sugarcane mosaic and sorghum mosaic virus strains and development of RT-PCR-based RFLPs for strain discrimination. *Phytopathology*, **87**, 932–939.

83 Bailey, R.A. & Fox, P.H. (1987) A preliminary report on the effect of sugarcane mosaic virus on the yield of sugarcane varieties NCo376 and N12. *Proceedings of the South African Sugar Technologists Association*, **61**, 1–4.

84 Zummo, N. & Charpentier, L.J. (1964) Vector–virus relationship of sugarcane mosaic virus. 1 – Transmission of sugarcane mosaic by the brick-red sowthistle aphid (*Dactynotus ambrosiae* Thos). *Plant Disease Reporter*, **48**, 636–639.

85 Harborne, K.M. (1988) Population dynamics of the main aphid vectors of sugarcane mosaic virus in Natal. *Proceedings of the South African Sugar Technologists Association*, **62**, 195–198.

86 Dean, J.L. (1971) Systemic host assay of sugarcane mosaic virus. *Phytopathology*, **61**, 526–531.

87 Bailey, R.A. & Fox, P.H. (1980) The susceptibility of varieties to mosaic and the effect of planting date on mosaic incidence in South Africa. *Proceedings of the South African Sugar Technologists Association*, **54**, 161–167.

88 Lockhart, B.E. & Cronjé, C.P.R. (2000) Yellow leaf syndrome. In: *A Guide to Sugarcane Diseases* (eds P. Rott, R. A. Bailey, J. C. Comstock, B. J. Croft & A. S. Saumtally), pp. 291–295. CIRAD and International Society of Sugar Cane Technologists, Montpellier.

89 Scagliusi, S.M. & Lockhart, B.E.L (2000) Transmission, characterization and serology of a luteovirus associated with yellow leaf syndrome of sugarcane. *Phytopathology*, **90**, 120–124.

90 Cronjé, C.P.R., Tymon, A.M., Jones, P. & Bailey, R.A. (1998) Association of a phytoplasma with a yellow leaf syndrome of sugarcane in Africa. *Annals of Applied Biology*, **133**, 177–186.

91 Cronjé, C.P.R. & Bailey, R.A. (1999) Association of phytoplasmas with yellow leaf syndrome of sugarcane. *Proceedings of the International Society of Sugar Cane Technologists*, **23**, 373–380.

92 Grisham, M.P., Pan, Y.B., Legendre, B.L., Godshall, M.A. & Eggleston, G. (2001) Effect of sugarcane yellow leaf virus on sugarcane yield and juice quality. *Proceedings of the International Society of Sugar Cane Technologists*, **24**, 434–437.

93 Egan, B.T., Ryan, C.C. & Francki, R.I.B. (1989) Fiji disease. In: *Diseases of Sugarcane – Major Diseases* (eds C. Ricaud, B. T. Egan, A. G. Gillaspie & C. G. Hughes), pp. 267–287. Elsevier, Amsterdam.

94 Comstock, J.C. & Lockhart, B.E.L. (1990) Widespread occurrence of sugarcane bacilliform virus in US sugarcane germplasm collections. *Plant Disease*, **74**, 530.

95 Viswanathan, R. (2000) Grassy shoot. In: *A Guide to Sugarcane Diseases* (eds P. Rott, R. A. Bailey, J. C. Comstock, B. J. Croft & A. S. Saumtally), pp. 215–220. CIRAD and International Society of Sugar Cane Technologists, Montpellier.

96 Chen, C.T. & Kusalwong, A. (2000) White leaf. In: *A Guide to Sugarcane Diseases* (eds P. Rott, R. A. Bailey, J. C. Comstock, B. J. Croft & A. S. Saumtally), pp. 231–236. CIRAD and International Society of Sugar Cane Technologists, Montpellier.

97 Suma, S. & Jones, P. (2000) Ramu stunt. In: *A Guide to Sugarcane Diseases* (eds P. Rott, R. A. Bailey, J. C. Comstock, B. J. Croft & A. S. Saumtally), pp. 226–230. CIRAD and International Society of Sugar Cane Technologists, Montpellier.

Chapter 4
Pests of Sugarcane

Graeme Leslie

INTRODUCTION

A large variety of insects feed on sugarcane. Many are only occasional feeders, but in most regions where this crop is grown insect pests are a significant factor in the economics of sugarcane production. Pemberton & Williams[1] noted that most pests of sugarcane are local species that have moved onto sugarcane from the habitat surrounding where the crop is cultivated. However, some, for example the lepidopterous stalk borer *Chilo sacchariphagus* (Bojer), have been spread into new regions accidentally from their area of origin as a result of the activities of man.

Pests of sugarcane can be grouped according to various criteria such as geographic distribution, taxonomic grouping, severity of damage caused, and feeding habit. Here, pests have been grouped by feeding habit as follows:

- pests that feed on stalks of sugarcane – stalk borers;
- pests that feed on the subterranean parts of stools – soil pests;
- pests that feed on the plant sap – sap feeders; and
- pests that feed on the leaves of sugarcane – leaf feeders.

The biology, damage, distribution and approaches to control of the major pests for each of these groups are examined. Within each of these groups, pest status may vary from region to region; what may be a minor pest in one region, may be a major pest in another. There are also several vertebrate pests of sugarcane including rats, pigs, monkeys and birds. Of these, rats are probably the most serious, and these and other pests in this group are also considered in this chapter.

STALK BORERS

Most pests in this group are lepidopterans, and there are a few coleopteran species of significance. In some regions stalk borers are the major pests of sugarcane and significant research effort is focused on them. Table 4.1 lists the major stalk borers of sugarcane in various parts of the world. The major groups of borers include many pyralids such as *Diatraea* spp., *Chilo* spp. and *Eldana saccharina* (Walker). Agrotid pest species are mainly found in the genus *Sesamia*.

Biology

The biology of these pests is essentially similar. Oviposition occurs on or in the crop. For example, in Louisiana *Diatraea saccharalis* (Fabr.) lays its eggs on the upper surface of the green leaves towards the midrib. *Eoreuma loftini* (Dyer) and *Eldana saccharina* lay their eggs on or behind folds in the dead leaf material attached to stalks[2,3]. After hatching, neonate larvae disperse in the crop to the preferred point of entry. Some produce silk threads by which they drop from leaves and may be wind dispersed. In the case of *D. saccharalis*, the larvae move from the green leaves to the nodes around the growing point, as does *E. loftini*. Species such as *E. saccharina* can disperse up stalks but eventually bore into the lower parts of stalks. This borer can cross the nodal plate of stalks as it feeds internally and can damage several internodes as it develops. Larvae undergo a number of moults

Table 4.1 A list of the more important stalk borers of sugarcane.

Pest species	Region
Lepidoptera	
Argyroploce schistaceana (Sn.)	Indonesia, Taiwan
Castniomera licus (Drury)	Guyana
Chilo sacchariphagus (Bojr.)	Far East, Mascerine islands
Chilo infuscatellus (Snellen)	Australasia
Chilo agamemnon (Blezynski)	Egypt
Diatraea saccharalis (Fabricius)	North and South America
Diatraea flavipennella (Box)	Brazil
Diatraea indigenella (D. & H.)	Brazil
Diatraea rosa (Heinrich)	Venezuela
Diatraea considerata (Heinrich)	Mexico
Diatraea grandiosella (Dyar)	Mexico
Eldana saccharina (Walker)	Africa
Eoreuma loftini (Dyar)	USA, Mexico
Scirpohaga excerptalis (Walker)	South-east Asia, India
Scirpophaga nivella (Fr.)	South-east Asia, India, Indonesia
Sesamia grisescens (Walker)	Papua New Guinea
Sesamia cretica (Lederer)	Africa
Tryporyza nivell intacta (Sn)	South-east Asia
Coleoptera	
Metamasius hemipterus (L.)	Florida
Rhabdoscelus obscurus (Boisd)	Australia, Papua New Guinea

(seven or more), development time being dependent on factors such as temperature and food quality. Pupation occurs within a cocoon or without, either in the soil or in the bored stalk, and adult emergence can occur within two to three weeks of pupation depending on climate and other environmental factors. In South Africa, one generation of the borer *E. saccharina* (egg to adult) develops in approximately eight weeks, while in Louisiana one generation of *D. saccharalis* takes 30 days. Some species can over-winter, for example in Louisiana *D. saccharalis* over-winters in the stubble of harvested fields or in wild hosts such as *Sorghum halipense* (W. H. White, personal communication).

Damage

The damage caused differs between species. Some attack young sugarcane killing the shoots (e.g. *Chilo agamemnon* Bles.) while others attack the top of stalks (e.g. *D. saccharalis)*, killing the growing point, which often results in side shooting. Still others, like *E. saccharina*, bore into the lower parts of stalks, which can cause a severe loss of purity because the sucrose content is greatest in the lower nodes of stalks. Losses in sucrose content are attributed primarily to the actions of fungal pathogens, particularly the red rot fungus (*Glomerella tucumanensis*) which produces the red discoloration often associated with borer damage to stalks. There have been many assessments made of the loss that borers cause to sugarcane, and the estimates are broadly similar. The approaches adopted attempt to relate damaged or bored stalk internodes with various parameters of yield, including such measures as purity, sucrose yield, crop yield, or combinations of these.

In Swaziland, King[4] showed that the borer *E. saccharina* caused an average loss of recoverable sucrose of 1% for every 1% of internodes bored. Rajabalee[5] reported that *C. sacchariphagus* caused a loss of 0.69% in sugar for every 1% of internodes bored. Simlarly in Indonesia, Kuniata[6] reported that losses of about 10% sugar loss from 20% bored internodes could be attributed to *C. sacchariphagus*. Kira and El-Sherif[7] in their study on *Chilo agamamenon*, estimated that with a 1% infestation, a loss of sugar yield of between 0.65% and 0.67% was produced. In Taiwan, Cheng[8] reported that an average infestation level of 8.9% by borers resulted in a reduction of between 19.3% and 43.2% of available sucrose. This gave a loss of 457 kg sucrose/ha. In Texas, Meagher *et al.*[9] estimated that losses owing to *E. loftini* were 0.1083 t sucrose/ha for every internode bored.

Damage caused by beetle borers has also been examined, e.g. losses caused by the weevil borer *Rhabdoscelus obscurus* (Boisd.) in Fiji[10]. They showed that 2% stalk damage for length and weight caused an average loss of 1.5% in pure obtainable sugar and a 3.35% loss in purity. In Papua New Guinea, this borer causes damage of < 2–8% of stalks bored. It was, however, difficult to separate damage caused by this borer and that caused by *Sesamia grisescens* (Walker) and *Chilo terrenellus* (Pagenstecher)[6].

Distribution

Lepidopterous sugarcane borers have been recovered from sugarcane in all regions where this crop is cultivated. The genus *Diatraea*, however, is restricted to the New World while the genus *Chilo* is found only in the Old World. Bleszynski[11] lists 32 species of the genus *Diatraea* associated with sugarcane. Most are minor pests, but some are serious and the more important of these are included in the list given in Table 4.1. In North America, the borers *D. saccharalis* and *E. loftini* are the most serious pests of sugarcane in Florida, Louisiana and Texas. *Diatraea saccharalis*, *D. grandiosella* (Dyar) and *D. considerata* (Heinrich) are considered to be important pests of sugarcane in Mexico in addition to *E. loftini*. Generally, species of *Diatraea* are the most serious pests of sugarcane in South America. In Venezuela for example, five species are found in cane, of which *Diatraea rosa* (Heinrich) is the most common. In Brazil, *D. flavipennela* (Box) and *D. indiginella* (D&H) are the most important, while in the Cuaca valley of Colombia, *D. saccharalis* and *D. indiginella* are the most important pest species.

Leslie[12] lists 15 lepidopterous species that are associated with sugarcane in Africa. Of these, only five are considered to be of any economic consequence, i.e. *E. saccharina*, *C. sacchariphagus*, *C. agamemnon*, *Sesamia cretica* (Lederer) and *S. calamistis* (Hampson). Rajabalee[5] lists 12 species of the genus *Chilo* associated with sugarcane worldwide, although not all can be considered serious pests of sugarcane. Kuniata[6] lists 12 lepidopterous and four coleopterous species as being associated with damage to sugarcane in the Far East. The more important lepidopteran pests include *Sesamia grisescens* (Walker), *C. sacchariphagus*, *Scirpophaga excerptalis* (Walker), *S. nivella* (Fr.) and *Argyroploce schistaceana* (Sn). Also in that region, Samoedi[13] states that the borer *Tryporyza nivella intacta* (Sn) is the most serious pest of sugarcane in Java and Sumatra. In the Indian subcontinent, species of *Chilo* considered to be economically important include *C. infuscatellus* (Snellen), and *C. sacchariphagus*. Coleopterous pests of sugarcane also include the weevil borers *Metamasius hemipterus* (L.) in Florida[14], and *Rhabdoscelus obscurus* (Boisd.) in Fiji[10], Australia and Papua New Guinea[6].

Control

Biological control is an important strategy against lepidopteran stalk borers. Additionally, cultural practices and host plant resistance are also approaches that have met with success. Insecticides have been used in some cases; but generally they are not widely used, often because of cost or because of their possible impact on the efficacy of natural enemies.

Biological control

Biological control is extensively used in the sugarcane growing regions of South America. In Brazil, the tachinid larval parasitoids, *Metagonistylum minense* (Tns.) and *Paratheresia claripalpis* (Wulp.) and the braconid *Cotesia flavipes* (Cameron) have been routinely released for the control of *D. saccharalis*[15]. Since 1988, parasitoid releases have reduced the infestation intensity from as high as 10% to an average in 1994 to about 3%[16]. Similarly in Venezuela, Salazar[17] reported that the *Diatraea* spp. occurring there were no longer considered of consequence because of good biological control. This has been achieved initially by releasing the larval parasitoid *M. minense*. Later, *C. flavipes* was released providing more effective control. Just how effective this approach has been can be seen in the levels of damage. Salazar[17] observed that 16% infestation was recorded in 1947 and in 1996 this was only 2%.

In Colombia, Gomez[18] reported that the artificially reared larval parasitoids *M. minense* and *P. claripalpis* have been effective against *D. saccharalis* and *D. indigenella*. Egg parasitoids have also been released. Both *Trichogramma pretiosum* (Riley) and *T. exiguum* (Pi. Pl. and O.) have been released; however, no field recoveries have been made of *T. pretiosum*. This was ascribed to its poor adaptation to *Diatraea* spp. as hosts. However *T. exiguum* has been recovered from the field and contributed to control[18]. An adequate supply of healthy host material is often a problem in mass rearing parasitoids. It is therefore of interest to note, as an aside, this author commented that parasitoid rearing had become more efficient with the elimination of a protozoan parasite from the laboratory cultures of the host insect.

In Mexico, biological control is one of several strategies adopted for the control of its borer complex, which comprises three species of *Diatraea* as well as *E. loftini* [19]. The indigenous parasitoid, *Allorhogas pyralophagus* (Marsh), has limited impact, but releases of *M. minense* have had some influence on damage. Releases are, however, restricted to October–February as high temperatures for the rest of the year prevent further releases. Florez[20] reported that *T. pretiosum* and *C. flavipes* are also being released against borers in Mexico.

In North America, White & Regan[21] reviewed the introduction of parasitoids in Louisiana for the control of *D. saccharalis*. Early experience involving the release of both egg and larval parasitoids was not sufficiently promising, despite establishment having been demonstrated. Recent attempts have focused on two species of *Cotesia*, viz: *C. flavipes* and *C. chilonus* (Ishii). Although these parasitoids have not yet become established, levels of parasitism by *C. flavipes* and *C. chilonus* were as high as 15% and 55%, respectively.

There are many borers of sugarcane in the Far East and Australasian region, and the use of biological control against these has been reviewed by Conlong[22]. Quoting several sources, he lists three major borers from Taiwan and seven from the Australasian region, and their parasitoids. Releasing the egg parasitoid *T. chilonus* reduced infestations of the borer *Argyroploce schistaceana* (Sn.) in Taiwan. Other examples include > 80% parasitism of late instar larvae of the borer *S. grisescens* in Papua New Guinea by *C. flavipes*, and the effective parasitism of *Chilo infuscatellus* (Sn.) by *Trichogramma* sp. in Indonesia.

An extensive biological control programme has been implemented against *E. saccharina* in South Africa. Conlong[23] lists those parasitoids tested against the different life stages of this borer. While in many cases successful laboratory rearing has been achieved and field recoveries made; however, their impact on crop damage has not been clear. Currently, the tachinid parasitoid *Sturmiopsis parasitica* (Curr.) has been released and recoveries have been made[24]. Similarly, a large programme

was developed in Mauritius against *C. sacchariphagus*. Since 1939, 30 egg, larval and pupal parasitoids were introduced into Mauritius against this borer. However, only *Xanthopimpla stemmator* (Thun.) and *Trichospilus diatraeae* (C&M) became established, and neither had an impact on the borer[25]. M. M. Embaby (unpublished data) in his study on the control of *C. agamemnon* on sugarcane in Egypt found that the egg parasitoid *Trichogramma evanescens* (Westw.) contributed significantly to the control of this borer. A reduction in the infestation level of between 50% and 60% was achieved at a release rate of 20 000–30 000 per feddan (1 feddan = 0.42 ha).

Spaull[26] investigated the use of nematodes as possible biocontrol agents of stalk borers, and *Heterorhabditis* sp. was examined for the control of *E. saccharina* larvae. Spaull showed that up to 40% larval mortality was achieved when a nematode suspension was sprayed on sugarcane at a rate of 11 000 million infective stage *Heterorhabditis* in 7400 L of water/ha. More larvae were killed when the nematodes were applied to the cane during the late afternoon than just before sunrise or at midday. However, desiccation of the nematodes in the field was a problem, and in a later study Spaull showed that the addition of the water thickener, Methocel J75, to the nematode suspension, increased mean larval mortality from 20% to 33% [27].

Insecticides

In Louisiana, insecticides are the most effective control measure used against *D. saccharalis* [28]. Routinely, the insecticides used are esfenvalerate and cyfluthrin at 420–630 g and 148 g a.i./ha, respectively. Applications are only recommended when an infestation exceeds the threshold value of 5% stalks infested with live larvae. Most applications are by air at a rate of approximately 19 L/ha (W. H. White, personal communication). In Texas, Meagher *et al.* [9] examined the use of insecticides against *E. loftini*, and the efficacy of three insecticides (monocrotophos, azinphosmethyl and cyfluthrin) was tested. Insecticides were applied weekly for 7, 4 and 1 months in the different treatments. Damage was least where the greatest number of treatments was applied. However, while damage could be shown to be significantly reduced by the treatments, there was not a corresponding increase in sucrose yield associated with any treatments.

On the East coast of Mexico, monocrotophos and trichlorfon have been used[29]; however, it was found that secondary pests such as lace bugs and the yellow flea became problematic. In the La Primavera mill (western Mexico), monocrotophos has been used against *D. grandiosella*, *D. considerata*, *D. saccharalis* and *E. loftini*. Use is now made of β-cyfluthrin and λ-cypermethrin as a direct application to the young crop when necessary. In Indonesia, *C. sacchariphagus* and *C. auricilius* have both been controlled by use of insecticides; however, the costs of treatment are apparently prohibitive[30]. In Africa, there is an extensive programme of field-testing of insecticides against *E. saccharina*. The approach has been to apply insecticides over the period of peak numbers of *E. saccharina* moths. The rationale behind this approach is that, over such a period, neonate larvae (the most exposed stage of this pest) will encounter a toxic treated surface as they disperse. Insecticides tested include synthetic pyrethroids (deltamethrin and cypermethrin) as well as the insect growth regulator, flufenoxuron. Despite some encouraging results, no treatment is used commercially; but recommendations exist for treating seed material (see later). Similarly, in Egypt, Awadallah *et al.*[31] tested chlordane and carbofuran against *Sesamia cretica*, but with no clear effect on damage. However, earlier Hassanien & El-Naggar[32] tested isobenzan, trichlorphon, heptachlor and endrin against *C. agamemnon*. Rates used per feddan were 2–3 kg; 1–1.5 kg; 12 kg; and 12 kg, respectively. The best results were obtained from isobenzan and endrin. Pandey *et al.*[33] reported on the use of carbofuran against *Scirpophaga excerptalis* (Walker) in India. They found that carbofuran, applied at a rate of 30 kg/ha, and in conjunction with inundative releases of the egg parasitoid *Trichogramma* sp., reduced borer incidence by nearly 13%. Also, Singla & Duhra[34] assessed the efficacy of five insecticides against the stalk borer *C. auricilius*. Two sprays of cypermethrin at 100 g a.i./ha proved better than other insecticides. Sprays applied in July gave better results compared to those applied in September.

Insecticides do not seem to be widely used for pest control problems in the Far East. However, there have been a few limited attempts, and the results from Papua New Guinea are discussed by Kuniata[6]. *Sesamia grisescens* is a serious pest in Papua New Guinea, and attempts have been made to control it using carbofuran. However, this only aggravated the cicada problem, and treatments were therefore discontinued. However, in 2000, Kuniata[35] reported the successful use of lambda-cyhalothrin and permethrin against this borer. At rates of 25 g and 250 g a.i./ha, sugarcane yields in sprayed plots were 57% and 64% greater, respectively, than those of the unsprayed plots. Some use has been made of chlorpyrifos as a treatment for seedcane against the weevil borer *Rhabdoscelus obscurus*, and was found to be effective. In Indonesia, the insecticides monocrotophos and methidathion have been found to be effective against various *Chilo* spp. However, their applications using conventional techniques pose problems, and are therefore rarely used.

Host plant resistance

Host plant resistance is an important component of any strategy aimed at reducing the economic impact of crop pests. It is based on three factors: antibiosis, antixenosis, and tolerance[36]. Antibiosis is based on the plant inhibiting the development of the feeding insect, while antixenosis acts by influencing adult and/or larval behaviour on the surface of the host plant. Most studies have focused on antibiosis, although Sosa[37] has shown that pubescence in sugarcane can be important. In his study he showed that leaf pubescence adversely affected oviposition and neonate larval movement of *D. saccharalis* in tests conducted with clones that were and were not pubescent. In Louisiana, an established programme for evaluating varieties against *D. saccharalis* has been developed based on rating of varieties in the field. Rating varieties, where a number of measures are taken can be difficult. In a study on varietal resistance to *D. saccharalis*, White[38] showed that cluster analysis was a useful procedure that allowed the use of multiple variables in assessing varietal resistance. Using such procedures, a more reliable estimate of varietal resistance may be obtained. In this study, assessments were based on visual damage and percentage of internodes bored. Other possible parameters include larval and pupal numbers, weight and pupal sex. Leslie & Keeping[39] showed that using measures such as these, both principal component analysis and cluster analysis could be useful in helping to evaluate varieties for resistance to the borer *E. saccharina*. The above procedures apply to field-based or large-scale shade-house trials. However, Sosa[40] reported on a simple laboratory-based procedure for evaluating resistance. In his study, third instar larvae were introduced into stalks of sugarcane via modified microcentrifuge tubes. Evaluation was based on the length of larval borings 10 days after inoculation. Results from this preliminary study were in general agreement with varietal evaluations based on other methods.

In Egypt, Allam & Abou Dooh[41] showed that there was a range of resistance in the 26 commercial varieties they screened against *Chilo* spp. They concluded that resistance was linked to fibre content, i.e. high-fibre varieties (15%) were more resistant than low-fibre ones (11.3%). As high-fibre stalks pose a problem for milling, however, resistance based on characteristics other than fibre would be more practical. In South Africa, Keeping & Leslie[42] examined the role of self-trashing (or self-stripping) stalks as a factor in crop resistance to *E. saccharina*. Because this pest selects dead leaf material on which to oviposit, any factor, such as self-trashing, that may possibly reduce availability or suitability of this material for oviposition would be a useful characteristic to select for. In their study, several self-trashing and non-self-trashing varieties were tested in field trials. The results showed, however, that there were no clear differences in the levels of damage in the varieties tested.

In their study on varietal resistance to *D. saccharalis*, Bessin *et al.*[43] noted that evaluation of the percentage of bored internodes measures the cultivar resistance only until penetration by young larvae into the stalk. By using the emergence holes of the sugarcane borer as a seasonal record of adult emergence, they developed a relative survival index to estimate the proportion of larvae inside the stalk that survived to adulthood. From this they developed a moth production index that serves as a measure of area-wide suppression (or enhancement) of borer populations. Their results showed that differences exist between the two methods (assessing bored internodes and moth production), and that the resistance estimate based on moth production should be considered for inclusion when assessing varieties.

Another development in the field of host plant resistance is the application of biotechnology to improve crop resistance to pest attack. The strategy aims to insert a gene (or genes) that codes for a toxin into the genome of sugarcane plants or into a bacterial species that may colonise the plant surface. In cases where the insertion is successful, and the expression of the gene is at a sufficiently high level, insects feeding on such plants either die or develop slowly, thereby reducing the damage caused. The South African study by Herrera *et al.*[44] is an example. A cryIA(c) gene was cloned from a native *Bacillus thuringiensis* strain showing activity against *E. saccharina*. The gene was introduced into an isolate of *Pseudomonas fluorescens*, capable of colonising sugarcane. Glasshouse trials indicated that sugarcane treated with the modified *P. fluorescens* was more resistant to *E. saccharina* damage than untreated sugarcane. Inserting toxin-coding genes into sugarcane itself has been reported in Cuba. Arencibia *et al.*[45] reported the successful expression of a truncated gene encoding for the active region of the *B. thuringiensis* δ-endotoxin in sugarcane. Despite low expression of the gene, the transgenic plants showed significant larvicidal activity against *D. saccharalis*. This approach to pest control shows great promise, and, as long as strategies recommended for reducing the build-up of pest resistance are implemented, transgenic crops will become increasingly relevant in pest control strategies.

Management practices

There are several crop management practices that can influence borer infestations in sugarcane including:

- altering planting dates;
- burning crops at harvest;
- collecting dead shoots;

- early harvesting;
- reduced fertiliser use;
- flooding;
- pest-free seedcane; and
- field monitoring.

All of these approaches have been applied to borer control in various regions. For example, selecting planting date was shown by Amin[46] to reduce damage to the crop caused by S. *cretica* in the Sudan. Similarly, M. M. Embaby (unpublished data) reported that choice of planting date, cutting out infested shoots and flooding the stubble of infested fields after burning the trash were effective against C. *agamemnon* in Egypt. Moreover, Kuniata[6] showed that crops planted towards the end of the dry season in Papua New Guinea tended to be more heavily attacked by *Sesamia grisescens*, so this practice is now avoided. The planting of a trap crop of maize was shown by De Charmoy[25] to influence infestations of S. *calamistis* in Mauritius. This approach has recently been re-examined in a modified form and could have considerable value in crop pest management[47]. It showed that inter-cropping maize with plants that repel ovipositing stemborer moths could reduce the infestation of maize by stemborers. Conversely, planting a plant species attractive to stemborers also resulted in an increase in maize yield.

Early harvesting and reduced fertiliser use have, amongst other practices, reduced the impact of E. *saccharina* on sugarcane in South Africa[48]. However, no clear effect of burning the crop at harvest has been proven, unless the crop is severely infested and has suffered drought stress. Under these conditions, burning the crop and careful attention to field hygiene can reduce damage in the ratooning crop. Using non-infested seed material when planting is an important aspect of crop hygiene that can help reduce the possibility of a pest becoming established in a crop. Where clean seedcane is not available, the seed material can be treated to reduce the survival of any borer material present. In South Africa, seedcane can be dipped in the insecticide phoxim, at a rate of 2 mL/L water for 10 min before planting. Trials with the fumigant methyl bromide at rates of 10–30 g/m^3 for 4 h are also promising. Results show that at the rates tested, larval

mortality was close to 100%, while germination was no different from that found in the control treatment[49]. Alternatively, it has been shown that heat-treating seedcane (at 50°C for 30 min) can be effective against E. *saccharina* larvae in the stalks. In their study on the giant lepidopteran borer *Castniomera licus* (Drury) in Guyana, Duke & Eastwood[50] reported that flash flooding fields for 48 h 2–3 weeks after harvest is the most common procedure used against this pest. Early harvesting and cutting out larvae and pupae (where labour was available) were also effective.

Monitoring methods

While there are many practices that may influence the incidence of pests in sugarcane, it is of critical importance to develop methods that allow assessments to be made of pest populations and the damage they cause. This then allows treatments to be recommended and their effect determined. What follows is not intended as a review of pest monitoring methods; rather it is a simple outline for general guidance.

Dent[36] divides monitoring into three basic categories: (a) general surveys; (b) field-based surveys; (c) fixed position monitoring. All are based on obtaining a representative sample from the study area. General surveys aim to provide an overview of the geographical distribution of a pest over seasons and may help in the process of predicting outbreaks and identifying migrating pests. Field-based surveys aim to provide the farmer with local information on which they can act. In most such surveys, samples are collected according to a predetermined pattern designed to make the sample as representative as possible. For borers, assessment may be made of measurements such as the percentage of stalks/shoots bored or killed, the percentage of internodes bored and the numbers of larvae in a sample. The percentage of stalks bored provides an indication of the extent of an infestation, while the percentage of internodes bored gives an indication of the intensity of damage. Counts of larval numbers indicate the population pressure in the field. For soil pests, assessment is often based on numbers per unit of soil sampled, e.g. sampling for white grubs. For sap feeders other methods are required. Grimes & des Vignes[51] compared three methods for estimating the populations of the froghopper *Aeneolamia varia saccharina* (Dist.) in Trinidad. They showed that, of the methods tested, the use of sticky traps was the most effective in their study.

Fixed position monitoring comes in various forms and includes pheromone, light and sticky trapping methods. All have been used in sugarcane. For example, in Texas Shaver *et al.*[52] developed a lure for monitoring field populations of the borer *E. loftini*. In Mexico, pheromone traps are being used to monitor *D. considerata*. No threshold values have been determined, but when numbers trapped are large (i.e. 37–60 moths/ha) insecticides are applied. Pheromone trapping of *C. saccchariphagus* has been conducted in Mauritius[5] to assess the feasibility of mating disruption. Results showed that this was indeed possible, but no association with reduced infestations could be shown. Light traps have also been used. In South Africa, they have been used to monitor the seasonal and regional trends in moth numbers of *E. saccharina* and other sugarcane pests[53]. In Zimbabwe, ultraviolet light traps have used as an attempt to trap out *Heteronychus licas* Klug[54].

Whatever the monitoring method used, the sample size taken and the frequency of sampling depend on the level of precision required. Southwood[55] provides various techniques that may be considered when planning a sampling programme for a pest and evaluating the results. Additionally, the purpose of the study must be clearly determined and goals defined.

SOIL PESTS

Soil pests form a diverse grouping of insects that feed on the subterranean parts of sugarcane. They include Coleoptera, Isoptera, Hompotera, cicada and nematode species. Of these, probably the most serious insect pest group is the Coleoptera, specifically the group of scarabaeoid beetles termed white grubs. Because many pest species are included in this grouping, it is difficult to generalise about their biology. Nevertheless, an outline of their biology

is given. The more important genera are listed in Table 4.2.

Damage

The pests in the group feed on the roots of the sugarcane plant or the planted sugarcane sett itself. This results in poor crop development because of root damage or damage to developing shoots. Additionally, because of the weakened root system, stools can easily be blown over, resulting in stool death.

Damage caused by white grubs depends on various factors such as the size and number of grubs present as well as the age and condition of the crop attacked. For the larger species of white grub, such as *Antitrogus consanguineus* (Blackburn), one or two grubs may cause serious damage. For smaller species, such as *Schizonycha affinis* Boh., larger numbers would need to be present before economic damage occurred.

In Florida, *Ligyrus subtropicus* (Blatchley) can be very destructive and Sosa[56] reported that grub numbers of 12/m row reduced cane t/ha by 28% and sucrose tonnes by 39%. In Australia, Allsopp *et al.*[57] quantified the damage caused by *A. consanguineus* larvae. In first ratoon crops, one larva reduced sucrose yield by between 0.61 and 0.63 t/ha.

Termite damage may also be severe, and Mohyuddin[58] recorded that in India termite damage caused a 4.5% reduction in sugar yield (mainly owing to *Microtermes obesi* Holmgren). In Pakistan, this termite species caused cumulative damage in a crop of 34.8%.

In discussing losses attributable to nematodes, Spaull & Cadet[59] commented that an estimated global annual loss in yield of cane was 15%. This exceeded all previous estimates of loss (ranging from 0.2% to 11%) and indicates the potential benefit that may be obtainable from controlling this group of pests.

Biology

As mentioned earlier, white grubs are arguably the most serious insect pests of sugarcane. Most pest species belong to the superfamily the

Family	Genus	Probable number of species	Region
Melolonthidae	*Astenopholis*	1	South Africa
	Cochliotis	1	Tanzania
	Dermolepida	1	Australia
	Eulophida	1	Burkina Faso
	Holotrichia	3	India
	Hoplochelus	1	Réunion
	Hypopholis	1	South Africa
	Lepidiota	13	Australia, India, Far East
	Leucopholis	1	India
	Phyllophaga	1	Mauritius
	Rhopaea	1	Australia
	Schizonycha	1	South Africa
Scarabaeidae	*Antitrogus*	4	Australia
	Eutheola	1	Venezuela
	Ligyrus	4	Venezuela, Brazil, Peru, Guyana
	Phyllophaga	2	Venezuela, Colombia, Peru
Dynastidae	*Alissonotum*	2	Mauritius, South-east Asia
	Heteronychus	2	Throughout Africa
Rutelidae	*Anomola*	1	South Africa
	Adoretus	1	South Africa
Cerambycidae	*Migdolus*	1	Brazil
	Dorysthenes	1	South-east Asia

Table 4.2 Genera and probable number of species of important whitegrub pests of sugarcane.

Scarabaeoidea and the melolonthid, rutelid and dynastid families. The general biology of white grub species is similar. Most species have a 1-year life cycle although some have a 2-year cycle, e.g. some members of the genus *Lepidiota* in Australia[60]. Adults emerge from the soil during summer, mate and disperse. They may roost in surrounding vegetation. Mated females oviposit in the soil around the base stools of sugarcane. There are generally three larval instars. The first and probably the second instar feed on the organic matter in the soil. It is the last instar that does the most damage to the crop by feeding on the sugarcane roots. Typically grubs inhabit the top 300 mm of soil around and under the stool where root density is highest. When fully developed, the grubs move deeper into the soil to pupate. Pupation occurs about 7 months after eclosion in the 1-year species and later in the 2-year species. The pupal period lasts about 2–3 weeks. Adults emerge during summer and disperse. They may roost and mate in trees surrounding fields, females returning to fields to oviposit.

Distribution

Rajabalee[5] lists 10 species of white grubs associated with sugarcane in the African region. Of these, the most widely distributed is *H. licas*. Ferrer[61] lists 18 species associated with sugarcane in South America, while Samoedi *et al.*[62] list 19 species attacking sugarcane in Australia. Mohyuddin[58] stated that there were no recorded white grub pests of sugarcane in Pakistan, but lists five species considered to be major pests in India. Although Charernsom & Suasa-ard[63] list over 80 species associated with sugarcane in south-east Asia, they state that only *Lepidiota stigma* F., *Alissonotum impressicolle* (Arrow) and *Dorysthenes buqueti* (Guerin) are of any consequence. A list of the genera recorded by the authors cited in this section is shown in Table 4.2.

Control

Insecticides

In Australia, the use of insecticides for the control of white grubs is widespread. Currently two insecticides are recommended, i.e. ethoprophos applied at a rate of 25 kg product/ha, and a slow release formulation of chlorpyrifos (SuSCon Blue). Chlorpyrifos is applied at a rate of between 14 and 21 kg/ha depending on the species being controlled. It is only recommended for use at planting, and correct placement of the granules in the furrow is important. The granules must be placed below the surface (about 100 mm) either side of the cane row or alternatively down the centre of the cane row. A tractor-drawn applicator with coulter discs that has a delivery tube behind has been developed for such application[60]. Although expensive (A$250/ha in 1993), this slow-release formulation provides control for up to 3 years. Ethoprophos can be applied to plant or ratooning sugarcane. As with SuSCon Blue, placement is important as the insecticide requires water to move in the soil zone where grubs are most active.

In South America, chlorinated hydrocarbons (heptachlor and endosulfan) have been successfully used though alternatives are being sought[61]. For example, in Brazil *Migdolus fryanus* (Westwood) is controlled by using heptachlor or endosulfan at planting. Various rates were tested (6–12 L/ha for endosulfan and 1.2–2.3 kg a.i./ha for heptachlor). Total production from four ratoons showed a 25% increase in yield (cane t/ha) when endosulfan was applied to the furrow[64]. Mohyuddin[58] quotes several sources on the use of insecticides in India. Larval control has been achieved using HCH, quinalphos and isophenphos as a soil treatment. Adult control has been achieved with variable success by treating the adult roosting sites with carbaryl and monocrotophos at the time of peak adult emergence. In Thailand, the insecticide carbofuran (3% granules) at 1 kg a.i./ha controlled grubs of *Lepidiota stigma* F. This was most effective if applied when newly hatched grubs were abundant[63]. Also chlordane or heptachlor applied at planting at a rate of 4–5 kg/ha was effective against *Dorysthenes buqueti* Guerin when applied in the furrows.

Agronomic practices

As well as insecticide applications, there are several agronomic practices that can help to reduce white grub damage. Allsopp et al.[60] reported from Australia that deep ploughing at the correct time (when grubs are most abundant and are in the top soil layers) killed many grubs. This is also recommended by Rajabalee[5] for the species in Africa. Ward & Cook[65] showed that harvesting and planting dates had a significant effect on crop damage by the greyback grub *Dermolepida albohirtum* (Waterhouse). Sugarcane planted early (i.e. between March and June) was more likely to be attacked than cane planted later the same season. Similarly, cane harvested early was more likely to be attacked than cane harvested later during the same season (early harvested cane was taller at the time of beetle oviposition). They deduced that this was a result of differences in sugarcane height at the time of beetle oviposition.

As with borers, host plant resistance can play a role. Allsopp & Mcghie[66] reported that snowdrop and wheatgerm lectins were found to be insecticidal and growth inhibiting dietary proteins for larvae of *Antitrogus parvulus* Britton. The value of such toxins is the possible inclusion of the gene(s) that code for them into the genome of sugarcane. Effective expression of such genes can result in the toxin being present in the roots of the sugarcane plant at sufficiently high levels to influence grub survival. It seems possible that the rooting habit of varieties would influence the degree of tolerance to white grub damage. This and similar aspects have been studied by Allsopp et al.[67]. They showed that there was variation in the tolerance of varieties to white grub feeding. Differences in top and root yields as well as top and stubble yields were examined. While differences between varieties were observed for each of these measures, no variety showed good resistance for all of them. As mentioned earlier, light traps have been used to control white grub adults. This approach, as well as collecting adults from the trees on which they feed, destroying such trees and the collection of grubs have all been attempted, but with little success[67].

Biological control

Much work has been done on the use of the fungal pathogen *Metarhizium anisopliae* (Metschnikoff) Sorokin. In Australia, Allsopp *et al.*[68] showed that a reduction of 42% in larval numbers was obtained 69 days after injecting the soil around sugarcane plants with a condial solution (7.5×10^{10} conidia/plant). They commented, however, that the pathogen took a long time to kill the grubs, which could have had time to damage the crop before dying. In Réunion, formulations of two strains of *Beauveria brongnartii* (Sacc. & Petch) have been a useful agent in the control of *Hoplochelus marginalis*[5]. Although a number of parasitoids of white grubs have been recorded[22], none has yet been used commercially in the suppression of these sugarcane pests.

Termites

These social insects are a ubiquitous pest of sugarcane in many parts of the world.

Biology

Winged alates disperse after the first summer rains. Once a new colony is established, differential castes are produced comprising workers and soldiers. The queen is sedentary and is usually located in the centre of the nest well below ground. It is the worker castes that forage for food. Colonies may be some distance away from the site of crop damage and can thus be difficult to locate and control. Table 4.3 lists those genera associated with damage to sugarcane as given by the authors cited.

Damage and control

Damage is caused by the workers feeding on the plant tissue and can result in stool or shoot death. This damage can be particularly severe in periods of low rainfall. Control options are generally based on protecting the planted setts by dipping them in an insecticide solution before planting. Organochlorine insecticides have been used in many industries. For example, in Brazil aldrin 5% and heptachlor 5% are recommended[15].

Table 4.3 A list of termite genera (number of species) associated with sugarcane.

Genus	Region
Amitermes (1)	Africa
Coptotermes (1)	
Macrotermes (2)	
Odontotermes (1)	
Psuedocanthotermes (1)	
Cornitermes (1?)	South America
Heterotermes (1)	Ferrer[61]; Anon[69]
Nasutitermes (1)	
Neocapritermes (2)	
Procarnitermes (1?)	
Syntermes (1)	
Coptotermes (1)	India, Pakistan
Eremotermes (1)	Mohyuddin[58]
Macrotermes (1)	
Microtremes (3)	
Microcerotermes (1)	
Odontotermes (10)	
Trinervitermes (1)	
Capritermes (1)	South-east Asia
Coptotermes (2)	Charernsom &
Heterotermes (1)	Suasa-ard[63]
Microcerotermes (1)	
Macrotermes (1)	
Nasutitermes (1)	
Odontotermes (3)	
Reticulitermes (1)	

More recently, imidacloprid has been used in bait, and was found to be successful at a rate of 0.01% concentration in controlling *Heterotermes tenuis* (Hargen). Also in Brazil, control of the serious termite pest *Cornitermes cumulans* (Koll.) has been achieved by thermo-fogging, e.g. chlorpyrifos or permethrin mixed either with diesel oil or mineral oil was tested. Highest mortality (90%) came from the chlorpyrifos (50g a.i./L) plus mineral oil treatment[69]. In Zimbabwe, imidacloprid was successfully used at a rate of 0.5 L product/ha at planting against the *Microtermes* sp. damaging cane[70]. In India, Mrig & Chaudhary[71] evaluated soil-applied insecticides for termite control. (Included in this study were trials against the root borer *Emmalocera depressella* (Swinh.), which had recently become a serious pest in subtropical India.) They showed that two applications of aldrin, quinalphos, chlordane and heptachlor gave effective termite control. However, timing of application seemed to be important for maximum effect. In another study in India, Madan *et al.*[72] evaluated five insecticides for the control of termites in sugarcane at planting. Germination was significantly greater in the treated than in the control plots, and heptachlor and chlorpyrifos were the most effective insecticides for improving crop yield.

Other approaches to termite control include avoiding crop stress by providing adequate irrigation where possible, digging out mounds and removing the queen (for mound-building species only), and the use of pathogens. For example, Milner *et al.*[73] isolated a strain of *Metarhizium anisoplae* that was effective against termite pest genera occurring in Australia, i.e. *Mastotermes* and *Coptotermes*. The isolate is currently being field-tested.

The biology of other soil insect pests

Cicadas (Cicadidae)

Cicadas are recognised as pests of sugarcane in Madagascar, Papua New Guinea and Australia. Eggs are laid on the midrib of dry leaves. On hatching, nymphs fall to the ground, penetrate the soil surface and start feeding on roots. Final stage nymphs move out of the soil and up the stalks to complete development to the adult.

In Madagascar, *Yanga guttulata* (Sign.) is the problem species. Control has been achieved by the use of aldrin, which in combination with rotavating reduces population by 77%[5]. In Australia, *Cicadetta crucifera* (Ashton), *Parnkalla muelleri* (Distant) and *Cicadetta* sp. (the brown, yellow and green cicadas, respectively) are the recognised pest species. Damage is most serious in older ratoon crops, and, where heavy infestations occur, the crop can fail to ratoon. Control is achieved by using rotation planting designed to break the cycle of migration and re-infestation[60].

Margarodes scale or earth pearls (Margarodidae)

This insect group is associated with sugarcane damage in Africa, Mauritius and Australia. At least two species are associated with sugarcane damage in Zimbabwe, i.e. *Margarodes salisburiensis* (Hall) and *M. peringueyi* (Brain)[74]. Similarly, *Eumargarodes laingi* (Jakubski) and *Promargarodes australis* (Jakubski) are associated with sugarcane in Australia[60]. Species in South Africa have yet to be identified.

Biology

Females move to the soil surface to mate, after which bundles of eggs are laid in the soil and covered with wax threads. On eclosion, the mobile nymphs disperse in the soil and attach themselves to roots. Over a period of weeks the nymphs gradually become encased in the typical cyst (pearl) that is seen when digging up a damaged stool. Feeding is by means of a tube extended from the cyst. The cysts may remain viable for years and can be of either sex. Females can reproduce parthenogenetically.

Damage and control

The nymphs feeding on the roots of the crop cause damage, crop development is slow, and in severe infestations ratoon failure can occur. In Australia, control practices comprise fallowing fields, fumigating fields with methyl bromide and, during the spring–summer months, cleaning machinery before moving between fields[60]. In Zimbabwe, Cackett[75] conducted a study on the possible use of oxamyl for margarodid control. He found that, in a pot trial, cyst populations were reduced when oxamyl was used at a rate of 24 L/ha. The conclusion drawn was that, while this treatment may reduce populations, the reduction was not sufficiently great to justify the cost of the treatment. Additionally, a pilot study showed that margarodids only affected crop growth when the crop suffered moisture and fertiliser stress. Cackett[75] concluded that, when such stress was absent, the effect of margarodids on crop development was minimal.

Nematodes

Nematodes cause significant crop loss in many

sugar industries; consequently mention must be made of this group as pests of sugarcane. More than 275 species from 48 genera of endoparasitic and ectoparasitic nematodes have been recorded from the roots and/or rhizosphere of sugarcane. The genera *Pratylenchus*, *Helicotylenchus* and *Tylenchorhynchus* are widespread. Several genera are locally common including *Meloidogyne*, *Xiphinema*, *Hoploliamus* and *Paratrichodorus*[59]. In most countries sugarcane is cultivated in soils with a clay content of > 5%. In such soils nematodes seem to have little effect on crop growth; however, in sandy soils damage can be severe. Inverting the soil and mixing the sandy topsoil with the clay subsoil, should such a layer be present, can reduce nematode damage. Control using nematicides is effective and Spaull & Cadet[59] list seven countries where nematicides have been successfully used. Aldicarb is the most widely used nematicide and is effective at rates between 2.25 and 4 kg a.i./ha.

SAP FEEDERS

Sap feeders belong to the Homoptera, a suborder of the larger group the Hemiptera. Species of the group are well represented in sugarcane with members of at least 19 families being associated with the crop. While many species of sap feeders are associated with sugarcane, only a few can be considered serious economic pests. These are listed in Table 4.4. Other species may be periodically troublesome for various reasons, including climatic and agronomic factors.

Biology

This diverse group of pests shows much variation on a basic pattern of biology; however, the general biology is as follows. The eggs are laid most frequently in plant tissue (e.g. green leaves are selected by *Aleurolobus barodensis* (Mask), *Ceratovacuna lanigera* (Zehntner) and *Rhopalosiphum maidis* (Fitch)) and also in dried leaves or soil (e.g. *Perkinsiella saccharicida* Kirkaldy). Nymphs feed on the tissue where they were laid as eggs. Some species (e.g. *Aeneolamia* spp.) produce a frothy spittle mass, which protects the nymphs from desiccation and probably natural enemies as well. Some species have a 'crawler' stage (e.g. *Aulacaspis tegalensis* (Zehntner)), in which the mobile larva moves over the plant until it settles in a suitable location, usually behind a leaf sheath. Nymphs of other species walk or jump within the crop. While nymphs may move, only adults have developed wings and disperse by flight.

Damage

Typically this group of pests attacks the leaves, but stems and buds can also be damaged. Symptoms include wilting and yellowing of leaves as well as discoloration of the leaves and stems, the latter being brought about by the growth of fungi called 'black sooty mould' (*Capnodium* sp. and *Fumago sacchari*) on the honey-dew secreted by some species. Some are thought to transfer toxic saliva to the plant causing damage in addition to the physical removal of sap. Several pests in this category are vectors of sugarcane diseases, and for some this is the main reason they are considered as pests.

Losses caused by sap feeders can be considerable. Greathead[76] commented that at the height of a massive outbreak of *Aulacaspis tegalensis* (Zehntner) in Tanzania, losses of up to 25% overall were recorded. This was then considered the maximum possible. However, Reagan[77] later reported that three species of frog-hopper (i.e. *Aeneolamia postica* (Walker), *A. postica jugata* (Fowler) and *A. postica campecheana* (Fennah)) could cause severe damage to sugarcane in Mexico, and this might be so severe that a crop may possibly not be cultivated. Even with moderate population levels, yield losses > 50% have occurred. During 1987 in Colombia an outbreak of the aphid species *Sipha flava* (Forbes), caused losses of as much as 53% reduction in sugar t/ha (L. A. Gomez, personal communication). In India the infestation of *Pyrilla perpusilla* (Walker) during 1968–69 caused losses of about 60 000 t of sugar[78]. Also in India, the diaspidid *Melanaspis glomerata* (Green) was reported to cause up to 43% reduction in crop weight. If losses attributable to plant pathogens transmitted by sap suckers are considered as well, then this group of pests is of considerable economic importance.

Table 4.4 A list of the more important sap feeding pests of sugarcane.

Pest species	Region
Coccidae	
Saccharipulvinaria iceryi (Signoret)	Mauritius
Saccharipulvinaria elongata (Newstead)	Mauritius
Tingidae	
Leptodictia tabida (Herrich-Schaeffer)	USA
Aleyrodidae	
Aleurolobus barodensis (Mask.)	India, Pakistan
Diaspididae	
Melanaspis glomerata (Green)	India
Cercopidae	
Aeneolamia postica (Walker)	Mexico
A. postica jugata (Fowler)	Mexico, Belize
A. postica campecheana (Fennah)	Mexico
A. flavilatera (Urich)	Guyana
A. varia saccharina (Dist.)	Trinidad
Callitettix (Calitettix) versicolor (F.)	Myanmar, Thailand
Mahanarva postica (Stål)	Brazil
Mahanarva fimbriolata (Stål)	Brazil
Aphididae	
Ceratovacuna lanigera (Zehntner)	Indonesia, Malaysia, Philippines, Myanmar, Vietnam, China
Hysteroneura setariae (Thomas)	South Africa
Longiunguis sacchari (Zehntner)	Brazil
Melanaphis sacchari (Zehntner)	South and East Africa
Rhopalosiphum maidis (Fitch)	Brazil, Indonesia, Philippines, Thailand, South and East Africa
Rhopalosiphum padi (Linn)	South Africa, Mascarene Islands
Psuedococcidae	
Saccharicoccus sacchari (Cockerell)	Philippines, Indonesia, Kampuchea, Laos, China, Thailand, Vietnam, Brazil
Delphacidae	
Eumetopina flavipes (Muir)	Papua New Guinea
Perkinsiella saccharicida (Kirkaldy)	Indonesia, Malaysia, Philippines, Myanmar, Vietnam, USA, China
Saccharosdyne saccharivora (Westw.)	Jamaica
Diaspididae	
Aulacaspis tegalensis (Zehntner)	Indonesia, Philippines, Thailand
Lophopidae	
Pyrilla perpusilla (Walker)	Thailand, India, Pakistan
Colobathristidae	
Phaenacantha saccharicida (Karsch)	Indonesia, Thailand

Distribution and control

Carnegie[79] stated that, in Africa and surrounding islands, there are at least 20 species of sap feeders associated with damage to sugarcane. Periodically serious pest species include the coccids *Saccharipulvinaria iceryi* (Signoret) and *Saccharipulvinaria elongata* (Newstead) and the diaspid *A. tegalensis*, as well as the tropiduchid *Numicia viridis* (Muir). The aphids *Rhopalosiphum padi* (Linn), *Hysterneura setariae* (Thomas) and *Rhopalosiphum maidis* (Fitch) are important vectors of

the sugarcane mosaic virus in South Africa. There have been several outbreaks (*S. iceryi*) in Mauritius, the latest of which occurred in the mid-1970s. Because the outbreaks occurred simultaneously in widely separated localities, Williams[80] suggested that climatic factors were involved in influencing host plant susceptibility and predator populations. Control is restricted to encouraging predators and parasitoids, the most effective of which seemed to be coccinellid predators and aphelinid parasitoids.

In South Africa, a recent outbreak of *Pulvinaria saccharia* (De Lotto) has been a cause for concern.

Investigations revealed no obvious reason, so it is probable that, as suggested by Williams for *S. iceryi*[80], climatic factors played a significant role, and that the populations will eventually be under adequate biological control again.

The armoured scale, *A. tegalensis*, is capable of causing serious damage to both the leaves and stalks of sugarcane, but the stalk is the most seriously damaged[79]. Insecticidal control is not satisfactory, and host plant resistance and biological control are the preferred approaches. For example, Carnegie[79] commented that varieties having loose-leaf sheaths, under which crawlers tend to settle, allow access of predators and parasitoids.

Until recently, sap-feeding insects have been of minor importance in North America[77]. However, two new pests have now been introduced, i.e. the sugarcane lace bug *Leptodictia tabida* (Herrich-Schaeffer) and the sugar cane froghopper, *Perkinsiella saccharicida* (Kirkaldy). *Leptodictia tabida*, the sugarcane tingid, was first recorded in Florida in 1990[81].

The major concern about the delphacid *P. saccharicida* is its ability to act as a vector for the Fiji disease virus. First recorded in Haiwaii in 1903[82], *P. saccharicida* soon became a serious pest there. *Perkinsiella saccharicida* was first observed in Florida in 1982[83]. During the1980s, two predator species (*Tytthus mundulus* and *T. parviceps*) were released, and the latter became established, as did the egg parasitoid (*Anagarus* sp.). *Perkinsiella saccharicida* currently has little economic effect as peak populations coincide with crop maturity and harvest[77].

In South and Central America, there are several species of sap-feeders that are associated with damage to sugarcane. Reagan[77] stated that the froghoppers *A. postica*, *A. postica jugata* and *A. postica campecheana* were serious pests of cane cultivated in Mexico. Control remains primarily by insecticides, and has been by the use of organophosphates and organochlorines. However, resistance to insecticides has been developing, and now three applications of an insecticide are required for control. Carbamates are now being used as well as azodrin + cypermethrin. Alternatives are being considered, and the use of the pathogenic fungus *Metarhizium* sp. shows promise[84].

In Brazil, the cercopids *Mahanarva postica* (Stål) and *Mahanarva fimbriolata* (Stål) are considered to be the most important of the six species of sap-feeders attacking sugarcane[15]. Control is achieved by the pathogenic fungus *M. anisoplae*[15]. During 1987 in Colombia there was an outbreak of the aphid species *S. flava*, but there have been no further outbreaks of this pest, and it is managed by varietal choice and the use of perimicarb or malathion (L. A. Gomez, personal communication).

Mohyuddin[58] lists 49 species of sap-feeder that are associated with sugarcane in India and Pakistan. However, of these only the lophopid *P. perpusilla*, the aleyrodid *Aleurolobus barodensis* (Mask) and the scale insect *Melanaspis glomerata* (Green) are considered to be serious pests of sugarcane. Initially, control of *P. perpusilla* was achieved by using chlorinated hydrocarbon insecticdes. These were subsequently replaced with organophosphate and carbamate insecticides. In Pakistan, the introduction of the parasitoid *Epiricania melanoleuca* (Fletcher) in 1975 had, by 1977, given complete control in the affected areas in that country. A similar result was found in India. By conservation practices (such as not burning the trash after harvest) the efficacy of the egg parasitoid *Parachrysocharis javensis* (Crawford) was improved, and the use of this parasitoid in combination with the effect of *E. melanoleuca* provided control without recourse to insecticides.

The diaspidid *M. glomerata* has only been reported as a pest of sugarcane in India. When an infestation is heavy, the entire stalk is covered with scale, giving it a greyish-black appearance. Control has been achieved by dipping seedcane in an insecticide solution such as malathion and dimethoate. Cultural control practices include the selection of clean seed, detrashing, trash burning and stubble shaving[58]. Many parasitoids and predators have been recorded, and some introductions have been made and recoveries obtained[58].

Suasa-ard & Charernsom[85] list 49 sap-feeding pests of sugarcane in south-east Asia. Of these, ten are considered to be economically important in various countries in the region and are included in Table 4.4. Natural control is considered by Suasa-ard & Charernsom[85] to be the most effective control measure available. They list 32 species of effective natural enemies of the ten most serious pests of sugarcane in south-east Asia. However Liu *et al.*[86] state that judicious timing of insecticide application (50% dimethoate) effectively controlled *Ceratovacuna lanigera* (Zehntner). Other effective approaches to control reported by Liu *et al.*[86] include intercropping with legumes, removal of wilted leaves and altered planting dates. These were effective against *S. sacchari* and *P. saccharicida*. Suasa-ard & Charernsom[85] note that host plant resistance (in the form of loose trashing varieties) has also been shown to be effective in controlling sap feeders such as *S. sacchari* and *A. tegalensis*. Trash removal was also found to an effective measure against these species in Thailand.

Allsopp[87] lists ten species of sap feeders associated with sugarcane in Australia. None is particularly important, though some are considered to be vectors of Fiji disease virus and sugarcane mosaic virus (*P. saccharicida*, *P. vastatrix*, *P. vitiensis* and *R. maidis*). Applied control measures are not considered necessary, as natural control is generally sufficient. Resistant varieties are seen by Allsopp[87] to be the most promising approach to control of both these and *Eumetopina flavipes* Muir – a vector of the Ramu stunt phytoplasmas in Papua New Guinea.

The pest problems of the West Indies have been outlined in an extensive report[88]. Several species of sap feeders can cause serious problems in this region. *Aeneolamia postica jugata* (Fowler) is periodically a serious pest in Belize, and the froghopper *A. flavilatera* (Urich) has been a serious problem in Guyana, particularly where poor drainage occurs. Insecticidal control of both nymphs and adults has been achieved with benzene hexachloride. However, this practice has been discontinued in favour of using the pathogenic fungus *Metarhizium* sp. In Jamaica, the formerly serious pest *Saccharosdyne saccharivora* (Westwood) has been effectively controlled through an integrated pest management programme. Reduction in the area treated with insecticides (19 000 ha in 1967 to 2000 in the 1990s); coupled with the effect of the efficient parasitoid *Stenocranophilus quadratus* (Pierce) has allowed the judicious use of malathion or fenetrothion against the nymphs. In Trinidad, the most serious pest problem is the froghopper *Aeneolamia varia saccharina* (Dist.), and losses of up to 60% have been recorded. Control has been based on suppressing the first of four generations with aerially applied insecticides. Timing of the applications is critical, and an elaborate sampling system has been developed for this. Also being examined is the use of pathogen *Metarhizium* sp.

LEAF FEEDERS

Pests in this group are generally of minor importance compared to other groups. Nevertheless, they can cause periodic, serious defoliation and therefore are worthy of consideration. The most serious pests in this group are lepidopterans and orthopterans. Lepidoterans include those families listed in Table 4.5. The most serious are noctuids, i.e. *Spodoptera* spp. referred to as 'armyworms',

Table 4.5 A list of the more important leaf feeding pests of sugarcane. Number of species recorded in parentheses.

Pest	Region
Noctuidae	
Leucania spp. (2)	Australia
Mythimna spp. (6)	Africa, Mauritius, Australia
Spodoptera spp. (3)	Africa, Mauritius, Australia
Acrididae	
Chortoicetes terminifera (Walker)	Australia
Gastrimargus musicus (Fabricius)	Australia
Locustana pardalina (Walker)	Africa
Locusta migratoria (Linnaeus)	Africa, Australia
Nomadacris guttulosa (Walker)	Australia
Patanga septemfasciata (Serville)	Africa
Schistocerca gregaria (Forskäl)	Africa

and *Mythimna* spp. collectively called 'trash caterpillars'. Orthopteran pests (mainly acridids) include the seven species listed in Table 4.5. Various species of grasshoppers are comparatively minor pests of sugarcane.

Damage

These pests feed on the green leaves of the crop, often stripping leaves entirely, leaving only the midrib. With much of the photosynthetic area destroyed, crop development is retarded and production can be severely affected, particularly if repeated infestations occur in the same area. There is not much information on the losses caused by this group to sugarcane. While invasion by migrating swarms of locusts can cause serious damage, these seem to be infrequent occurrences. Similarly, damage caused by species of *Spodoptera* and *Mythimna* are only of sporadic concern. Ganeshan & Rajabalee[89] investigated the economics and management of these to genera in Mauritius. They showed that, while one artificial defoliation (clipping of leaves near the dewlap of young sugarcane) showed no effect on crop yield, there was a significant reduction in pol per cent cane after two defoliations. This shows that the effect of defoliators is dependent on the extent of removal of leaf area.

Distribution

Spodoptera spp. are widespread and have been associated with sugarcane damage in Africa (e.g. *Spodoptera exempta* (Walker)[90]); Malaysia (e.g. *Spodoptera litura* (F.)), Asia and Australasia (e.g. *Spodoptera* spp.[91]). *Spodoptera frugiperda* (J.E.S.) and the looper *Mocis latipes* (Guenée) are recorded as pests of sugarcane in Brazil[15]. *Caligo ilioneus* causes some foliar damage in Colombia.

In Australia, distinction is made between day feeding and night feeding armyworms. The latter comprises *Mythimna* and *Leucania* spp. and the former *S. exempta*. In Africa the term armyworm is generally restricted to *Spodoptera* spp., while *Mythimna* spp. are referred to a trash caterpillars. In Mauritius, Geneshan and Rajabalee[92] list four species of trash caterpillars (referred to as 'armyworms') attacking sugarcane there. Carnegie

et al.[90] list seven species of trash caterpillars associated with sugarcane in South Africa. Of these, *Mythimna phaea* (Hamps.) was considered to be the most important. As with the armyworms and trash caterpillars, locusts and grasshoppers are generally considered to be minor pests. However, during outbreak periods serious local damage can occur. Greathead[76] lists 18 species of locusts that have been known to feed on sugarcane in Africa, and in Australia, Allsopp *et al.*[60] list four species of locust associated with damage to sugarcane.

Biology and control

Spodoptera *spp.*

Eggs are laid in masses and covered with scales from the abdomen of the female. On hatching the larvae disperse and feed. They often occur in large numbers and move en masse from grassland or pasture to the cane fields as the grass is consumed – hence the term 'armyworm'. *Spodoptera exempta* larvae have two forms, and in the outbreak form the last two instars are dark[91]. Damage is sporadic and dependent on environmental factors that may or may not encourage population explosions. Larvae pupate in an earth cell in the soil, and the adults are night flying and can migrate over long distances. Generally applied control measures are not necessary; however, insecticides may be used where it is felt such treatments are warranted. In Australia, chlorpyrfos, permethrin and trichorophon may be used at 700, 200 and 900 mL product/ha, respectively.

Mythimna *spp.*

Eggs are laid on the host plant and on hatching the larvae disperse, but not as deliberately as *Spodoptera* larvae. They feed at night on the young shoots and rest under the dead leaf material in sugarcane fields during the day. In South Africa, this pest is most common at the beginning of the harvesting season (i.e. September/October) in fields normally trashed at harvest, and attacks the young developing shoots of the following ratoon. Pupation takes place in the soil. Control may be achieved by treating with an insecticide when larvae are small and damage is beginning to manifest itself. Ganeshan & Rajabalee[92] showed that thiodicarb at 187.5 g a.i./ha significantly reduced the number of live larvae recovered 15 days after treatment. However, it is often too late to treat once extensive damage has occurred as the many natural enemies present may be adversely affected by any insecticidal treatments.

Locusts and grasshoppers

Distribution

This group of pests occurs in all parts of the world, and can damage many crops besides sugarcane.

Biology

The general biology of locusts is as follows. Eggs are laid in the soil after the first summer rains. On hatching the nymphs (hoppers) disperse in large bands. At this stage treating the bands with insecticides easily controls the flightless hoppers. There are between six and seven instars depending on the sex and/or phase. In some species, hoppers exist in two phases, solitary and gregarious. Once mature, the adults of some species swarm and migrate over great distances.

Damage and control

In Africa and Australia[60], locust damage to sugarcane is not common. In Brazil, four species of grasshopper are recorded pests of sugarcane. These are *Staurorhactus longicornis* (G. Tos.), *Rhammatocerus pictus* (Bruner), *Chromacris speciosa* (Thumb.) and *Zoniopoda tarsata* (Serv.). All are considered to be minor pests[93]. In South Africa, recent outbreaks of grasshoppers have been controlled by the application of deltamethrin at a rate of between 20 and 30 g a.i./ha depending on the stage targeted. Greathead[76] states that since the withdrawal of dieldrin, organophosphates are being used. Synthetic pyrethroids are also being used, but are more expensive. Insect growth regulators are also effective, being promising for the control of hopper bands. Such treatments can be applied as ultra-low volume sprays to hopper bands or by air on to flying swarms.

VERTEBRATE PESTS OF SUGARCANE

Rodents

Of the vertebrates that attack sugarcane, rats are the most frequently cited cause of serious crop damage. Rats are recorded pests in South America, Hawaii, the Indian subcontinent, the Far East and Australia. Occasional damage can be attributed to rodents in other regions of the world where this crop is cultivated.

In Hawaii, four species of rat attack sugarcane of which *Rattus exulans* (Peale) is the most serious. *Holochilus sciureus* (= *Holochilus brasiliensis* (Desmarest)?) is a major pest in Central America, particularly in Guyana[94]. This rat is a major seasonal pest in Guyana, and considerable damage can occur. In Mexico, *Sigmodon hispidus* (Say) is considered a serious pest, and 200 000 ha need to be treated for rats, i.e. 35% of the total annual crop[95]. Lefebvre *et al.*[96] list three species of rats associated with sugarcane in Florida: *S. hispidus*, *Rattus rattus* L. and *Neofiber alleni* (True). In Australia, three species of rodent are problematic: the ground rat *Rattus sordidus* (Gould) and the climbing rats *Melomys burtoni* (Ramsay) and *Melomys cervinipes* (Gould). Of these, *R. sordidus* is the most serious pest[60]. In the Indian subcontinent, *Millardia meltada* (Gray), *Bandicota bengalensis* (Gray), *B. indica* and *Nesokia indica* (Gray) attack field crops including sugarcane, and in Tawain *Mus formosanus* (Kuroda) and *Apodemus agrarius* (Swin.) are common in the crop[97].

Damage

Damage is mainly to the basal nodes of stalks. This allows ingress of pathogens to the stalks that can cause serious yield loss. In addition, stalks may break at the point of damage, increasing crop loss. Some species, such as *N. indica*, feed below ground causing stool death. In Mexico, losses can be as high as 50–60% of the crop[95], while in Florida incidence of damage ranged from 4.5% to 38.6%[96].

In Barbados, losses of 13.7% of millable cane have been recorded[88].

Biology

Rodents are generally prolific breeders, and Macdonald & Fen[98] state that many species become sexually mature in 2–3 months. Females can produce litters of up to seven young after a 2–3 week gestation period. Additionally females can become pregnant soon after giving birth. Such breeding obviously requires favourable conditions, and it is the duration of such conditions that dictates the severity of an outbreak.

Control

Most rodenticides are administered as baits, and fall into two basic categories, i.e. fast-acting compounds (e.g. zinc phosphide and sodium fluoroacetate), and slow acting ones, mainly anticoagulants such as difenacoum and bromadiolone[99]. In Mexico, baits incorporating zinc phosphide are used but are being replaced with strychnine sulphate applied at a rate of 2–4 kg/ha[94]. In Hawaii, Sugihara *et al.*[100] tested various formulations of zinc phosphide baits for rat control in sugarcane fields. They showed that, while pre-baiting fields with non-toxic grain improved the performance of the zinc phosphide oat bait, substantial numbers of rats remained in all fields, regardless of rodenticide or whether fields were pre-baited with non-toxic grain. Absorption of moisture and physical degradation were thought to have reduced the acceptance of pellets, and they concluded that a more weather-resistant bait formulation was required. Allsopp *et al.*[60] suggested some non-chemical approaches to control, such as the mowing of grasslands and grass verges to reduce the cover that provides a habitat suitable for rodents. In addition, Smith[101] suggested clear cultivation, flooding land (where possible), chemosterilisation, and parasites and diseases.

Other vertebrate pests

Several other vertebrates are considered periodic pests of sugarcane. In South Africa, the vervet monkey *Cercopithicus aethiops* can cause considerable damage to sugarcane, particularly where the crop is cultivated in close proximity to natural bushlands. Wild pigs have also been periodically troublesome, and severe damage to ratooning cane has been noticed where pigs have dug up roots, apparently in search of white grub larvae. In Australia, wallabies *Macropus* spp., foxes *Vulpes vulpes*, possums *Dactylopsila trivirgata*, swamp hens *Porphyrio porphyrio* and cockatoos *Cacatua galerita* are all recognised as periodic pests of sugarcane. Control measures include hunting and the use of normal or electric fences[60].

REFERENCES

1 Pemberton, C.E. & Williams, J.R. (1969) Distribution, origins and spread of sugar cane pests. In: *Pests of Sugarcane* (eds J. R. Williams, J. R. Metcalfe, R. W. Mungomery & R. Mathes), pp. 1–9. Elsevier, Amsterdam.

2 van Leerdam, M.B., Johnson, K.J.R. & Smith, J.W. Jr (1984) Effects of substrate physical characteristics and orientation on oviposition by *Eoreuma loftini* (Lepidoptera: Pyralidae). *Environmental Entomology*, **13**(3), 800–802.

3 Leslie, G.W. (1990) The influence of dead leaf material on the oviposition behaviour of *Eldana saccharina* Walker (Lepidoptera: Pyralidae) in sugarcane. *Proceedings of the South African Sugar Technologists Association*, **64**, 100–102.

4 King, A.G. (1989) An assessment of the loss in sucrose yield caused by the stalk borer *Eldana saccharina* in Swaziland *Proceedings of the South African Sugar Technologists Association*, **63**, 197–201.

5 Rajabalee, A. (1990) Management of *Chilo* spp. on sugarcane with notes on mating disruption studies with the synthetic sex pheromone of *C. sacchariphagus* in Mauritius. *Insect Science and its Application*, **11**(4/5), 825–836.

6 Kuniata, L.S. (1994) Pest status, biology and effective control measures of sugar cane stalk borers in the Australian, Indonesian and Pacific Island sugar cane growing regions. In: *Biology, Pest Status and Control Measure Relationships of Sugarcane Insect Pests* (eds A. J. M. Carnegie & D. E. Conlong), pp. 83–96. Proceedings of the International Society of Sugar Cane Technologists Second Entomology Workshop, SASA Experiment Station.

7 Kira, M.T. & El-Sherif, H. (1971) Estimation of losses in cane and sugar yields caused by infestations of *Chilo agamemnon* Bles. *Proceedings of the International Society of Sugar Cane Technologists*, **14**, 427–428.

8 Cheng, W.Y. (1994) Sugarcane stem borers of Taiwan. In: *Biology, Pest Status and Control Measure Relationships*

of Sugarcane Insect Pests (eds A. J. M. Carnegie & D. E. Conlong), pp. 97–105. Proceedings of the International Society of Sugar Cane Technologists Second Entomology Workshop. SASA Experiment Station.

9 Meagher, R.L. Jr, Smith, J.W. Jr & Johnson, K.J.R. (1993) Insecticidal management of *Eoreuma loftini* (Lepidoptera: Pyralidae) on Texas sugarcane: A critical review. *Journal of Economic Entomology*, **87**(5), 1332–1344.

10 Tamanikaiyaroi, R., Johnson, S.S. & Wood, R.A. (1995) Loss in cane quality caused by the cane weevil (*Rhabdoscelus obscurus* Boisd.) in the Fiji sugar industry. *Proceedings of the International Society of Sugar Cane Technologists*, **22**(2), 599–604.

11 Bleszynski, S. (1969) The taxonomy of the crambine moth borers of sugar cane. In: *Pests of Sugar Cane* (eds J. R. Williams, J. R. Metcalf, R. W. Mungomery & R. Mathes), pp. 11–59. Elsevier, Amsterdam.

12 Leslie, G.W. (1994) Pest status, biology and effective control measures of sugar cane stalk borers in Africa and surrounding islands. In: *Biology, Pest Status and Control Measure Relationships of Sugarcane Insect Pests* (eds A. J. M. Carnegie & D. E. Conlong), pp. 61–70. Proceedings of the International Society of Sugar Cane Technologists Second Entomology Workshop. SASA Experiment Station.

13 Samoedi, D. (1995) Yield losses of commercial cane varieties due to infestation of white top moth borer, *Tryporyza nivell intacta* Sn. in Java. *Proceedings of the International Society of Sugar Cane Technologists*, **22**(2), 610–617.

14 Sosa, O. Jr, Shine, J.M. & Tai, P.Y.P. (1997) West Indian cane weevil (Coleoptera, Curculionidae) – a new pest of sugarcane in Florida. *Journal of Economic Entomology*, **90**(2), 634–638.

15 Anon (1982) *Guia das pricipais pragas da cana-de-açúcar no Brazil.* Instituto do Açúcar e do Alcool, Planalsucar, Brazil.

16 Anon (1997) *Centro de Tecnologia Copersucar Annual Report for 1996–1997.* Copersucar, Brazil.

17 Salazar, J. (1997) In: *Internal SASEX Report on the Third International Society of Sugar Cane Technologists Entomology Workshop*, Culian, Mexico (eds G. W. Leslie & M. G. Keeping). Copersucar, Brazil.

18 Gomez, L.A. (1995) Pre-Congress forum discussion papers. Forum A.1 Plant Health. *International Sugar Journal*, **97**(1161), 488.

19 Pantoja, N. (1997) In: *Report on the Third International Society of Sugar Cane Technologists Entomology Workshop*, Culian, Mexico (eds G.W. Leslie & M.G. Keeping). SASA Experiment Station Internal Report, Mount Edgecombe, Durban, South Africa.

20 Florez, S. (1997) In: *Report on the Third International Society of Sugar Cane Technologists Entomology Workshop*, Culian, Mexico (eds G. W. Leslie & M. G. Keeping). SASA Experiment Station Internal Report, Mount Edgecombe, Durban, South Africa.

21 White, W.H. & Regan, T.E. (1999) Biological control of the sugarcane borer with introduced parasites in Louisiana. *Sugar Journal*, **9**, 13–17.

22 Conlong, D.E. (1994) Biological control in sugarcane.

In: *Proceedings of the Third International Conference on Tropical Entomology* (ed. R. K. Saini), pp. 73–101. ICIPE Science Press, Nairobi, Kenya.

23 Conlong, D.E. (1994) A review and perspectives for the biological control of the African sugarcane stalkborer *Eldana saccharina* Walker (Lepidoptera: Pyralidae). *Agriculture Ecosystems & Environment*, **48**, 9–17.

24 Conlong, D.E. (1998) Field surveys. In: *Progress Report of the Entomology Department 1998–1999.* SASA Experiment Station Internal Report, Mount Edgecombe, Durban, South Africa.

25 Williams, J.R. (1983) The sugar cane stem borer (*Chilo sacchariphagus*) in Mauritius. *Revue Agricole et Sucriere de i'Ile Maurice*, **62**, 5–23.

26 Spaull, V.W. (1990) Field tests to control the pyralid, *Eldana saccharina*, with an entomogenous nematode, *Heterorhabditis* sp. *Proceedings of the South African Sugar Technologists Association*, **64**, 103–106.

27 Spaull, V.W. (1992) On the use of a methylcellulose polymer to increase the effectiveness of a *Heterorhabditis* species against the sugarcane stalk borer, *Eldana saccharina*. *Fundamental and Applied Nematology*, **15**(5), 457–461.

28 Bessin, R.T., Moser, E.B. & Reagan, T.E. (1990) Integration of control tactics for management of the sugarcane borer (Lepidoptera: Pyralidae) in Louisiana. *Journal of Economic Entomology*, **83**(4), 1563–1569.

29 Martinez, A. (1997) In: *Report on the Third International Society of Sugar Cane Technologists Entomology Workshop, Culiacan, Mexico* (eds G. W. Leslie & M. G. Keeping). SASA Experiment Station Internal Report, Mount Edgecombe, Durban, South Africa.

30 Boedijono, W.A. (1974) An attempt to control sugar-cane stem borers with the dipterous parasite *Diatraeophaga striatalis* (Towns). *Proceedings of the International Society of Sugar Cane Technologists*, **15**, 393–396.

31 Awadallah, W.H., El-Metwally, E.F. & Aly, F.A. (1982) The effect of soil insecticides on sugarcane growth and borer infestation. *Agricultural Research Review*, **58**(1), 107–111.

32 Hassanien, M.H. & El-Naggar, M.Z. (1972) Susceptibility of certain sugarcane plant varieties to borer infestation in Egypt. *Bulletin of the Entomological Society of Egypt*, **55**, 27–43.

33 Pandey, K.P., Singh, R.G. & Singh, S.B. (1997) Integrated control of top borer *Scirpophaga excerptalis* Wlk in Eastern U.P. *Indian Sugar*, **47**(7), 491–493.

34 Singla, M.L. & Duhra, M.S. (1992) Efficacy of some insecticides for the control of stalk borer, *Chilo auricilius* Ddgn. on sugarcane. *Journal of Insect Science*, **5**(1), 111–112.

35 Kuniata, L.S. (2000) Integrated management of sugarcane borers in Papua New Guinea. In: *Sugarcane Pest Management in the New Millennium* (eds P.G. Allsopp & W. Suasa-ard), pp. 37–50. *Proceedings of the International Society of Sugar Cane Technologists*, Fourth Sugarcane Entomology Workshop, Khon Kaen, Thailand.

36 Dent, D. (1991) Host plant resistance. In: *Insect Pest Management*, pp. 213–292. CAB International, Oxford.

37 Sosa, O. (1988) Pubescence in sugarcane as a plant resistance character affecting oviposition and mobility by

the sugarcane borer (Lepidoptera: Pyralidae). *Journal of Economic Entomology*, **81**(2), 663–667.

38 White, W.H. (1993) Cluster analysis for assessing sugarcane borer resistance in sugarcane line trials. *Field Crops Research*, **33**, 159–168.

39 Leslie, G. W. & Keeping, M. G. (1995) Approaches to the control of the pyralid sugarcane borer *Eldana saccharina* Walker. *Proceedings of the International Society of Sugar Cane Technologists*, **22**(2), 561–566.

40 Sosa, O. Jr (1995) Evaluation for resistance in sugarcane to the sugarcane borer *Diatraea saccharalis* (F.). *Proceedings of the International Society of Sugar Cane Technologists*, **22**(2), 594–598.

41 Allam, A.I. & Abou Dooh, A.M. (1995) Strategies for breeding varietal resistance to sugarcane stalk borer (*Chilo* spp). *Proceedings of the International Society of Sugar Cane Technologists*, **22**(2), 604–609.

42 Keeping, M.G. & Leslie, G.W. (1998) Are self-trashing varieties of sugarcane resistant to the stalk borer *Eldana saccharina* Walker (Lepidoptera: Pyralidae)? *Proceedings of the South African Sugar Technologists Association*, **73**, 100–101.

43 Bessin, R.T., Reagan, T.E. & Martin, F.A. (1990) A moth production index for evaluating sugarcane cultivars for resistance to the sugarcane borer (Lepidoptera: Pyralidae). *Journal of Economic Entomology*, **83**(1), 221–225.

44 Herrera, G., Snyman, S.J. & Thomson, J.A. (1994) Construction of a bioinsecticidal strain of *Pseudomonas fluorescens* active against the sugarcane borer, *Eldana saccharina*. *Applied Environmental Microbiology*, **60**(2), 682–690.

45 Arencibia, A., Vàzquez, R.I., Prieto, D. *et al.* (1997) Transgenic sugarcane plants resistant to stem borer attack. *Molecular Breeding*, **3**(4), 274–255.

46 Amin, E.M. (1988) Some aspects of the ecology of the stem borer *Sesamia cretica* Led. in sugarcane and sorghum in the Sudan. *Beiträge zur tropischen Landwirtschaft und Veterinärmedizin*, **26**(1), 33–37.

47 Khan, Z.R., Chiliswa, P., Ampong-Nyarko, K. *et al.* (1997) Utilisation of wild gramineous plants for management of cereal stemborers in Africa. *Insect Science and its Application*, **17**(1), 143–450.

48 Carnegie, A.J.M. (1981) Combating *Eldana saccharina* (Lepidoptera: Pyralidae): a progress report. *Proceedings of the South African Sugar Technologists Association*, **55**, 116–119.

49 Kuppen, J.P. & Leslie, G.W. (1999) The use of the fumigant methyl bromide in the control of the sugarcane borer *Eldana saccharina* Walker (Lepidoptera: Pyralidae) in seedcane. *Proceedings of the Twelfth Entomological Congress of the Entomological Society of Southern Africa*.

50 Duke, N.H. & Eastwood D. (1997) Production losses in sugarcane attacked by the giant borer *Castniomera licus* (Drury) Lepidoptera: Castniidae in Guyana. *Proceedings of the West Indies Sugar Technologists Association*, **26**, 169–176.

51 Grimes, N. & des Vignes, W.G. (1997) Froghopper population estimation – a preliminary comparison of three different traps and traditional observer counts. *Proceedings of the West Indies Sugar Technologists Association*, **26**, 177–179.

52 Shaver, T.N., Brown, H.E. & Hendricks, D.E. (1990) Development of pheromone lure for monitoring field populations of *Eoreuma loftini* (Lepidoptera: Pyralidae). *Journal of Chemical Ecology*, **16**(8), 2393–2400.

53 Carnegie, A.J.M. & Leslie, G.W. (1990) *Eldana saccharina* (Lepidoptera: Pyralidae): ten years of light trapping. *Proceedings of the South African Sugar Technologists Association*, **64**, 107–110.

54 Musikavanhu, F. (1996) Black maize beetle control strategy at Triangle, Zimbabwe. *Proceedings of the South African Sugar Technologists Association*, **70**, 18–21.

55 Southwood, T.R.E. (1975) *Ecological Methods*. Chapman & Hall, London.

56 Sosa, O. Jr (1994) Soil pests of sugarcane in North America. In: *Biology, Pest Status and Control Measure Relationships of Sugarcane Insect Pests* (eds A. J. M. Carnegie & D. E. Conlong), pp. 13–14. Proceedings of the International Society of Sugar Cane Technologists Second Entomology Workshop. SASA Experiment Station, Mount Edgecombe, Durban.

57 Allsopp, P.G., Bull, R.M. & McGill, N.G. (1991) Effect of *Antitrogus consanguineus* (Blackburn) (Coleoptera: Scarabaeidae) infestations on the sugarcane yield in Australia. *Crop Protection*, **10**, 205–208.

58 Mohyuddin, A.I. (1994) Pest status and control measures for soil pests of sugar cane in India and Pakistan. In: *Biology, Pest Status and Control Measure Relationships of Sugarcane Insect Pests* (eds A. J. M. Carnegie & D. E. Conlong), pp. 139–145. Proceedings of the International Society of Sugar Cane Technologists Second Entomology Workshop. SASA Experiment Station, Mount Edgecombe, Durban.

59 Spaull, V.W. & Cadet, P. (1990) Nematode parasites of sugarcane. In: *Plant Parasitic Nematodes in Subtropical and Tropical Agriculture* (eds M. Luc, R. A Sikora & J. Bridge), pp. 461–491. CAB International, Oxford.

60 Allsopp, P.G., Chandler, K.J., Samson, P.R. & Story, P.G. (1993) *Pests of Australian Sugarcane*. Bureau of Sugar Experiment Stations, Brisbane, Australia.

61 Ferrer, W.F. (1994) Soil-inhabiting insect pests and nematodes that damage sugar cane in South America: biology and control measures. In: *Biology, Pest Status and Control Measure Relationships of Sugarcane Insect Pests* (eds A. J. M. Carnegie & D. E. Conlong), pp. 15–22. Proceedings of the International Society of Sugar Cane Technologists Second Entomology Workshop. SASA Experiment Station, Mount Edgecombe, Durban.

62 Samoedi, D., Allsopp, P.G. & Kuniata, L. (1994) Pest status, biology and control measures for soil pests of sugar cane in the Australian, Indonesian and Papua New Guinea regions. In: *Biology, Pest Status and Control Measure Relationships of Sugarcane Insect Pests* (eds A. J. M. Carnegie & D. E. Conlong), pp. 27–41. Proceedings of the International Society of Sugar Cane Technologists Second Entomology Workshop. SASA Experiment Station, Mount Edgecombe, Durban.

63 Charernsom, K. & Suasa-ard, D.W. (1994) Pest status, biology and control measures for soil pest of sugar cane in South East Asia. In: *Biology, Pest Status and Control*

Measure Relationships of Sugarcane Insect Pests (eds A. J. M. Carnegie & D. E. Conlong), pp. 53–60. Proceedings of the International Society of Sugar Cane Technologists Second Entomology Workshop. SASA Experiment Station, Mount Edgecombe, Durban.

64 Anon (1993) *Centro de Tecnologia Copersucar Annual Report for 1992–1993.* Copersucar, Brazil.

65 Ward, A.L. & Cook, I.M. (1996) Effect of planting and harvesting date on greyback canegrub damage to sugarcane grown in the Burdekin river area. In: *Sugarcane: Research towards Efficient and Sustainable Production* (eds J.R. Wilson, Hogarth, D.M., Campbell, J.A. & Garside, A.L.), pp. 226–227. CSIRO Division of Tropical Crops and Pastures, Brisbane.

66 Allsopp, P.G. & Mcghie, T.K. (1996) Snowdrop and wheatgerm lectins and avidin as antimetabolites for the control of sugarcane whitegrubs. *Entomologia Experimentalis et Applicata,* 80 (2), 409–414.

67 Allsopp, P.G., McGhie, T.K., Smith, G.R., Ford, R. & Cox, M.C. (1995) Progress in the development of cane varieties with resistance to canegrubs. *Proceedings of the Australian Sugar Technologists Association,* 17, 97–105.

68 Allsopp, P.G., McGill, N.G., Licastro, K.A. & Milner, R.J. (1994) Control of larvae of *Antitrogus consanguineus* (Blackburn) (Coleoptera: Scarabaeidae) by injection of *Metarhizium anisopliae* conidia into soil. *Journal of the Australian Entomological Socety,* 33 (3), 199–210.

69 Anon (1994) *Centro de Tecnologia Copersucar Annual Report for 1993–1994.* Copersucar, Brazil.

70 Clowes, M.St.J. & Breakwell, W.L. (eds) (1998) *Zimbabwe Sugarcane Production Manual.* Zimbabwe Sugar Association Experiment Station, Chiredzi, Zimbabwe.

71 Mrig, K.K. & Chaudhary, J.P. (1991) Evaluation of some soil insecticides for the control of root borer and termites during monsoon and post monsoon season in sugarcane ratoons. *Indian Sugar* 41 (4), 25–30.

72 Madan, Y.P., Maan Singh & Singh, M. (1998) Evaluation of some soil insecticides for termites and shoot borer control in sugarcane. *Indian Sugar* 48 (7), 515–518.

73 Milner, R.J., Lomer, C.J. & Prior, C. (1992) Selection and characterization of strains of *Metarhizium anisopliae* for control of soil insects in Australia. In: *Biological Control of Locusts and Grasshoppers* (eds R. J. Milner, C. J. Lomer & C. Prior). CAB International/University of Arizona Press.

74 Cackett, K.E. (1983) Pearl scale life studies. *Zimbabwe Sugar Association Experiment Station Report* No. 3500/11.

75 Cackett, K.E. (1984) Control of pearl scale with vydate. *Zimbabwe Sugar Association Experiment Station Report* No. 3500/15.

76 Greathead, D.J. (1994) Locusts (Acridoidea) as pests of sugar cane in Africa. In: *Biology, Pest Status and Control Measure Relationships of Sugarcane Insect Pests* (eds A. J. M. Carnegie & D. E. Conlong), pp. 155–161. Proceedings of the International Society of Sugar Cane Technologists Second Entomology Workshop. SASA Experiment Station, Mount Edgecombe, Durban.

77 Reagan, T.E. (1994) Pest status, biology and control measures of sugar cane sap suckers in the North American sugar cane growing regions. In: *Biology, Pest Status and Control Measure Relationships of Sugarcane Insect Pests* (eds A. J. M. Carnegie & D. E. Conlong), pp. 123–127. Proceedings of the International Society of Sugar Cane Technologists Second Entomology Workshop. SASA Experiment Station, Mount Edgecombe, Durban.

78 Gupta, S.C. & Gupta, A.P. (1969) The loss in sugar production in Western Uttar Pradesh due to *Pyrilla* in the 1968–69 season. *Indian Sugar* 19, 159–167.

79 Carnegie, A.J.M. (1994) Sugar cane sucking insects (Homoptera), Africa and close islands. In: *Biology, Pest Status and Control Measure Relationships of Sugarcane Insect Pests* (eds A. J. M. Carnegie & D. E. Conlong), pp.107–112. Proceedings of the International Society of Sugar Cane Technologists Second Entomology Workshop. SASA Experiment Station, Mount Edgecombe, Durban.

80 Williams, J.R. (1980) The biology of the soft scale insect *Pulvinaria iceryi* (Sign). *Proceedings of the International Society of Sugar Cane Technologists,* 17 (2), 1843–1854.

81 Hall, D.G. (1990) The sugarcane lace bug, an insect pest new to Florida. *Sugar Journal,* 53 (5), 10–11.

82 Sosa, O. Jr, Cherry, R.H. & Nguyen, R. (1986) Seasonal abundance and temperature sensitivity of sugarcane delphacid (Homoptera: Delphacidae). *Environmental Entomology,* 15, 1100–1103.

83 Sosa, O. Jr (1985) The sugarcane delphacid, *Perkinsiella saccharicida* (Homoptera: Delphacidae), a sugarcane pest new to North America detected in Florida. *Florida Entomologist* 68 (2), 357–360.

84 Hernandez, J. (1997) In: *Internal SASEX Report on the Third International Society of Sugar Cane Technologists Entomology Workshop,* Culian, Mexico (eds G. W. Leslie & M. G. Keeping). SASA Experiment Station Internal Report, Mount Edgecombe, South Africa.

85 Suasa-ard, D.W. & Charernsom, K. (1994) Pest status, biology and effective control measures for sugar cane sap suckers in South East Asia. In: *Biology, Pest Status and Control Measure Relationships of Sugarcane Insect Pests* (eds A. J. M. Carnegie & D. E. Conlong), pp. 147–152. Proceedings of the International Society of Sugar Cane Technologists Second Entomology Workshop. SASA Experiment Station, Mount Edgecombe, South Africa.

86 Liu, Z.C., Sun, Y.R., Warg, Z.Y. & Lie, G.F. (1987) The role of biological control in integrated management of sugarcane insect pests. *Biocontrol News and Information,* 8 (4), 311.

87 Allsopp, P.G. (1994) Australian sugar cane sap suckers. In: *Biology, Pest Status and Control Measure Relationships of Sugarcane Insect Pests* (eds A. J. M. Carnegie & D. E. Conlong), pp. 133–138. Proceedings of the International Society of Sugar Cane Technologists Second Entomology Workshop. SASA Experiment Station, Mount Edgecombe, South Africa.

88 Anon (1997) *Sugar Association of the Caribbean Handbook 1961–1996.* Sugar Association of the Caribbean Technologists Committee, Belize.

89 Ganeshan, S. & Rajabalee, A. (1995) Sugarcane leaf-eating caterpillars: their bionomics and management.

Proceedings of the International Society of Sugar Cane Technologists, **22** (2), 575–580.

90 Carnegie, A.J.M., Dick, J. & Harris R.H.G. (1974) Insect and nematodes of South African sugarcane. *Entomology Memoir Department of Agriculture Technology Service Republic of South Afroica*, **39**, 1–19.

91 Hill, D.S. (1983) *Agricultural Insect Pests of the Tropics and their Control*, 2nd edn. Cambridge University Press, Cambridge.

92 Ganeshan, S. & Rajabalee, A. (1996) The *Mythimna* spp. (Lepidoptera: Noctuidae) complex on sugarcane in Mauritius. *Proceedings of the South African Sugar Technologists Association* **70**, 15–17.

93 Anon (1977) *Guia das pricipais pragas da cana-de-açúcar no Brazil*. Instituto do Açúcar e do Alcool, Planalsucar, Brazil.

94 Dasart, L. & Victorine, C. (1977) Strategies for rodent control in the Guyana Sugar Industry. *Proceedings of the West Indies Sugar Technologists Association*, **26**, 191–196.

95 Leslie, G.W. & Keeping, M.G. (1997) *Internal SASEX Report on the Third International Society of Sugar Cane Technologists Entomology Workshop, Culian, Mexico*. SASA Experiment Station Internal Report, Mount Edgecombe, South Africa.

96 Lefebvre, L.W., Ingram, C.R. & Yang, C. (1978) Assessment of rat damage to Florida sugarcane in 1975. *Proceedings of the American Sugar Technologists Association*, **7**, 75–80.

97 Wood, B.J. (1994) Rodents in agriculture and forestry. In: *Rodent Pests and Their Control* (eds A. P. Buckle & R. H. Smith), pp. 45–83. CAB International, Oxford.

98 Macdonald, D.W. & Fenn, M.G.P. (1994) The natural history of rodents: preadaptions to pestilence. In: *Rodent Pests and Their Control* (eds A. P. Buckle & R. H. Smith), pp. 1–21. CAB International, Oxford.

99 Buckle, A.P. (1994) Rodent control methods: chemicals. In: *Rodent Pests and Their Control* (eds A. P. Buckle & R. H. Smith), pp. 127–160. CAB International, Oxford.

100 Sugihara, R.T., Tobin, M.E. & Koehler, A.E. (1995) Zinc phosphide baits and prebaiting for controlling rats in Hawaiian sugarcane. *Journal of Wildlife Management*, **59** (4), 882–889.

101 Smith, R.H. (1994) Rodent control methods: non-chemical and non-lethal chemical. In: *Rodent Pests and Their Control* (eds A. P. Buckle & R. H. Smith), pp. 109–126. CAB International, Oxford.

Chapter 5
Sugarcane Agriculture

R. D. Ellis and R. E. Merry

INTRODUCTION

Sugarcane agriculture practices are influenced by many factors, including climate, landform, soil composition and structure, irrigation and drainage requirements, varieties, pests and diseases, management and labour skill availability, and harvesting methods. Many of these interact also in different ways and are often interdependent. As a consequence, different practices have developed in the sugar producing countries to satisfy local conditions. This chapter aims to set out the general principles that are followed in most places, but also highlights specific practices in some regions that are of interest.

For many years, land was cleared by axe and saw, and much of the land cultivated to sugarcane in the Caribbean, North and South America, Asia and Australia was initially cleared in this manner. It was a tedious and highly labour intensive method. There was a rapid transformation when bulldozers and tractor-drawn or mounted rippers were developed. These work quickly and require few but highly skilled operators and maintenance crews. Trees are uprooted, pushed into windrows with the remaining vegetation, and burnt when dry. The cleared area can be land-formed to provide surface drainage or given a precise grade to facilitate irrigation.

When preparatory cultivation was carried out by hand, using hoes or forks or animal-drawn implements, the heavier soils could be worked only when they were too moist to allow a satisfactory tilth to be obtained. This changed around 1900 when ploughshares carried on cables and drawn across the field by steam engines were introduced

into several countries. In the 1930s, steam ploughs were replaced to some extent by gyro-tillers. Both gave way, a few years later, to crawler and wheeled tractors. The first tractors were driven by petrol or paraffin engines and later by diesel engines. They draw implements mounted on tool bars controlled by hydraulic linkages, and thereby eliminate wasteful headlands.

SYSTEMS OF CULTIVATION

Although several different agricultural systems were developed when cane was first planted on a large scale, generally under rainfed conditions, they had two things in common:

- they were devised to meet local climatic conditions; and
- they were dependent upon an abundant supply of cheap labour.

In dry areas, in the absence of irrigation, soil and moisture conservation were of supreme importance, whilst in wet low-lying areas adequate drainage was the main requirement.

Soil conservation and field layouts

Soil conservation aims to protect the soils on which sugarcane is grown in order to produce consistently high returns for as long as possible. All field layouts should aim to achieve this and the tools that are used by planners are based on the Universal Soil Loss Equation[1]. This defines the expected soil loss under any conditions and is expressed as follows:

soil loss (A) =
rainfall erosivity factor (R) × soil erodibility factor (K) × topographic factor (LS) × crop management factor (C) × practices (P).

- *Soil loss* (A) is the total soil lost from any set of conditions and will have a different value for each soil type. It is measured in tonnes/annum. The factors that influence it need to be managed carefully where they are particularly limiting.
- The elements that determine the *rainfall erosivity factor* (R) are the quantity of rain falling in a particular storm, which gives a broad estimate of damage, and its intensity. The interactive forces that take place in a rainfall event are raindrop splash and run-off of water. A mulch cover or trash, and sensible row and drainage gradients can mitigate the effect of these on soil loss.
- The *soil erodibility factor* (K) is dependent on soil constituent composition (sand, silt and clay percentage), soil structure, organic matter content and soil permeability. Poorly structured, permeable, sandy soils with low organic matter content are more erodible than structured clay soils with moderate to high organic matter content.
- The *topographic factor* (LS) is a function of steepness and length of slope. As these increase so does the amount of water available to cause separation and transport of soil particles.
- The *crop management factor* (C) has different values determined by various practices such as trashing, cane burning, strip cropping and minimum tillage, as well as timing of operations, e.g. planting in relation to expectation of high intensity rainfall. These practices are also important in conserving soil moisture during the crop growth phase.
- The *practices factor* (P) is mitigated by conservation practices such as contour row planting and the use of graded terrace banks.

The crop management and practices factors are the most controllable factors by the farmer; but mechanical conservation works are essential to control water flowing over the land before it causes serious damage, and conduct it at a safe velocity off the land to an area where it can do no harm. These works include:

- storm drains situated on higher ground above the land to divert storm water before it enters the field;
- terraces within arable land which collect surface run-off and discharge into a stabilised waterway; and
- waterways which may be natural or constructed depressions with a stable vegetative cover.

The most vulnerable time when the erosion hazard is greatest is during the plough out or bare fallow period, so these should be minimised or restricted to a period when high intensity rainfall events are unlikely.

Drainage of soils is often as important as soil conservation for maintaining crop productivity. Because of its close association with the irrigation of sugarcane, this subject is more fully dealt with later in this chapter.

The impact of soil and moisture conservation practices on different cultivation systems is described in the following sections.

Row cropping

The most common system of cultivation is row cropping, where the sugarcane is planted in rows, either on the flat or on ridges.

Crops planted on the flat normally have no deep furrows or high ridges. This method is widely used for mechanical planting, and is also suitable for mechanical harvesting. The shallow furrows or interrows are more vulnerable to sheet erosion, unless protective measures are taken. Normally, the rows are laid out on the level contour. In fields of a regular configuration, short rows are eliminated, but changes in row direction will often be required where the landform alters in the field. Rows of sugarcane planted on the flat provide dense vegetative barriers limiting soil loss and run-off, but significant losses can take place between the stools, especially in older ratoons.

Sugarcane is grown on ridges for several reasons:

- to prevent sheet erosion and channel run-off water into conservation structures;
- to provide greater depth of suitable soil for cold and poorly drained soils;

- to facilitate irrigation, especially furrow irrigation and, to a certain extent, drip irrigation;
- to give better control of certain pests (i.e. soil-inhabiting larvae) that prefer the moister conditions which occur in the furrow;
- to suit machinery operations as the wheels run in the furrow and are less likely to damage stools, provided wheel spacings are compatible with the inter-row width; and
- to lay cane on after hand cutting for the operation of grab loaders and provide a 'face' for chopper harvesting.

It is not always possible to plant sugarcane on ridges, especially where furrow irrigation is practised, as poor germination results because of inadequate soil moisture contact with the cane sett. In this case, 'middle busting', which involves splitting the inter-row ridge after germination and placing the soil around and on top of the emerged shoots in the furrow, is necessary to obtain the desired ridge. This can be damaging and sets back the growth of the crop, unless it is done carefully and before the crop is knee-high.

Ridge height is variable, usually settling to between 0.15 and 0.25 m in ratoon crops, and between 0.5 and 0.8 m wide. After ratooning, the ridges may be reconstructed to satisfy any of the above criteria for which they are required.

Crop row lengths vary and should be determined by the requirements of good conservation. Rows are commonly a maximum of 200 m on light textured soils and up to 400 m on heavy soils. Very long rows do occur in suitable soil and slope conditions, and rows of 500–1000 m are used in the very flat, vertisol soils in the Ord River development in Western Australia and Nakambala in Zambia. Longer rows suit machinery operations better, but have severe soil conservation limitations. They are unsatisfactory where operations are highly labour intensive, because the management of labour is difficult.

Row gradients are determined by soil classification, land form, uniformity of slope, irrigation method and row length. Typical gradients are rarely steeper than 2.0–2.5%, except on very short rows; they are often flatter where conditions permit.

Row spacing is decided on the basis of response to yield for different spacing and management factors. Most experimental work has shown that, where moisture stress is not severe, sugarcane yield increases as the distance between the rows decreases, within certain limits. In South Africa[1], for example, there is a 3% increase in the plant crop for every 300 mm decrease in row spacing from 2.0 m to 0.6 m. In practice, row spacing of 1.0 m is as close as field equipment will conveniently allow, but row spacing tends to be between 1.5 and 1.8 m wide in highly mechanised operations. In cooler, slower growing conditions, on steep land where a quick canopy is required, on erodible soils, and where varieties with erect leaves are planted, closer rower spacing is more suitable. Where growing conditions are better, or the soils are shallow or the rainfall is low, slightly wider spacing is preferred. In good growing conditions, where leaf canopy development is rapid, irrigation is practised, and operations are highly mechanised, the widest row spacing is most suitable.

In wider spaced rows, sugarcane is planted in double or triple rows. This is often referred to as tramline or pineapple planting. This row configuration has added advantages with cane harvesting, as the inter-rows are about 1.8 m wide. It results in a reduction in traffic damage to the cane row and compaction near the cane row, as well as fewer passes of harvesting machines. Slower canopy cover is, however, a disadvantage to the wide spacing as it affects the effectiveness of weed control.

The yield increase from closer row spacing is relatively small and there are increased costs involved, such as greater seedcane requirement and longer planting time needed, and more rows to be weeded and treated.

Cambered beds

Adequate drainage of low-lying flat land in areas of high rainfall is achieved in many countries by growing cane on cambered beds separated by deep drains. The beds are of varying width, most frequently 6–7 m wide and the drains 0.6 wide and 0.45 m deep. The camber is maintained by mouldboard ploughs or discs travelling along the beds and turning the furrow slices towards the centre (gathering). If the camber becomes too pronounced the mouldboard or disc is replaced by chisel tines in this first operation. The full sequence of cultivation operations on soils, which are mainly heavy clays, is:

- plough or chisel to uproot old cane stools to break the soil into large clods;
- harrow and re-harrow, if necessary, after an interval of 10 days to produce good tilth;
- ditch using a suitable implement to re-open the drains; and
- make planting furrows 1.5 m apart.

All cultivation is done in dry weather. Where surface irrigation is practised, cane is laid in the furrows, running along and not across the cambers, covered with soil by breaking the banks, and the cambered shape of the beds restored. Sprinkler irrigated camber beds are easier to irrigate, and good surface drainage results in well-grown cane. In non-irrigated areas it is necessary to wait for the first showers of the season before fields can be planted. Cross sections of a typical cambered bed, before and after planting, are shown in Fig. 5.1.

In theory, the chisels create bands of cultivated soil 0.45 m deep of the same curvature as the cambered surfaces. Excess rain or overhead irrigation percolates through this layer and, because of the camber, enters the drains through their sides. In practice there can be considerable soil erosion, especially at the sides of the beds, and frequent cleaning of drains is necessary. Because of this, and also because the outside rows of cane are 0.6 m

further apart than the centre rows, overall growth can be restricted. Nevertheless, cambered beds can be successful in providing reasonable drainage on heavy clay soils, and are in use in East and South Africa in these conditions.

The system in Guyana

In Guyana, sugarcane is grown on narrow strips of land along the coast. The cultivated area is within 13 km of the Atlantic Ocean and much of it is below sea level. The mean annual rainfall is 2340 mm, and there are two distinct dry seasons. Heavy clay, saline soils are found near the coast but they decline in salinity inland. More distant from the coast and the rivers these heavy clays are replaced by highly acidic peat soils. Beyond the cultivated area are vast tracts, known as swamp reservoirs, in which water from the interior is collected in conservancies for use in transport and irrigation on the estates during the dry season.

It is only by a complex system of drains, dykes and canals that cane farming is possible. Dutch colonial engineers reclaimed the low-lying areas for cultivation. A wide, well built sea wall protects the coastal land from inundation, and substantial banks safeguard the riverside fields. Drainage water is discharged into the rivers and sea by pumps or, where possible, by sluice gates which are opened at low tide. The drainage canals (sidelines) pass along the ends of the fields and receive water from the in-field drains. At the opposite ends, and at a higher level, are the transport

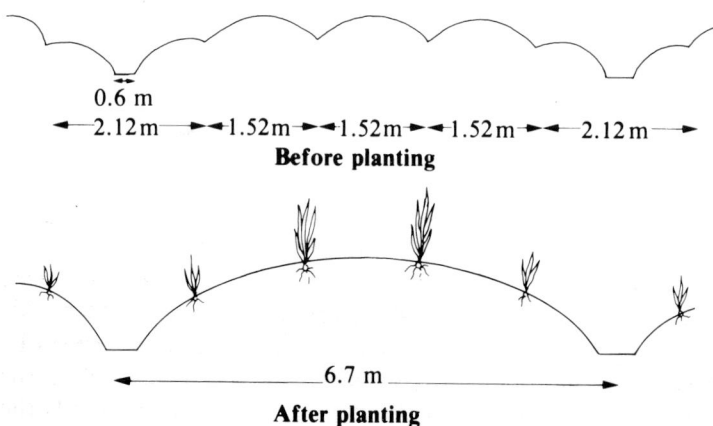

Fig. 5.1 A cross-section of a typical cambered bed.

and irrigation canals (middle-walks) whose off-shoots (cross-canals) establish field boundaries. An 11.3 m wide dam bed surrounds each field, which prevents water from the higher-level canals flowing into the field, except when required, and irrigation water from the field into the drainage canal. Within the field the cambered beds of varying width (usually 7.3 m from centre to centre) are separated by drains. The beds run from either middle-walk to sideline (English layout) or from cross-canal to cross-canal (Dutch layout). Cane is planted in rows 1.8 m apart, which run across the beds. Typical English and Dutch layouts are illustrated in Fig. 5.2.

The typical sequence of cultivation operations carried out when a field is replanted is as follows:

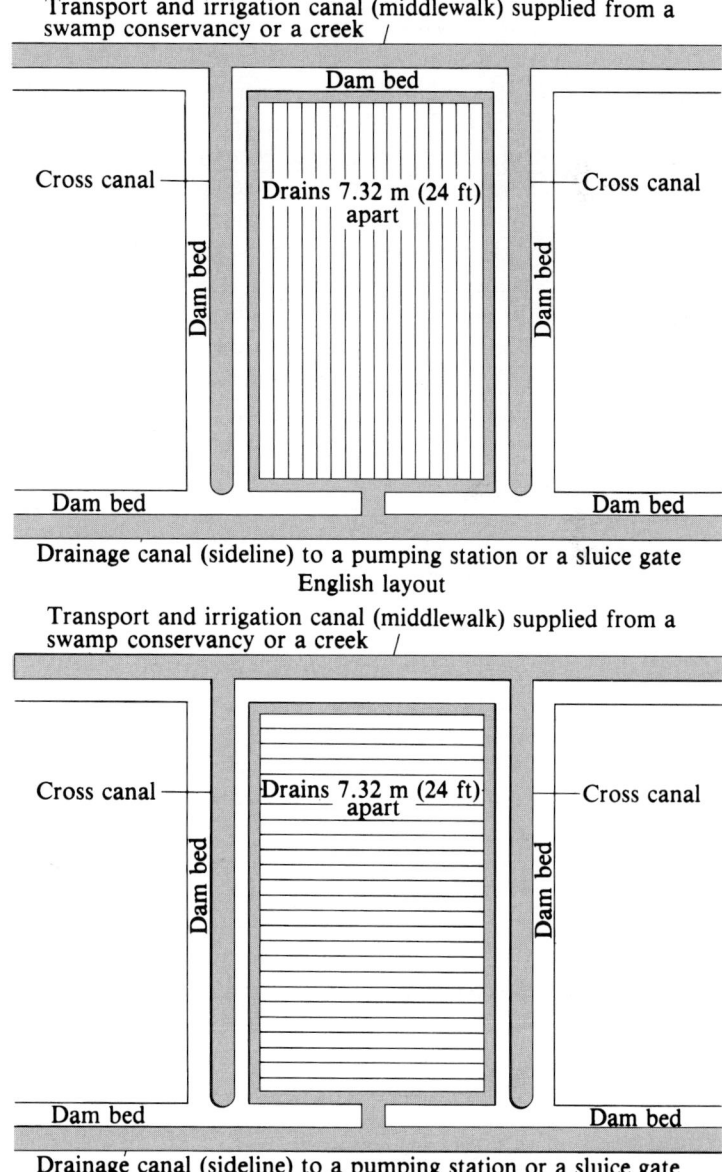

Fig. 5.2 The layout of cambered beds in Guyana.

- plough with discs to uproot the old cane stools and restore the camber of the beds;
- harrow with discs to improve soil tilth;
- re-open the drains with a mechanical digger;
- move the soil from the drains to the centre of the beds to complete the camber; and
- harrow with tines or chisels to break up clods and smooth the surface of the bed.

The field may then be submerged to a depth of 0.3–0.45 m for a period varying from 3 to 6 months (flood fallowing), after which the water is drained off and the sugarcane planted. Flood fallowing improves the texture of the soil, which becomes more friable, eliminates dry land weeds and increases soil nitrogen content. It has been suggested that a layer of ferrous ions, developed by the reducing conditions brought about by flood fallowing, which becomes oxidised to the ferric state when the water is removed, protects the soil crumbs, and that this is responsible for the improvement in tilth. Flood fallowing causes an increase in yield of 40% over a 3- or 4-year crop cycle, although this increase has to compensate for the loss of one crop in the cycle.

Florida and Mozambique

Difficulties similar to those of Guyana occur elsewhere, e.g. in Florida and Mozambique.

In Florida, sugarcane is grown on land that has been reclaimed from the Everglade swamps by the installation of an extensive and well-managed drainage system. The 'muck' soil that has been recovered has a high organic matter content, in some places in excess of 60%, and its fertiliser requirement for successful crop production is unusually low, if not unique.

The difficulties of combining flood protection with the provision of an adequate drainage system have also been faced in Mozambique, where cane is grown on vertisols in the deltas of the Zambezi and Komati rivers protected by high embankments constructed around the entire cane production area. The problems of field drainage on very flat soils have not been entirely resolved, although on one estate all the drainage water is pumped out of the enclosed land within the embankment back into the river.

Louisiana banks

In Louisiana, the successful growth of sugarcane on low-lying flat land, with a high water table and under heavy rainfall, was achieved by the development of a system of ridge and furrow cultivation in fields formed in the shape of turtlebacks. The ridges on which the cane is grown are 0.45 m high, and are spaced at intervals of 1.8 m. Each of the furrows acts as a drain. Water flows from the furrows into slightly deeper quarter drains, 20 m or more apart, which are at right angles to the ridges. It is then discharged at the sides of the field into deeper ditches that run parallel with the ridges aided by the turtleback shape. The field drains lead into area drainage canals (Fig. 5.3).

The great merit of the Louisiana bank system (as the ridges and furrows are called) is that it allows all stages of sugarcane production to be mechanised. Wheeled tractors that straddle the cane rows and draw suitable implements can carry out all field operations. A unique feature of Louisiana bank system is the use of high-clearance tractors. They haul discs in fields of established cane to increase the height of eroded banks to the required level, and destroy weeds at the same time. This and other cultivation operations fill the quarter drains that must be quickly re-opened.

Cultivation on sloping land

For sloping land, the emphasis is on the prevention of soil loss and soil moisture conservation. Design of layouts is simplified now that aerial photography and contour maps are readily available. The major considerations in the spacing of structures are slope and soil type[1]. The structures that need to be incorporated into designs for sloping land are roads, waterways, and terraces.

The design of the waterway is crucial for the safe discharge of storm water. The USA Soil Conservation Service design is the most appropriate method and takes into account the expectation of peak flows in any situation. The steeper the ground, the closer the terrace spacing must be to control soil and water losses. Also the more erodible the soil, the closer the terraces must be to reduce the flow length of run-off water. Management factors

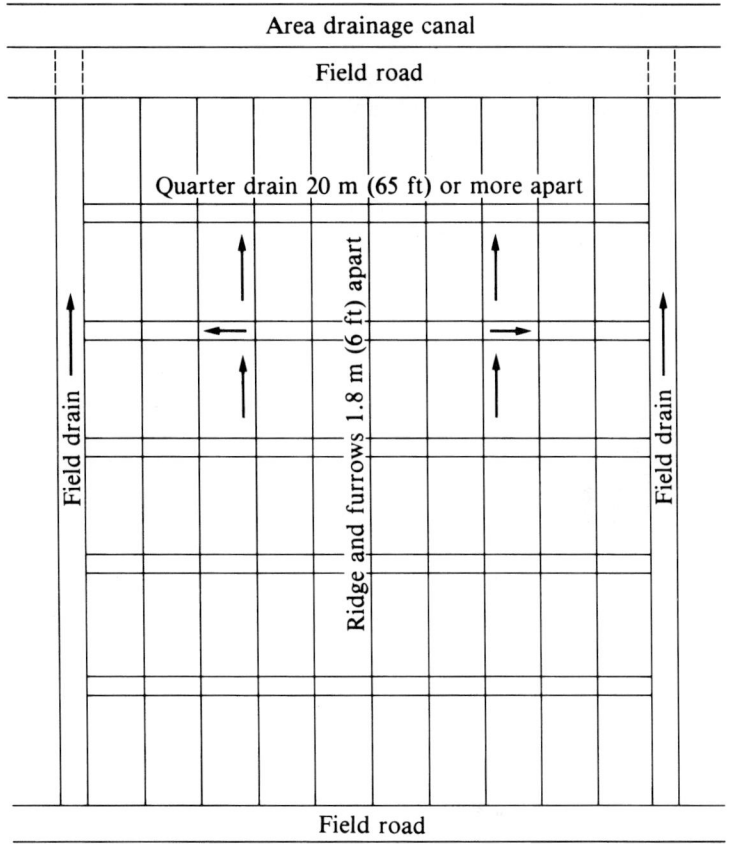

Fig. 5.3 A modified Louisiana bank layout.

⟶ Indicates direction of flow of water

play an important part in the spacing of terraces, and these include minimum tillage, strip cropping, burning, trashing, and time of planting. Nomographs are now available that take these factors into consideration when designing layouts for cultivation on sloping land.

Typical examples of field layouts on sloping land include:

- *Sloping ridge.* This has a crest road running down the centre with natural and/or constructed waterways on either side. Terraces are at right angles to the crest road, parallel to each other and discharge water from the crest into the waterways. Crop rows are parallel to the terraces. Terraces are useful for discharging excess water and for loading cane.
- *Dome.* This has a road across the top of the hill, with terraces encircling the dome, discharging into constructed grassed waterways running down the slope.
- *Saddle.* The crest road links the saddle from high point to high point. Terraces encircle the saddle and discharge into waterways leading off the lower point of the saddle.

The construction of conservation structures needs careful attention. Terraces on slopes of up to 15% can be made using a reversible disc or mouldboard plough, moving soil upwards if possible. On steeper slopes, a blade terracer or bulldozer is more suitable. Waterways are made using reversible ploughs, bulldozers or dam scoops. Field-edge drains are best constructed with graders. Immediately after construction, waterways should be grassed with *Cynodon* or *Stenotaphrum* species and watered to produce a grass cover before the rains.

YIELD

One of the most significant indications of the success of a farmer is the productivity of his land expressed as the yield of cane or sucrose per hectare per annum. The yield per hectare harvested, though often quoted, is not valuable as an indicator unless related to the age of the cane at harvest. For example, a cane yield of 170 t/ha when the crop is harvested at 24 months indicates lower productivity than 90 t/ha if annual harvesting is practised. In annual cropping, age at harvest may vary between 9 and 14 months, depending on the season of harvest and whether the fields are plant or ratoon cane. Dividing by the age in months only gives a crude comparative figure, as growth is limited in winter, and a 14-month plant crop harvested late in the season cannot be directly compared with a 10-month crop harvested early in the season.

Cane age at harvest is very variable around the world. Most irrigated estates harvest annually at about 12 months, as do high-rainfall productive industries such as those in Australia and Colombia. The more unsuitable the climatic potential, the higher is the age at harvest, e.g. with dry land cane in South Africa being harvested at 15–18 months and in Kenya at 18–22 months.

Various attempts have been made to improve the expression of yield in terms of age. Sweet & Patel[2] suggested the COTCHM method (Corrected Tonnes Cane per Hectare per Month). From an analysis of a large database of yields of variety Nco376 at Triangle in Zimbabwe and Simunye in Swaziland, Sweet & Patel produced correction factors for cane yields for cane harvested in each month of the season at different ages, referenced to the yield attained in April at the beginning of the season. The method is very useful for correcting field yields to a common base, to determine whether they are underperforming in relation to the age and season of harvest, and it has particular application in the decision making process of replanting.

Very useful work is being done in Swaziland to determine the potential yield of sugarcane based on the use of the CANEGRO climatic model. The potential yield is adjusted by specific estate corrections based on mean historical yield achievements on each soil type, for each ratoon, the season of harvest, variety, and irrigation method[3]. The method is used for an assessment of a manager's performance and in the forecasting of cane yields.

Farmers generally place greater emphasis on cane yield as a measure of performance. This is because it is usually more tangible than sucrose yield and more controllable by them. This is particularly true when the farmer is not paid on the basis of sucrose production nor given any feedback by the miller on the sucrose levels in cane. There is no incentive in these circumstances for the farmer to deliver fresh, high sucrose cane to the mill. This situation pertains in many mills in Asia, Africa, and South America. Where, however, sucrose is measured and paid for, growers use many techniques to increase their sucrose percentage and express their yield in terms of it. These include delivery of fresh cane < 72 h after burning or cutting, good topping practices, drying-off and ripening. These techniques result in higher sucrose payments to producers and improved extraction and recoveries by the factories. Particular examples of this are found in Australia and Southern Africa.

RATOONS AND RATOONING

After a newly planted cane field has been harvested, the bud and root primordia of the stool develop when ecological conditions are favourable and produce a stubble or ratoon crop. As the new shoots grow and develop roots, the old roots die and decompose. Thus each successive crop is sustained by water and nutrients absorbed by its own new root system. The soil loses its structure with time, however, and becomes compacted by wheeled equipment. The tilth created by land preparation is lost, the efficiency of drainage declines, soil salinity and sodicity problems are exacerbated, the stool is damaged by harvesting equipment (especially with combine harvesters), and pests and diseases cause more damage. Consequently, other factors being constant, the formation of an adequate root system becomes increasingly difficult for each successive ratoon, plant populations decline, and cane yields are reduced to a point where it becomes economic to plough out the crop and replant.

Well structured, freely draining soils may produce many ratoons before this stage is reached, whereas poorly structured, sodic soils require frequent replanting. In Swaziland over 20 ratoons have been produced on the best, structured, freely drained clay loam soils under irrigated conditions; whereas poorly drained, sodic, duplex soils are replanted after four or five ratoons. Intermediate quality soils are replanted after 8–10 years. In Australia, it is rare for more than four ratoons to be harvested, whilst the maximum is usually two for smallholder fields in Kenya.

The establishment cost of plant cane is very much higher than the cultivation cost of ratoon cane, a difference which has been increased by the widespread cultivation of modern vigorous hybrid varieties, and the increasing realisation that, in many instances, practices like subsoiling can cause a loss in yield, particularly when irrigation is not applied promptly after the cultivation is completed. The high cost of establishing plant cane should be accounted as a capital investment.

In some countries the maximum number of ratoons permitted to be grown, or the proportion of the assigned area planted with cane harvested each year, used to be determined by statute or by an inflexible system of crop rotation. This was the case in Queensland, Barbados, Java, Taiwan and Kenya, but is less rigorously applied now because of a realisation of the economic benefit to be obtained from ratooning. In deciding the size of the replanting programme, and therefore the area to be ratooned, the cane farmer considers many factors, of which the most important are:

- the balance between the lower yield of cheaper old ratoon cane and the higher yield of more expensive younger cane, taking special note of the effect of output on fixed charges;
- the condition of the sugar market, bearing in mind the need to fulfil quotas fixed by international agreement and to take up the shortfalls of other producers;
- the social consequences, in some countries, of having little or no work for field employees in the period immediately preceding the start of milling operations if the area in young plant cane is small;

- technical considerations which include:
  (a) multiplication of a new, promising variety with potential for higher yield or disease resistance;
  (b) control of soil inhabiting pests which are usually greater in older ratoons;
  (c) improvement of field layouts and changes in irrigation methods;
  (d) rectifying damage caused to fields by harvesting practices;
  (e) improvement of surface and sub-surface drainage;
  (f) measures to reduce soil salinity, sodicity or rectify soil pH; and
  (g) declining stool population.

It is preferable to have a relatively constant annual replant programme, especially on large estates, and this should be evaluated regularly, based on the above criteria. If a three-year replant plan is drawn up, adequate preparation for the replant in each field can take place that includes arranging for suitable harvest dates, and undertaking drainage, soil amelioration and irrigation investigations. When deciding the rate of annual replant, the decision should be based on sound economic principles. Determining the Net Present Value (NPV) of various ratoons is the most suitable method and a number of models have been developed which do this. The information required for these models is:

- an estimation of yield for plant crops and each ratoon on each class of soil, which can be obtained from historical yield data and preferably adjusted for the season of harvest, to eliminate bias particularly from low yielding fields at the end of a harvest season;
- the estimated sucrose percentage, in the case of a grower, or of sugar extracted in the case of a miller;
- the cost of plough out and replant, and the annual cultivation cost of plant and ratoon cane, expressed as cost per hectare, and the harvesting and haulage cost per tonne;
- an estimate of the sucrose or sugar price that can be expected; and
- an estimate of the discount value of money, usually the rate at which money can be borrowed.

The highest NPV of any ratoon is the ratoon at which most benefit will accrue from replanting.

LAND PREPARATION

The principal objectives of land preparation are:

- to destroy the previous crop;
- to remove roots and stones to improve harvesting conditions, and volunteer re-growth, which may transmit smut and ratoon stunting disease (RSD) to the next crop;
- to break up existing plough pans;
- to install suitable surface and sub-surface drainage and field layouts;
- to ameliorate soil conditions by the addition of gypsum, lime or mill mud; and
- to provide an adequate tilth for germination and growth of the sugarcane crop.

In order to achieve these objectives a range of activities take place, depending on the results that are required and the cost of operating different machinery in each farmer's circumstances. The timing of land preparation operations is important. It is preferable if the fields to be replanted are harvested early in the dry season, when rainfall is least expected and the growth potential is lowest, thus minimising potential loss of cane. The fallow period should be as short as possible, consistent with the operations that have to be carried out although, if the previous crop had high levels of RSD, a longer fallow period is desirable. Ideally, the field should be replanted in time for the germination and early tillering phase to be completed by the time of rapid growth potential in early summer. Fields that are to be fallowed for autumn planting with a green manure crop are an exception, and will normally be harvested late in the harvesting season.

In many instances the methods are unique to each farmer, and the description that follows aims to illustrate those that are generally used. Each operation is not necessarily done in the order given nor are they all always done.

- *Environmental considerations.* There is an increasing realisation that sugarcane farmers must be more pro-active in managing their environ-

ment. This has resulted in reluctance to develop new land. Where new land is to be developed, this should be done more sensitively, i.e. by only opening land that can be profitably cultivated, excluding poor soils, riverine vegetation, rocky areas and places of archaeological interest. Farmers are increasingly required to undertake Environmental Impact Assessments and to abide by the recommendations. This has a twofold effect (i.e. reducing areas developed and undertaking practices that will result in more productive use of existing land under cultivation) that has a significant impact on land preparation methods.

- *Stool destruction.* This may be carried out using reversible or one-way mouldboard ploughs, which are set shallow to turn over the existing stool and expose it to desiccation. Alternatively, heavy disc harrows can be used which chop up the stool. These are effective in dry conditions but less so than mouldboard ploughs when the fields are wet. Disc ploughs can also be used but they tend to turn in the old stool, allowing it to re-grow. Any volunteer stools should be removed by hand before the next operation commences, especially if the previous variety is susceptible to smut or RSD.
- *Ploughing.* The object of this operation is to cultivate the soil to depth in order to break any soil pans that occur as a result of compaction and to mix the topsoil. It may be done with conventional or reversible ploughs or heavy Rome disc ploughs.
- *Harrowing.* This is done to produce a good surface tilth or seed bed and to allow the following operations to proceed on a smooth surface, by breaking up soil clods from other land preparation operations and to incorporate ameliorants. Two harrowings may be required, the final one to produce the planting tilth. The operation is done with disc harrows, chisels or tines.
- *Ripping.* This operation is done to shatter deep pans, break up compacted soil and to improve rooting depth. It is usually done using crawler or high-powered, wheeled tractors with tined rippers. It is an expensive operation but very effective, particularly in heavy clay or duplex soils.
- *Levelling.* In newly developed or uneven ratoon fields, graders or dump-scoop levelling may be

required to fill in hollows and create terraces and waterways. In replanted fields this may only be required in bad spots. Land planing is important to smooth out high spots and fill low spots, particularly in furrow-irrigated fields, and in fields with very shallow slopes. In sophisticated operations, a laser instrument controls land planing.

- *Sub-surface drainage.* The land preparation operation is the most suitable time to install sub-surface drainage works to reduce water tables, salinity and sodicity. The need for this should have been planned by investigations conducted before the replanting decision is taken. The spacing of drains, depth and size of the pipe are determined by these investigations. Drainage pipe is normally slotted PVC pipe but mole drains may also be effective. Drainage pipe is usually installed by trenching the soil and laying the pipe in a sand envelope. Laser controlled trenching and drainage pipe installation machines are available and are used in large-scale sophisticated operations. Inspection and cleaning boxes must also be built.

- *Soil ameliorants.* These are best added before the ripping or the final harrowing in order to incorporate them satisfactorily into the soil. Gypsum is applied to reduce soil sodicity, particularly in sodic soils. Lime should also be incorporated to reduce high soil pH levels. Both these materials are applied using tractor drawn spreaders. Mill mud is used to improve the organic matter status of the soil and soil structure at rates of about 100 t/ha and, at lower rates, provide a source of phosphate fertiliser. It is often necessary to dispose of this material on factory estates and, even though it is very bulky, it has a high value particularly in virgin soils, which are often very low in phosphorus. Mill mud is applied using tractor drawn spreaders or by spreading dumps in the field with graders. On estates with distilleries, 'stillage' or 'dunder' can be applied that provides an adequate level of potassium and some nitrogen. It is applied with tractor drawn bulk tankers through nozzles on a boom or, later, through the sprinkler or furrow irrigation system.

- *Rock removal.* Rocks in fields are a serious hindrance to mechanical operations and can be very destructive if they are gathered at harvesting and end up going through the mill. Strenuous efforts to remove them at replanting and after harvest are worthwhile. In Mauritius, which is of volcanic origin, the land is strewn with basalt rocks and these have been piled up into high heaps in the field. When this land is replanted, subsoilers drawn by crawler tractors, pass through to expose the stones. These are removed by bulldozers or manually and stacked in pyramids or as walls between rows of cane.

- *Cover cropping and fallowing.* This technique is used in many countries:
    (a) to provide a break between crops of sugar cane because of time or machinery constraints on replanting;
    (b) to improve soil conditions especially the organic matter status; and
    (c) to control resistant weeds or to prevent the build-up of diseases and pests.

Common cover crops are usually legumes, such as cowpeas, velvet beans or sunn-hemp. The cover crop is planted after late harvested cane has been ploughed out and following preliminary land preparation operations. It is optimally turned in at the time of maximum vegetative production when in the early flowering stage of growth, and allowed to decompose before final land preparation is completed. In some instances a commercial crop such as cotton replaces legumes.

- *Minimum tillage.* This technique is now quite well established. Stool destruction is effected by the application of glyphosate at 8 to 10 litres/ha or with fluazifop at 6 litres/ha. Fluazifop should not be applied less than 8 weeks before replanting to avoid poor germination. The chemicals should be applied to actively growing cane more than 400 mm high for optimum stool eradication. Good coverage of the foliage is important. A narrow band of about 200 mm of fine tilth is produced between the old cane rows into which the cane setts are planted. This is achieved with discs in light soils or with rotary hoes in heavier soils. The advantages of minimum tillage are a reduction in soil erosion on steep lands because of the continued presence of the old, dead crop, and the small area of soil disturbed. Soil structure, organic matter, moisture and nutrients are preserved; the volunteer population is reduced; costs are generally lower; and a yield increase can be expected on light and medium textured soils. The disadvantages are that planting is delayed because of the requirement for the previous crop to be actively growing before it can be sprayed with the non-selective herbicide; new layouts and land levelling cannot be undertaken, and the incorporation of ameliorants is difficult.

- *Planting furrows.* The preparation of these is the last step in land preparation. They are made by the passage of V-shaped double mouldboards, tines or discs. Care is required to ensure that they are a constant distance apart and of uniform depth, i.e. approximately 100 mm. The spacing of the rows has been discussed earlier. A fine seedbed is important for even germination, as a cloddy bed will result in poor contact of the seedcane with the moist soil.

SEEDCANE PRODUCTION

The quality of seedcane used in planting is fundamental to the production of high yielding, healthy plant cane and that of subsequent ratoons. Quality is determined by the freedom from diseases and pests, varietal purity and germination ability. Many diseases are transmitted through infected seedcane (e.g. smut, RSD, mosaic, yellow leaf syndrome (YLS), leaf scald, chlorotic streak and streak), and many of them can cause severe yield losses. Varietal purity ensures that the desired variety is grown and the seedcane is not contaminated with unwanted varieties susceptible to disease. Young, well-grown seedcane is more vigorous than old cane, and germinates better giving early, rapid establishment, and uniform growth. All these requirements can only be satisfied by seedcane produced in managed nurseries, and not from commercial fields.

The seedcane nursery should preferably be isolated from other sugarcane, on good soils, and in a frost-free area. Seedcane should be 8 to 10 months old for optimum vigour when planted commercially in warm regions, and 12 to 15 months in cooler areas. The planting of the nursery should

be planned to supply seedcane of this age. Two-stage nurseries are ideal. The primary or first-stage nursery should be grown in isolation following a fallow break. Often this is a co-operative nursery supplying a number of growers, and managed intensively as such. Introductions of new varieties to an industry or region will usually be placed in this first-stage nursery. On large estates, a secondary or second-stage nursery will bulk up the seedcane before planting commercially. In both types of nursery, the level of management and control of diseases and pests should be intensive.

The size of the nursery depends on the expected yield of seedcane, the extent of the area to be re-planted, and the rate of seedcane in tonnes/hectare to be planted commercially. Low population varieties will require a greater area of the seedcane nursery. It is better to over-estimate the nursery size in case of mishaps or increased demand The nursery should be isolated and clearly marked, fallow for a period of three months, and the land thoroughly cultivated and free of volunteers. Seedcane should be selected from another registered nursery, be vigorous and true to type. The primary nursery seedcane should be heat treated to eliminate RSD, and dipped in fungicide afterwards for disease control. Heat treatment also has the effect of suppressing apical dominance, thus eliminating the need to cut stalks into sets at planting. After planting, it is very important to monitor the crop for diseases, dig out all infected stalks, particularly those with smut or other recognisable systemic diseases, and those not true to type.

In Australia, South Africa, Swaziland, and Zimbabwe very high standards are set for the production of seedcane. In Swaziland seedcane nurseries must be registered with the Extension Services who are responsible for inspection and certification. No nursery will be certified for use unless it meets the following criteria:

- smut and mosaic infection less than 0.1%;
- freedom from RSD and YLS;
- *Eldana saccharina* infection less than 4 per 100 stalks;
- freedom from off types, and less than 0.1% of other varieties;

- seedcane may not be moved from one grower to another without a certificate and movement permit; and
- severely lodged and over-aged cane may also not be certified.

Seedcane for commercial planting should be young plant cane from defined nurseries, virtually free of diseases, undamaged by pests and lodging, and a pure varietal stand. First ratoon cane can be used in certain circumstances if it meets these criteria, particularly if it is a newly released variety where only small quantities are available for planting.

PLANTING

The time of planting in the year has a significant impact on the yield achieved in the plant cane crop. In general, the earlier in the planting period it can be done, the higher the yield will be. This is because the emergent cane has a longer growth period, and is better able to attain full tillering and leaf canopy, before the commencement of rapid stalk elongation with the onset of warmer conditions. There is also a greater likelihood of rainfall. This was demonstrated in Zimbabwe[4] where the conditions for the onset of rapid stalk elongation were the attainment of a mean temperature of 18.5°C, and the production of five unfurled leaves on the primary tillers. Late planted cane did not achieve this until well into the growing season, and had less chance of achieving satisfactory yields. Other studies in Swaziland showed that there was a decline in yield of plant cane of about 1.25 t/ha for each week of delay in planting from July onwards. Similar results have been reported from other countries.

Irrigation farmers have much more control of their planting programme than do producers of rainfed cane. The latter have to delay their planting until the onset of the rains, and often have to rush to complete it before high rainfall and wet soil conditions prevent any further planting. Even then, they have to accept that some plantings may be failures because only intermittent or light rainfall is received. Provided land preparation is complete

and sufficient equipment and labour are available, planting can be successful, especially as in many tropical countries the timing of the onset of the wet season can be predicted with reasonable accuracy, and heavy showers often precede its arrival. As long as water penetrates the soil to a depth of 100 mm, cane stalks can be laid in the furrows and covered with soil. The exposed soil may dry out, but that in contact with the cane sett will be moist, and germination will usually be acceptable. Compressing the soil above the cane sett to ensure better soil and seedcane contact will improve germination.

Seedcane in most countries is still planted by hand; but, where labour is expensive, mechanical planting is practised, e.g. Australia and the USA. For manual planting, the seed rate ranges from 5 to 10 t seedcane/ha. This averages about 7 t/ha for the conventional one-and-a-half stick overlap method. Wholestalk seedcane is normally cut in the nursery and stacked in heaps. The trash is normally left on to protect the buds from damage during transit, and to prevent them from drying out before planting. At this stage lodged or pest infected seedcane should discarded if possible. The seedcane is then transported to the commercial field on cane trailers, and may be dumped in the field for distribution by the labour, or on the field edge for transfer to trailers that travel through the field where the planting gang place it in the furrows. The stalks are cut into setts on the field edge and carried into the field for planting, or cut *in situ* in the furrow. The seedcane is cut into varying lengths – normally into three or four-budded setts of no more than 450 mm in length. This is done with cane knife or machete. Sometimes guillotines are used when the setts are prepared at the field edge. This is to reduce the apical dominance that occurs in the whole stalk. When the top bud of a stem grows, hormone-like substances or auxins retard the development of the lower buds to an increasing degree from top to bottom of the stalk. Therefore, if whole stalks are planted, there are large gaps in the rows of young shoots that arise from them. These gaps must be filled with extra seedcane, which results in a significant wastage of time and money as well as planting material. However, if the stalks are cut into three-budded setts before being covered, all the buds germinate to form continuous rows of

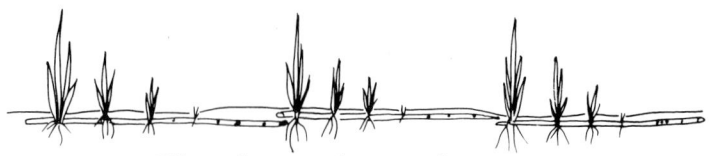

Where the stems have not been cut

Where the stems have been cut

Fig. 5.4   The effect of top dominance on germination.

uniform growth, and supplies are often unnecessary (Fig. 5.4).

The cutting knives should be dipped in a disinfectant (e.g. a 5–15% Lysol solution or 0.1% Mirrol or Roccal solutions) when each row has been planted to minimise the spread of RSD. In some countries (e.g. Zimbabwe), setts are dipped into a fungicide before planting as a preventative measure against smut disease[5].

The setts are then covered with soil to about 100 mm by breaking the banks by hand using hoes or by discs or mouldboard covering implements. In irrigated fields, the planted seedcane should be irrigated as soon as possible, particularly if it is a shy germinating variety. In rain fed areas, the cane should be planted as near as possible to expected rainfall.

Multi-cropping is practised in some countries, e.g. Africa, Asia, Fiji and Mauritius. This entails inter-planting food crops in the inter-row with the sugarcane. These crops are usually fast growing such as vegetables, sugar beans, soya beans, groundnuts, and maize, which are planted at the same time as the primary crop of cane. These food crops are harvested within 90 days of sowing, before the sugarcane has attained full canopy. In high rainfall areas, they are unlikely to place undue water stress on the cane, and the shading effect is limited during the early development of that crop. A substantial amount of research has demonstrated that the net biomass production of

the multi-cropping is measurably greater than that produced by sugarcane alone. The production of some food is very important to the farmers who do this, because often most of their land is devoted to the main cash crop sugarcane.

The implements used in machine planting are of varying complexity, and are usually hauled by wheeled tractors. In the simplest, the setts, which have been previously cut by means of chopper harvester, are carried in bins and fed by chute into furrows and then covered with soil by discs. Others have attachments which open the furrows, chop whole stalks into setts, place the setts in the furrows, apply bands of fertiliser parallel with and on each side of the row, and finally consolidate the soil with light rollers. In Australia, where it is necessary to take precautions against pineapple disease (*Ceratocystis paradoxa*) in susceptible varieties, the machines may be fitted with a series of nozzles by which the setts are sprayed with a fungicide before being planted. Alternatively, the machines may have tanks containing the fungicide through which the setts must pass. The planting rate of machine planters tends be higher than hand planting at about 8 t/ha. A typical dual row planter is shown in Fig. 5.5.

RATOON MANAGEMENT

The good management of ratoons is very important

Fig. 5.5 Dual row cane planter.

to sustain productive crops over a long period. The highest priority in ratoon maintenance is to repair a damaged field's infrastructure, which may include roads, waterways, drainage boxes, open drains and sub-surface drains. The window of opportunity for this is often short, because of the rapid growth of ratoons, and because of the high demand on machinery at a time when land preparation and planting is often receiving a higher priority.

There are some differences in ratoon management between rain fed and irrigated sugarcane, particularly over the issue of whether or not to burn the trash after harvest. In rain fed cane, one of the aims should be to maintain a trash mulch[1] in order to increase cane yield by:

- retention of soil moisture;
- reduction of loss of water from the soil by evaporation, surface water run-off, soil erosion and surface capping;
- improvement in the soil's receipt of rainfall, the aggregation of soil particles and the consequent increase in air-pore space;
- effective suppression of weed growth;
- elimination of the hazardous burning operation and environmental pollution;
- reduction in the likelihood of soil compaction; and
- reduction in the amount of applied P fixed in the soil when fertiliser is applied over the mulch.

The disadvantages are that:

- there is a lower cutter output;
- very few mechanical harvesters can cut green cane efficiently with a consequent reduction in payloads and an increase in extraneous matter, resulting in lower mill throughput and sucrose extraction;
- considerable quantities of cane may be lost at harvest under the trash blanket;
- ratoon regeneration is delayed particularly during cool weather;
- in stressed cane the trash blanket may exacerbate insect damage (e.g. various species of trash worms and *Eldana saccharina*) in the regenerating ratoon; and
- a heavy trash blanket may suppress cane yields.

Consequently, although the arguments for a trash blanket are convincing, it has been recommended for only limited use in coastal areas, on steep slopes and erodible soils, in close proximity to urban areas and neighbouring main roads to avoid fire hazards, and where the thermal inversion layer is low to prevent air pollution.

Despite the difficulties of trash mulching, it is practised to an extent in many non-irrigated industries for the advantages outlined above. For example it is used extensively by the smallholders in Kenya because of the high annual rainfall of

2000 mm recorded in the sugarcane area. Another country that practises trash mulching is Colombia, where environmental pressure groups forced the industry to commit itself to abandoning cane burning by the year 2000. This has resulted in the development of effective green cane harvesters and self-trashing varieties. Australia is also making strides towards a green cane harvesting policy with non-irrigated cane.

Most sugarcane industries burn the cane trash either before or after harvesting for the ease of management of ratoon crops, especially where irrigation is practised, although there is some compromise when the residual trash may be scattered over the rows or gathered into windrows every 5–8 rows. However, in furrow-irrigated fields the trash is a severe impediment to the even flow of water down the furrows. Consequently, it is usually burnt to avoid this problem.

The ripping of inter-rows is practised in some countries to improve infiltration and break up compacted layers in the soil but there is not very strong evidence that these benefits are achieved. If harvesting has caused some damage to the inter-row, this can be ameliorated by a ripping operation but there is also the danger of damaging the residual root system that the crop depends on for the first few weeks after harvest. Re-shaping of collapsed ridges after harvest is probably more useful, especially in furrow-irrigated fields and in shallow soils to provide greater root depth in the immediate vicinity of the stool.

Gap filling of ratoons may sometimes be necessary. If it is required, the operation should be done as soon as possible after harvest, and in irrigated fields with the first irrigation. Gaps are only worth filling if they exceed at least 1 m, and if there is a corresponding gap in an adjacent row. Cane can compensate quite well for missing stools, and gap filling should only be undertaken if the reason for the missing plants has been identified and corrected. Otherwise, the agent that caused the gaps may also damage the supplied cane. The most effective form of gap filling is with emerged setts that have been pre-germinated in a nursery. This form of planting material is known as 'speedlings'. Other ratoon management operations are fertilising and weed control.

WEED CONTROL IN SUGARCANE

Weeds have been described as plants that are out of place, and there is certainly no place for them in efficient sugarcane growing. When they are allowed to grow without restraint, as in abandoned fields, they quickly smother and destroy sugarcane. Weeds affect the crop in a number of ways:

- they compete with sugarcane for water, nutrients, light and space, and have an effect on germination;
- they also harbour diseases and pests; and
- they can excrete damaging chemicals into the soil.

They are most harmful when the crop is young, and more injurious to plant cane than ratoons; however, they are relatively unimportant when the crop is fully canopied. The effect of weeds on sugarcane yield has been extensively documented. In Hawaii and Trinidad, yields have been increased by up to four times by weeding, and in South Africa yields have been doubled by effective weed control.

Weed species are either i) annuals, which live for one year and set seed in that year, or ii) biennials, which live for two years setting seed in the second year, or iii) perennials, which live for more than three years and produce seed each year. They may be dicotyledons (broadleaf weeds) or monocotyledons (grasses or sedges). Weeds are successful in crops because their seeds remain viable for a long time, they possess effective methods of dispersal, are able to produce large quantities of seed, have few predators, and they are hardy.

Weed control can be obtained by a number of methods:

- *Preventative*. Keeping field verges free of weeds and irrigation canal verges cut short to prevent the seeding and distribution of weeds. Land preparation also aims to control weeds in the subsequent crop.
- *Cultural*. Using light and water to control weeds by:
  (a) ensuring good, even germination and avoiding gaps in the cane row;
  (b) obtaining early full canopy to shade out weeds;

(c) using close row spacing;

(d) having a thick trash blanket in ratoon crop;

(e) using a break crop and crop rotation to alter the dominant weed spectrum; and

(f) effective irrigation management, e.g. limiting the wetted area with drip or furrow irrigation.

- *Mechanical.* Tractor-drawn cultivators are most common, but mule, horse, cattle or buffalo-drawn cultivators are also used on small farms. Weeding can include other operations such as middle busting or ridge re-building. However, the cane can be damaged and fresh weed seeds germinate. Only weeds in the inter-row are controlled, and the method is unsuitable where there are open drains or steep slopes.

- *Hand weeding.* This includes:

(a) pulling weeds out by hand from the row, especially large grass species like *Sorghum* spp., *Panicum* spp., and *Rottboelia* spp., and the trailing broadleaf weeds of the *Cucurbitaceae* family and *Commelina* spp.;

(b) hand hoeing, which is very common on small farms, but is only really effective when weed are small and conditions dry, and as a follow up to chemical weed control.

- *Chemical (herbicides).* This is an expensive method, but very effective when the correct chemical is chosen and applied at the right time. Herbicides control weeds in the row, can give a long period of control, treat large areas rapidly, and have low labour and machinery requirements.

Chemical weed control

Chemical weed control was first attempted in the 1920s and 1930s, but the substances used were very dangerous, e.g. sodium chlorate (inflammable) and sodium arsenate (high mammalian toxicity). As a consequence they were little used. Then came the development of the hormone type herbicides (e.g. 2,4-D and MCPA), and with them the beginning of a new era in weed control. These herbicides were the first used to destroy weeds in established cane fields. As their use was limited to non-woody broad-leaved weeds, oil and pentachlorophenol (PCP) were frequently added to improve the efficacy of 2,4-D and MCPA as contact herbicides. This technique, however, only killed established weeds, and it is the newly germinating plant cane that suffers most from competition. In Hawaii a method was developed for spraying 2,4-D immediately after planting. A single treatment controlled the weeds until the cane canopied, then the leaf canopy restricted further weed growth.

With the development of the triazine herbicides (e.g. atrazine, ametryn, metribuzin and hexazinone) and the ureas (e.g. diuron) that were relatively insoluble, herbicides could be sprayed after planting the setts but before the weed seeds germinated. Consequently, a much wider range of weeds could be controlled including many annual grasses.

It is also possible to apply these herbicides after the cane has germinated and on ratoon crops. Where the cane trash is left as mulch, this restricts weed growth in the mulched areas; however, creeping weeds and some strong growing grasses can often grow through the mulch. Dalapon was the first effective grass herbicide to be developed, but it damaged many varieties of cane and it could not be sprayed over the crop. Paraquat will knock back the grasses, but it also scorches cane severely if sprayed on the green leaves. However, this effect is normally temporary. Paraquat, and more recently glyphosate, has been widely used to clean up a grass infestation before the plant cane germinates or the ratoon cane develops after cutting. Asulam has been used to kill grass weeds selectively, particularly *Sorghum*, *Digitaria* and *Rottboelia*. It is also used in mixtures with actril to control a range of grasses and broad-leaved weeds in established cane.

More recent herbicides that have been released include the acetanilides (e.g. alachlor and metolachlor), which are very effective pre-emergent grass herbicides. The dinitroanilines (e.g. trifluralin and pendimethalin) are incorporated into the soil before planting and give effective control of sedges as well as grass weeds. Acetochlor is another new herbicide that is particularly effective against annual grasses and a range of broadleaf weeds. Fluazifop-*p*-butyl is a grass herbicide that also kills cane at herbicide rates, and should not be applied directly to the crop at these rates. Sedges are often difficult to control, but halosulfuron is a new herbicide particularly effective against this group of weeds.

Successful chemical weed control requires knowledge of:

- the products available;
- their mode of action and selectivity against the target weed spectrum;
- the point of uptake of the herbicide by the plant;
- the clay and organic content of the soil; and
- the stage of growth when the herbicide is most effective.

These stages are pre-emergent, early post-emergent, post-emergent, or late post-emergent. Normally younger weeds are easier to control with an herbicide. In addition, the choice of herbicide will be influenced by the period of control required and the cost of the product, expressed as unit cost per ha per week.

Combinations of herbicides are most commonly used in order to control the usual weed spectrum of grasses and broadleaf weeds. A number of combinations have been used in various countries depending on their success under local conditions, and on the chemical being registered in that country for particular use. Single herbicides are often used for specific problems, and for the control of difficult weeds, e.g. the sedges (*Cyperus esculentus* and *C. rotundus*), grasses (*Panicum* spp., *Sorghum* spp., *Rottboelia cochinchinensis*, *Paspalum* spp., *Cynodon* spp., and *Digitaria* spp.), and broadleaf weeds – particularly vines.

The methods by which herbicides can be applied vary, e.g. aerially, tractor-mounted boom sprayers, knapsack sprayers (either lever operated or motorised) or controlled droplet, low volume applicators. The choice of method depends on many factors including the weed spectrum, stage of growth of the crop, size of operation, accessibility of the field and ground conditions, and the relative cost of the application method. Surfactants are often added to herbicides to improve the effectiveness of uptake of the herbicide by the weed. They decrease the surface tension of the spray solution on the weed foliage allowing a better wetting and sticking action.

Herbicides are toxic, and safety practices are essential to avoid danger to people, animals, non-target plants and the environment. All herbicides undergo rigorous testing before registration by controlling bodies in each country. Therefore, users should follow the directions for application on the label and understand the recommendations for which the herbicide is designed, the precautions that must be taken in their use, as well as the remedies in the case of poisoning.

Most major sugarcane growing countries issue recommendations for the use of herbicides. An example of the recommendations developed for South Africa is summarised in Table 5.1[6]. This table illustrates the flexibility available to sugarcane growers, and the wide range of mixtures that can be used to grow the crop whilst controlling weeds. Note that the chemical name is given in Table 5.1 rather than the registered proprietary name, because different proprietary names are sometimes registered in other countries.

Some aspects of herbicide management are given below:

- *Damage to cane or phytotoxicity.*
  (a) Herbicides sprayed before the cane emerges are far less damaging than those sprayed at post emergence.
  (b) Older cane is more susceptible to damage from chemicals than younger cane.
  (c) Post-emergence applications should always be directed so as to spray as little of the cane's leaf area as possible, e.g. use drop-arms and flood jets in the inter-row.
  (d) Poorly grown cane suffering from drainage problems (i.e. 'wet feet'), nematode damage or nutrient deficiency, is more susceptible to damage than well-grown cane.
  (e) Some varieties appear to be more susceptible to damage then others. When these varieties are grown, good pre-emergence sprays should preferably be used, or particular care should be taken to direct post-emergence sprays away from the can foliage.
  (f) Hot, humid conditions increase the likelihood of damage to cane.
- *Ratoon cane.* Ratoon cane develops ground cover more rapidly than plant cane and suffers

Table 5.1 Herbicide recommendations in South Africa 1997.

Herbicide	Rate (L/kg)/ha*	Weeds controlled†	Weeks control	Clay content %	Comment
Pre-emergence (short term)					
MCPA	7	B/L	5		Requires moist soil conditions
Pre-emergence (long term)					
Alachlor + MCPA	(5–6) + 4	B/L, G, YWG	8	All	Requires moist soil conditions
Alachlor + Atrazine	(5–6) + (2–6)	B/L, G, YWG	8	All	
Alachlor + Ametryn	(5–6) + (2–3)	B/L, G, YWG	8	All	
Alachlor + Diuron	(5–6) + (2–4)	B/L, G, YWG	8	All	
Metolachlor + Ametryn	(1–1.6) + (2–3)	B/L, G, YWG	9	All	
Acetochlor + Ametryn	(2–3) + (2–3)	B/L, G, YWG	9	All	
Acetochlor + Diuron	(2–3) + (2.5–3)	B/L, G, YWG	9	All	
Acetochlor + Atrazine	(2–3) + (2–6)	B/L, G, YWG	9	All	
Imazethapyr + Acetochlor + Atrazine	(0.75–1) + (2–3) + (3–4)	B/L, G, YWG	10		Imazethapyr very damaging when in contact with cane foliage
Metazachlor + Atrazine	(1.5–2) + (2–3)	B/L, G, YWG	9	> 10	
Metazachlor + Ametryn	(1.5–2) + (3–4)	B/L, G, YWG	9	> 10	
Metazachlor + Diuron	(1.5–2) + 3	B/L, G, YWG	9	> 10	
Thiazopyr	(1–4)	G	16	All	
Thiazopyr + Acetochlor	(1–3) + 2.5	G, YWG	16	All	
Thiazopyr + Diuron	(1–3) + 2.5	B/L, G	16	All	
Thiazopyr + Acetochlor + Diuron	(1–3) + 1.5 + 2.5	B/L, G, YWG	16	All	
Hexazinone	(0.6–1)	B/L, G, YWG	12	> 5 %	
Pre-emergence (incorporated)					Dry soil on application before planting followed by moist conditions
EPTC	3–7	G, YWG, PWG	8	All	Soil incorporated
Pre- to early post-emergence (long term)					
Alachlor + Atrazine + Paraquat	6 + (2–6) + 1	G, YWG, PWG	8	All	
Alachlor + Ametryn + Paraquat	(5–6) + (3–5) + 1.5	B/L, G, YWG	8	All	
Alachlor + Diuron + Paraquat	(5–6) + (2.5–3) + 1.5	B/L, G, YWG	9	All	
Alachlor + Ametryn + Surfactant	(5–6) + 6 + 0.6	B/L, G, YWG	8	All	

Table 5.1 (Continued.)

Herbicide	Rate (L/kg)/ha*	Weeds controlled†	Weeks control	Clay content %	Comment
Metolachlor + Ametryn + MCPA	(1–1.6) + (4–5) + 3.5	B/L, G, YWG	9	All	
Metolachlor + Ametryn + Paraquat	(1–1.6) + (2–3) + 1.5	B/L, G, YWG	9	All	
Metolachlor + Ametryn + Surfactant	(1–1.6) + 6 + 0.6	B/L, G, YWG	9	All	
Metolachlor + Diuron + Paraquat	(1–1.6) + (2–2.5) + 1.5	B/L, G, YWG	8	All	
Acetochlor + Ametryn + Paraquat	(2–3) + 4 +(1–1.5)	B/L, G, YWG	9	All	
Acetochlor + Diuron + Paraquat	(2–3) + (2.5–3) + 1.5	B/L, G, YWG	9	All	
Acetochlor + Atrazine + Paraquat	(2–3) + (2–6) + (1–1.5)	B/L, G, YWG	9	All	
Acetochlor + Ametryn + Surfactant	(2–3) + 6 + 0.6	B/L, G, YWG	9	All	
Metazachlor + Ametryn + Paraquat	(1.5–2) + 3 + 1	B/L, G, YWG	9	> 10	
Metazachlor + Diuron + Paraquat	(1.5–2) + 3 + 1	B/L, G, YWG	9	> 10	
Metribuzin + Diuron	3 + 2	B/L, G, YWG	12	6–35	
Metribuzin + Diuron + Paraquat	3 + 2 + 1	B/L, G, YWG	12	6–35	
Hexazinone + Diuron	0.8 + 1.5	B/L, G, YWG	12	> 5	Ratoon crops only
Sulcotrione + Atrazine (Proprietary mixture)	1.6–3.6	B/L, G, YWG	8	All	
Chlorimuron-ethyl + Metribuzin (Proprietary)	0.8–1	B/L, G, YWG, PWG	12	> 7	
Tebuthiuron + Diuron	(2–2.5) + 2.5	B/L, G, YWG	9	8–50	
Tebuthiuron + Ametryn	(2–2.5) + 4	B/L, G, YWG	9	8–50	
Post-emergence (long term)					More effective when applied to moist soil
Acetochlor + Diuron + Oxytril	(2–3) + (2.5–3) + 1.25	B/L, G, YWG	9	All	
Alachlor + Diuron + Oxytril	6+ 2.5 + 1.25	B/L, G, YWG	8	All	
Alachlor + Ametryn + Oxytril	(5–6) + (3–5) + 1.25	B/L, G, YWG	8	All	
Metribuzin + Diuron	3 + 2	B/L, G, YWG	12	6–35	
Metribuzin + Diuron + Oxytril	2.9 + 2.5 + 1.25	B/L, G, YWG	12	6–35	
Metribuzin + Ametryn	3 + 3	B/L, G, YWG	12	6–35	
Metribuzin + Ametryn + Paraquat	3 + 3 + 1	B/L, G, YWG	12	6–35	
Hexazinone + Diuron	0.8 + 1.5	B/L, G, YWG	12	> 5	Ratoon crops only
Hexazinone + Diuron + Oxytril	(0.6–1.2) + (1–2) + 1.5	B/L, G, YWG	12	> 5	Ratoon crops only
Hexazinone + Ametryn	(0.6–0.7) + (3–4)	B/L, G, YWG	12	> 5	Ratoon crops only

Herbicide combination	Rate	Weeds	No.	Stage	Comments
Metolachlor + Metribuzin + Paraquat	(1–1.6) + 2 + 1.5	B/L, G, YWG	12	All	
Spotaxe (Proprietary) + surfactant	2.5 + 0.3	B/L	8	All	
Spotaxe + Diuron	2 + 2.5 + 0.3	B/L, G	10	All	Moist soil conditions are preferable
Post-emergence (short term)					
Ametryn + surfactant	8 + 0.6	B/L, G	6	All	Satisfactory in dry conditions
Ametryn + MCPA + surfactant	(4–5) + 3.5 + 0.6	B/L, G, YWG	6	All	
Ametryn + Oxytril	(4–5) + (1–1.25)	B/L, G, YWG	5	All	
Ametryn + MCPA + Oxytril	(4–5) + 3.5 + 0.5	B/L, G, YWG	6	All	
MCPA + surfactant	7 + 0.6	B/L	5	All	
MSMA	4 + 4 (split applications)	G, YWG, PWG	4	All	Effective control of PWG
Diuron + MCPA + surfactant	2.5 + 4 + 0.6	B/L, G, YWG, PWG	6	All	
Diuron + Oxytril	2.5 + (1–1.25)	B/L, G, YWG	5	All	
Diuron + MCPA + Oxytril	2.5 + 3 + 0.5	B/L, G, YWG	6	All	
Terbuthylazine + Bromoxynil (Proprietary)	2	B/L	6	All	
Halosulfuron + surfactant	(0.05 + 0.5) + (0.05 + 0.5) split applications	YWG, PWG	6	All	Effective control of PWG
Late post-emergence (short term)					
Paraquat + MCPA	3 + 4	B/L, G, YWG, PWG	5	All	Effective for both dry and moist conditions
MSMA	6	G, YWG, PWG	4	All	Useful for resistant tufted grasses
Ametryn + MSMA	3 + 3	B/L, G, YWG, PWG	5	All	Useful for resistant tufted grasses
Diuron + Paraquat	2 + 2.5	B/L, G, YWG, PWG	5	All	Useful for resistant tufted grasses
Diuron + MSMA	3 + 3	B/L, G, YWG, PWG	5	All	Useful for resistant tufted grasses
Treatments for special problems					Treatments should be applied with extreme caution
Glyphosate	6–8	B/L, G, YWG, PWG			Useful for resistant creeping grasses
Glyphosate	8–10	B/L, G, YWG, PWG			Eradication of actively growing cane
Fluazifop-p-butyl	6	B/L, G			Eradication of actively growing cane
Hexazinone + Diuron	0.6 + 2.5	B/L, G, YWG, PWG			
Sulfosate	4	B/L, G, YWG, PWG			Useful for resistant creeping grasses
Sulfosate	5.3–6.7	B/L, G, YWG, PWG			Rate for cane eradication

* Read the label for recommended rates, compatibility with other herbicides, mixing instructions, safety requirements, and phytotoxic effects on sugarcane.

† B/L, Broadleaf weeds; G, grasses; YWG, *Cyperus esculentus*; PWG, *Cyperus rotundus*.

Source: *South African Sugar Association Experiment Station Herbicide Guide, 1997*[6].

less from competition, particularly from *Cyperus* spp. that do not grow very tall. Cheaper, short-term treatments are often satisfactory, but care must be taken to prevent aggressive weeds becoming established and competing with successive ratoon crops, as these are difficult to control.

- *Field borders and railways*. Chemical weed control is also used to protect field borders, waterways, irrigation canals, main drains and railway tracks, not only to keep them clean but also to prevent them becoming a source of weed invasion into cane fields. Glyphosate is often used in these situations where the weeds have to be totally eliminated but, with waterways, a broadleaf herbicide is preferable in order to leave the grass unaffected.
- *Damage to crops in adjoining fields and gardens*. The herbicides used in sugarcane fields (particularly glyphosate, fluazifop-*p*-butyl, 2,4-D and MCPA) can be phytotoxic to other plantation and garden crops. The small droplets in the spray can drift with the wind, and may also volatilise and move downwind for quite a distance. Consequently, all herbicides should be used with great caution in boundary fields, and on roads and railways that pass through land not under the control of the cane grower. Sensitive areas should be sprayed only on calm days, or using non-volatile products when a favourable wind is blowing.

IRRIGATION AND DRAINAGE

Sugarcane is one of the heaviest users of water. Only rice and tree crops perhaps use more. Lysimeter studies[7] carried out in the 1960s determined an empirical yield/water use relationship, which roughly equates to 10 mm of water (crop evapotranspiration) producing a yield of 1.0 t cane/ha. A crop of sugarcane will require in the region of 1100–2000 mm of water depending on the local climatic factors and crop age.

This water demand is received either as rainfall or applied as irrigation, or a combination of both. Entirely rainfed cane areas require a consistent and reliable rainfall pattern that meets, or provides a good percentage of the crop's water demand for at least 9 months of the year. This is possible in such places as the traditional sugarcane areas of the Caribbean, the equatorial highlands of Kenya and Uganda, the coastal regions of South Africa and Queensland, and the tropical and subtropical countries of South East Asia. But cane production is always at the mercy of weather patterns, which El Niño can demonstrate to severe effect. Rainfed cane will never match the yield performance of irrigated cane, other than in exceptional conditions. Increasingly, therefore, cane farmers are evaluating the introduction of supplementary irrigation to boost production.

Over the past century, new sugarcane developments have been established very successfully in climatic regions where full irrigation is required to sustain the crop. The most extreme example is in the arid coastal area of Peru, where rainfall is virtually nil and the whole crop is grown under irrigation. In other irrigated sugarcane areas in South America, Africa and Australia, rainfall will account for typically 25–75% of the crop's water demand, and the remainder is covered by irrigation.

Irrigation water can be applied in many different forms and with varying degrees of efficiency. Older schemes tended to be laid out with capital and operating cost in mind rather than efficiency but with competing demands for water from other users and increasing water charges, irrigation farmers are now turning to more efficient systems.

While providing significant benefits irrigation can also create some serious problems; the chief one being the rise in groundwater table to within the root zone whereas under rainfed conditions this would never have occurred. The speed of water table rise is dependent on irrigation efficiency and soil porosity, and the earlier the cane farmer accepts the inevitable and plans for it the better. Many mature irrigation schemes now require conjunctive use of groundwater to maintain water tables below the root zone. If irrigators had recognised the problem earlier and applied mitigation measures, then costly groundwater pumping or loss of land through salinisation could have been delayed or avoided.

Irrigation and drainage are therefore two issues closely interrelated and inter-linked, but it
is beyond the scope of this book to discuss all the
principles and theory. There are already excellent
publications available[8–10] that cover the subject
well. Instead, the objective in this section has been
to provide an overview of the various irrigation and
drainage systems used in sugarcane cultivation and
to describe some of the methods that can be adopted to improve performance and efficiency.

Appropriate irrigation systems

Sugarcane can be grown under virtually any irrigation system, but because irrigated cane is mostly
grown on a ridge and furrow system, basin or
border strip irrigation is not usually appropriate.
Similarly, micro-jet sprinklers are more suited to
horticulture and tree crops rather than a row crop
such as sugarcane.

Table 5.2 defines the irrigation systems considered most appropriate for sugarcane. There
can only be two methods for delivering irrigation
water, i.e. by gravity or by pressure. With gravity
there is only the furrow irrigation system; but there
are numerous methods for delivering the water to
the head of the furrow. Pressure methods can be
categorised as overhead (spray) irrigation with
water delivered through nozzles, or drip irrigation
with water delivered through tubes and emitters.

Table 5.2 Suitable irrigation systems for sugarcane.

Delivery method	Primary category	Secondary category
Gravity	Furrow*	Feeder ditch
		Siphon
		Spile and drop spile
		Gated pipe
		Lay-flat fluming
		Surge
Pressure	Overhead irrigation	Centre pivot
		Linear move
		Boom irrigator
		Sprinkler
		Floppy
		Rain gun or cannon
	Drip irrigation†	Surface drip
		Subsurface drip

* Also known as surface or flood irrigation.
† Also known as trickle irrigation.

Furrow irrigation

Although complete statistics are not available to
prove it, furrow irrigation is the dominant system
in sugarcane farming. Surveys undertaken by
the International Commission on Irrigation and
Drainage (ICID) and the Food and Agricultural
Organisation (FAO) indicate that surface systems
account for > 80% of the total irrigated area worldwide.

Furrow irrigation of sugarcane is popular for
several reasons:

- water is applied by gravity without the need for
  power;
- wind has no effect on application efficiency;
- it is a simple and cheap system to install and
  operate;
- it can be adapted to a wide range of soil types,
  topography and irregular field boundaries; and
- it can use a variety of water delivery methods at
  the head of the furrow.

The furrow method can represent the very best
in irrigation practice; but unfortunately it can also
represent the very worst with efficiencies lower
than 30%. The factors that depress efficiency include low quality land preparation and levelling,
poorly maintained and leaking canals, tail water and
percolation losses. Labour productivity is also generally lower than in the other irrigation systems.

Best practice furrow irrigation

With the continuing pressure on water resources
and water use efficiency, furrow irrigation has to
perform better. Best practice furrow irrigation can
achieve efficiencies of 80% or more but it will need
to contain the following features:

- restricting furrow irrigation to clayey soils where
  deep percolation losses during irrigation applications are minimal (free draining soils would
  be acceptable if integrated with a groundwater
  recirculation system);
- optimising field layout and furrow direction,
  slope and shape to the local soil type and topography, which can be determined with the aid of
  field trials and specialist software such as sirmod
  ii[11];

- conveying water to the field edge via buried pipe or lined canal (clay, concrete or membrane lining) to reduce seepage losses – earth canals are only acceptable in self-sealing clay soils;
- distributing water along the field edge and furrow heads via gated pipe, fluming or lined canals to reduce seepage losses – earth canals are only acceptable in self-sealing clay soils;
- selecting a furrow flow rate that is optimal for the particular furrow configuration and soil characteristics. Flow rates are typically in the range of 0.5–8.0 L/s;
- collecting, storing and recycling tail water from furrow ends and canal spillways;
- following an appropriate irrigation scheduling system, which uses a climate-based soil moisture balance or soil instrumentation to determine the optimum timing for irrigation applications;
- applying soil-binding polymers in the irrigation water to improve water infiltration and to settle out sediment, if appropriate for the soil type;
- monitoring soil and water quality, and adopting a management plan to preserve soil structure and prevent the build-up of salinity/sodicity. Applications of soil ameliorants such as gypsum for alkaline sodic soil and lime for acidic sodic soils may be necessary; and
- retaining a trash blanket to preserve soil moisture, although this practice is not always suitable where there are long furrows on a shallow gradient as flows can be impeded.

Feeder ditch delivery system

Many methods have evolved for the delivering of water from the field edge to the head of the furrows. The most basic method is to form a feeder ditch that is dammed in sections of around 20 furrows to form a set. The sides of the feeder ditch are breached at intervals using a hoe or spade to let water flow into each furrow. This method is cheap to construct but is labour intensive, since the breaches have to be repaired before progressing to the next set. Furrow flow rates are low, and water distribution is irregular and very dependent on the individual irrigator's skill. Furrows located at the upstream end of the set tend to receive more water

than the downstream end. A further refinement of this method to reduce the hoeing operations is to run the head ditch alongside an earth tertiary canal. A single breach therefore only needs to be made in the tertiary canal bank to feed a section of head ditch, and the furrow outlets are left open all the time.

The feeder ditch method tends to be adopted by farmers with small parcels of land and short furrows, where only basic land levelling has to be done. The 'twig and main' system of Jamaica and the basin furrow system of Barahona in the Dominican Republic are variations of this method. The furrow lengths are only 20–50 m, and feeder channels are used to convey water to irrigation sets within the field as well as along the top edge of the field.

Siphon pipes

Siphon pipes provide a much better control of water to individual furrows, and will cover the full range of flow rates required. Siphons are constructed from polythene pipe curved into sections, and have diameters ranging from 25 to 75 mm. The ideal operating head (i.e. tertiary canal water level to furrow water level) is 50–300 mm. Figure 5.6 gives the head-discharge curves for various siphon sizes and Fig. 5.7 illustrates the siphon system in operation in Zambia.

Spile pipes

An alternative to siphons is spile pipes, which are placed into the wall of the tertiary canal to feed a launching bay serving a set of 10–20 furrows. A lever-flap gate on the inlet of the spile pipe controls the water flow, and the gate is operated either fully opened or fully closed. Spile diameters range from 150 to 300 mm, and the operating head is typically 50–500 mm.

Another version is the drop spile tertiary canal (Fig. 5.8), which has one or two spile pipes of 50 or 75 mm in diameter per furrow. The spile pipe is located in a drop section in the invert of the tertiary canal that is constructed from precast or in-situ concrete, or gunite (a sprayed concrete). This system is popular for furrow layouts where tertiary ca-

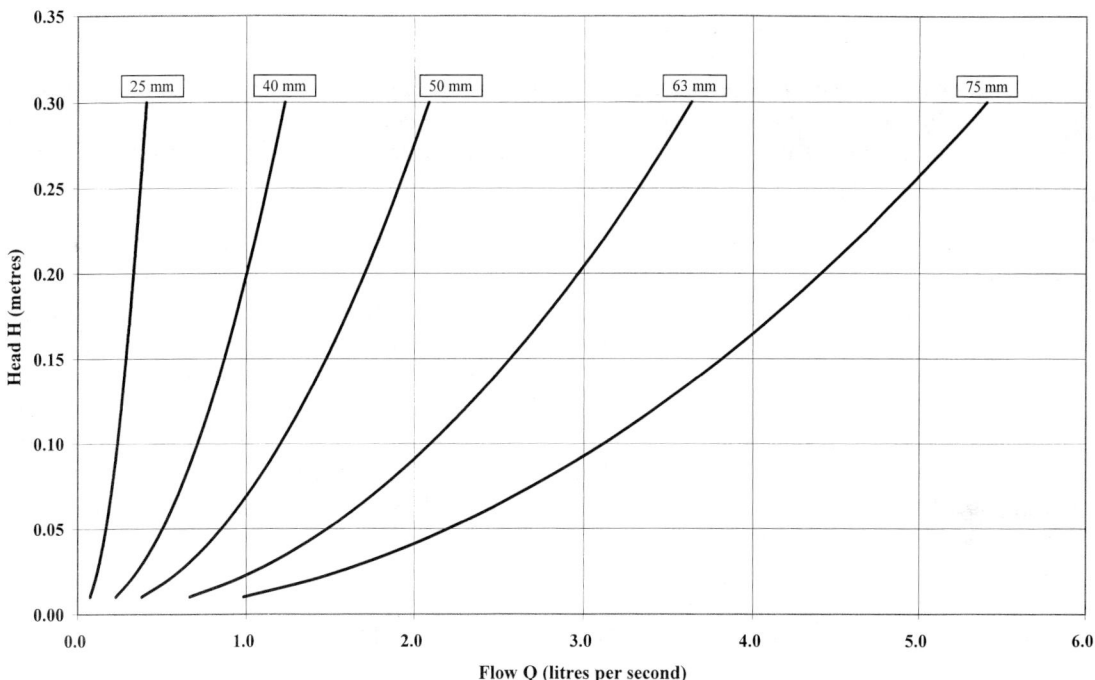

Fig. 5.6 Head discharge curves for siphons of 25 to 100 mm diameter.

Fig. 5.7 Siphon irrigation in Zambia.

nals are on steep gradients of 1–3%, since portable gates and stepped reaches are not required as in a siphon system. The canal is run directly down the slope, and the spile outlets are only opened where the furrows are being irrigated. Irrigators tend to prefer spiles because they are quicker and easier to operate than transporting and priming siphon pipes. Spile pipes and their close-fitting caps can easily be sourced locally from plastic-moulding companies.

Fig. 5.8 Drop spile canal system in Swaziland.

Fig. 5.9 Typical fluming system in Queensland, Australia. Source: C. E. Bartlett Pty Ltd, Australia.

Gated pipe and lay-flat fluming delivery systems

Tertiary canals for siphons and spiles act as a barrier to harvesting operations where a headland area is needed for harvesting equipment to turn. Gated pipe and plastic lay-flat fluming are alternative irrigation delivery systems that are dismantled at harvest to allow free access to cane fields for harvesting traffic. Figure 5.9 illustrates a lay-flat fluming system in Australia.

Gated pipe and fluming systems require a low head pressure system with 0.2–2.0 m residual head at the outlet nozzle and a buried pipe system is needed to distribute water to the hydrant valves that control flows to the gated pipe or lay-flat fluming. The operating head for the system is usually provided by a pump, high-level canal or storage dam. The installation cost is higher than for si-

phons or spiles; but to some farmers the operating benefits outweigh the extra cost.

Surge irrigation

An additional refinement on the gated pipe system is to install surge valves for pulsing the water flow to furrows. The valves are normally located at the supply hydrant in the centre of two branches of gated pipe or lay-flat fluming. The operating principle is to surge the water down the furrows on alternating irrigation sets at timed intervals (e.g. 30 min on and 30 min off) until the water reaches the end of the furrow. This is called the 'advance phase'. The time interval is then reduced to around half until completion of the irrigation cycle. This is called the 'cutback or soak phase'. The objective of the filling and emptying of the furrows in the advance phase is to prepare the soil for the cutback phase, by lowering its infiltration capacity and smoothing the furrow profile. The cutback phase then uses much lower average flow rates, and achieves a more uniform wetting of the soil along the furrow length.

Surge irrigation is most effective on free-draining alluvial soils where high percolation losses at the head of furrows are a problem, and where high furrow streams are needed to get water to reach the end of the furrows. The benefits of surge irrigation are water and labour savings, higher application efficiency and higher yield. Fertiliser can also be added and controlled through the surge valve. Surge irrigation techniques have mostly evolved in the USA, through the Colorado and Utah State University Cooperative Extension Services where research work on many different crops has been undertaken.

Polyacrylamides

Polyacrylamide (PAM) is a soil-binding polymer that can provide the following benefits:

- reduces soil erosion from the furrows;
- improves water infiltration; and
- reduces sediment-bound nutrients and pesticides being exported from farms and contaminating watercourses.

PAM is also commonly used as a settling and clarifying agent in food and sugar processing, and in water treatment. PAM does not pose any environmental threat, and is broken down by sunlight and cultivation.

PAM can be supplied as granules, tablets, aqueous or emulsified concentrates, and typical applications are 1–8 kg/1000 m^3 of water applied, depending on the soil type, furrow shape and flow rate. The chemical is mixed and injected at the head of the tertiary canal or pipeline during the advance phase only until the irrigation water has reached the end of the furrows. PAM is generally only used in the first irrigation after planting or inter-row cultivation when the risk of soil loss is greatest. If the soil conditions warrant it, then PAM can be used in alternate or other intermediate irrigation applications. For maximum effect, the PAM treated water must be applied to dry furrows without any previous pre-wetting.

The use of polyacrylamide soil conditioners is worth considering on heavy clay soils where water infiltration and lateral spread from the furrow profile is poor. The chemical is used to good effect in Australia and the USA, and there are many informative research and guideline reports[12–15] available. The Agricultural Research Service in Kimberley, Idaho, USA, has done most of the pioneering research.

Overhead irrigation

The second most popular form of sugarcane irrigation on a worldwide basis is overhead irrigation, where the water is applied in droplet form by nozzles positioned above the crop. Under this method, complete and uniform field coverage is achieved and water will infiltrate vertically into the soil. This is a markedly different wetting pattern from furrow irrigation, where only 40–60% of the soil surface is wetted, and water infiltration is both lateral and vertical.

There are broadly three categories of overhead nozzles:

- low-pressure – operating at 0.6–2.0 bar, with a typical throw radius of 3–12 m;
- medium-pressure – operating at 2.0–5.0 bar, with a typical throw radius of 12–30 m; and
- high-pressure – operating at 5.0–7.5 bar, with a throw radius of 30–65 m.

Low-pressure nozzles are generally plastic and mounted on a boom irrigation system such as a centre pivot, linear move or boom irrigator. This layout provides a more economic arrangement for mounting the closely spaced sprinkler nozzles compared to a fixed or movable riser pipe system. The exception to this is the Floppy™ sprinkler, which is mounted on a riser pipe or overhead wire system. However, the Floppy system operates on the borderline of low and medium pressure, and the sprinkler spacing for sugarcane is set at 12–15 m. The horticulture industry may opt for low-pressure sprinklers on a closely spaced riser and pipe system, but crop areas are compact, and in sugarcane this would not be economic.

Medium-pressure nozzles are generally constructed in brass or plastic and mounted in sprinkler bodies of plastic or cast bronze. For sugarcane, the traditional method of mounting the sprinkler is on a 3–4 m riser pipe connected to fixed or moveable pipe distribution system. Sprinkler spacing is commonly on an 18 × 18 m grid.

High-pressure nozzles are generally constructed in plastic, brass or stainless steel, and mounted in sprinkler-gun bodies of cast aluminium. These rain-gun sprinklers (or cannons) are usually mounted on trolleys towed by hose-reel irrigators (hard-hose or soft-hose types). In some cases, the sprinkler guns are mounted on tripods connected to a fixed or movable pipe distribution system. Tripod assemblies are heavy and require two people to move them into position.

The low- and medium-pressure systems are more expensive to install, but will have lower energy requirements and operating costs. These systems are more suited to fully irrigated farms where the annual irrigation requirement is 500 mm and above. The high-pressure systems with their high energy and operating costs are more suited to supplementary irrigated farms where the annual irrigation requirement is < 500 mm. Brief descriptions on the main overhead irrigation systems used by the sugarcane industry, and a detailed overview of sprinkler systems and equipment can be found in a specialist publication[16].

Boom irrigation systems

The three forms of boom irrigation system are the centre pivot irrigator, the linear move irrigator, and the boom irrigator. All forms use the same principle of a single high-level pipe boom to convey irrigation water to spray nozzles. The boom typically has pipe diameters of 80–250 mm and spans of 15–60 m, and is stiffened by a truss arrangement of rods and/or wires. The differences between the three forms are the manner in which water is picked up by the irrigator, and the way that they move. The early boom irrigators had conventional impact sprinklers mounted on top of the boom, but the modern trend is to install low-pressure nozzles, typically of 0.7–1.4 bar (10–20 psi), on drop tubes below the boom.

The centre pivot has a fixed water inlet at the centre, and the boom moves in a circle. A 50-ha size pivot would have seven main spans plus an overhang span at the outer end, giving an overall length of 400 m. The spans are supported on tower legs and large diameter wheels driven by small 1 kW electric motors. Sensors mounted on each tower control the movement and alignment of the pivot spans. The irrigation application rate can be adjusted by changing the speed at which the pivot rotates. A graduated set of sprinkler nozzles is required along the pivot to compensate for the varying speed of travel. Figure 5.10 shows a typical

Fig. 5.10 Centre pivot irrigator with low-pressure nozzles on drop tubes. Source: Valmont Irrigation, USA.

centre pivot irrigator in operation as it passes from mature cane through a fire break.

The longer the pivot span, the greater the irrigated area and the cheaper the system becomes. The spray rates at the pivot ends for areas >90 ha are, however, too high for most clayey soils and moderate slopes. Pivots >90 ha should only be considered for sandy soils on shallow slopes. The best pivot systems in sugarcane use a high profile pivot with the underside of the boom at least 4 m above ground level, compacted gravel wheel tracks to prevent rutting, and cane rows planted parallel to the circular wheel tracks.

Centre pivots have the simplest method of water pick-up and alignment control, and hence are the more popular boom irrigation system; but the circular field layout does provide a significant challenge to farm planning. Cornering systems exist but are not usually adopted by cane farmers. They would rather choose to leave the land fallow (if water rather than land is at a premium), or to install a separate irrigation system such as drip or solid set sprinkler for the land that would otherwise not be irrigated. Centre pivots are very suited to large-scale commercial farms where mechanised irrigation is required to save on labour. Computerised control is an option for multi-pivot schemes where full automation is required. It is feasible and generally more economic to apply fertiliser, herbicide, ripener and insecticide through the pivot, if the selected nozzle package is appropriate. Nevertheless, good flushing is required to prevent corrosion of the galvanised pipe work. Plastic liners are available for very corrosive irrigation waters such as diluted distillery effluent. In 2002, the cost of a 50-ha pivot (including delivery and erection but excluding the pump and mainline) would be approximately US$65 000, giving a unit cost of US$1300/ha. Smaller pivots would have a greater unit cost, while larger pivots would have a lower unit cost.

The linear move irrigator is more adaptable to conventional rectangular field layouts, since the spans move in a straight line down the field, and the sprinkler nozzles all have the same specification. Water is picked up from an open channel or hydrant and flexible hose system at the field edge. A two- or four-wheel end-cart carries the diesel pump for water pick-up, and a diesel generator for powering the drive system. Alignment control is achieved via an above-ground cable, a furrow wheel, or a buried cable and antenna pick-up.

Irrigation management is more complex with the linear system in that once a field is irrigated the irrigator has to dry-reverse back to the start, or to double-irrigate back over the same field. Alternatively, if the field layout is suitable, the linear irrigator can swing round to an adjacent field and continue irrigating back to the start.

Boom irrigators tend to be low profile and more suited to supplementary irrigation for establishing plant cane or boosting young ratoons. Some farmers may modify proprietary boom sprayers to give higher ground clearance for irrigating mature cane; but the topography needs to be flat to provide stability for the irrigator.

Boom irrigators can be fed from a hose reel machine or a buried pressure pipe system with hydrants and hoses. Typical boom lengths are 30–50 m and can provide an effective irrigation width of 45–72 m if fitted with end sprinklers. Dedicated lanes that are smooth and level across their width should be established for the irrigator trolleys to run along, otherwise grounding of the boom ends could occur. Boom irrigators are designed to be lightweight and foldable for ease of towing around farms and setting up into position for rapid irrigation applications.

Low energy precision application (LEPA) nozzles

LEPA is a concept developed in the USA for achieving high levels of water and energy efficiency in circular and linear boom irrigation systems. A LEPA system is a combination of good equipment specification to give a coefficient of uniformity (CU) greater than 94% and good soil-water management. One characteristic of the LEPA system is that nozzles are located in alternate crop rows and < 450 mm above the ground surface to reduce wind drift and evaporation loss. Drag-socks can be used as an alternative to spray nozzles so that irrigation water can be applied directly to the soil surface.

The LEPA system can be used in tall field crops such as maize; but in sugarcane problems could develop with the long drop tubes becoming entangled with the cane stalks laying across the inter-rows if the crop lodges. However, many of the principles of LEPA could be adopted on cane farms where they are appropriate.

Sprinkler systems

Sprinkler systems are hand-operated schemes and are categorised as portable, semi-permanent, or permanent systems. There are countless variations in the way the sprinklers can be mounted, laid out and moved; but some of the more common ones for sugarcane are listed in Table 5.3.

The dragline sprinkler system in its many variations of module configuration has become an extremely popular and versatile irrigation system in the irrigated cane industry of southern Africa. Figure 5.11 illustrates a typical dragline sprinkler

layout, and in 2002 the cost of infield equipment would be approximately US$1100/ha (excluding pump and main supply pipeline).

Floppy sprinkler

The Floppy sprinkler system is a relatively new overhead irrigation system invented and developed in South Africa. The innovative nozzle design is radically different from other conventional sprinkler systems and has a flexible silicone tube that snakes to-and-fro (see Fig. 5.12) while slowly rotating through 360° forming heavy droplets similar to rainfall. To operate effectively the Floppy sprinkler requires a minimum pressure of 2 bar. Each sprinkler is fitted with a diaphragm flow-control device so that the flow rate is constant irrespective of fluctuating pressure. The sprinkler heads can be mounted on riser pipes or on an overhead wire system.

Table 5.3 Types of sprinkler systems.

Category	System type	Description
Portable	Hand-move lateral lines	Sprinkler risers are connected directly to surface-laid aluminium pipes (laterals), and both are moved as irrigation progresses across a field. Aluminium main lines and diesel pumps are also moved to rotate irrigation around a farm. Low capital cost but very labour intensive. More appropriate for supplementary irrigation on young cane, as moving 9 m long pipe sections in standing cane is very difficult.
Semi-permanent	Hop-a-long	Sprinkler risers are located at every two or three outlets along the lateral so that the lateral pipe need only be moved half or a third the number of times as the sprinkler is moved. Sprinklers are hopped along the lateral in between lateral pipe moves. Main lines are permanently installed underground. Requires less labour than a portable system, but still awkward for moving lateral pipes through mature cane.
	Dragline (surface laterals)	Aluminium lateral pipes are laid out on the surface and left in place throughout the irrigation season. Sprinklers are mounted on portable tripods or riser stakes, and connected by flexible hose to the laterals. Sprinklers are moved to grid positions either side and along the lateral in a 'module', e.g. 3 x 4 positions on an 18 m grid covers a module of 54 m by 72 m = 0.3888 ha. Requires less labour than hop-a-long since no pipes are moved until harvest; but paths are needed through the cane to ease sprinkler moves.
	Dragline (buried laterals)	Lateral pipes and branches to sprinkler grid positions are permanently constructed from black polyethylene pipes 25 to 80 mm in diameter laid underground. Riser pipes and valve outlets feed water to the surface. Sprinklers are mounted on riser stakes and are connected by a short length of flexible hose to the valve outlet. Sprinkler rotation is the same as the surface lateral system. Only the sprinkler equipment needs to be removed at harvest; but valve outlets need protection from damage.
Permanent	Solid-set	A buried pipe network is constructed to supply each sprinkler position and a sprinkler riser assembly is installed at every position, only being temporarily removed at harvest. Sprinklers operate in groups and are controlled by hand-operated field valves. Capital cost is high; but labour requirement is low, and sprinkler paths are not required through the cane.

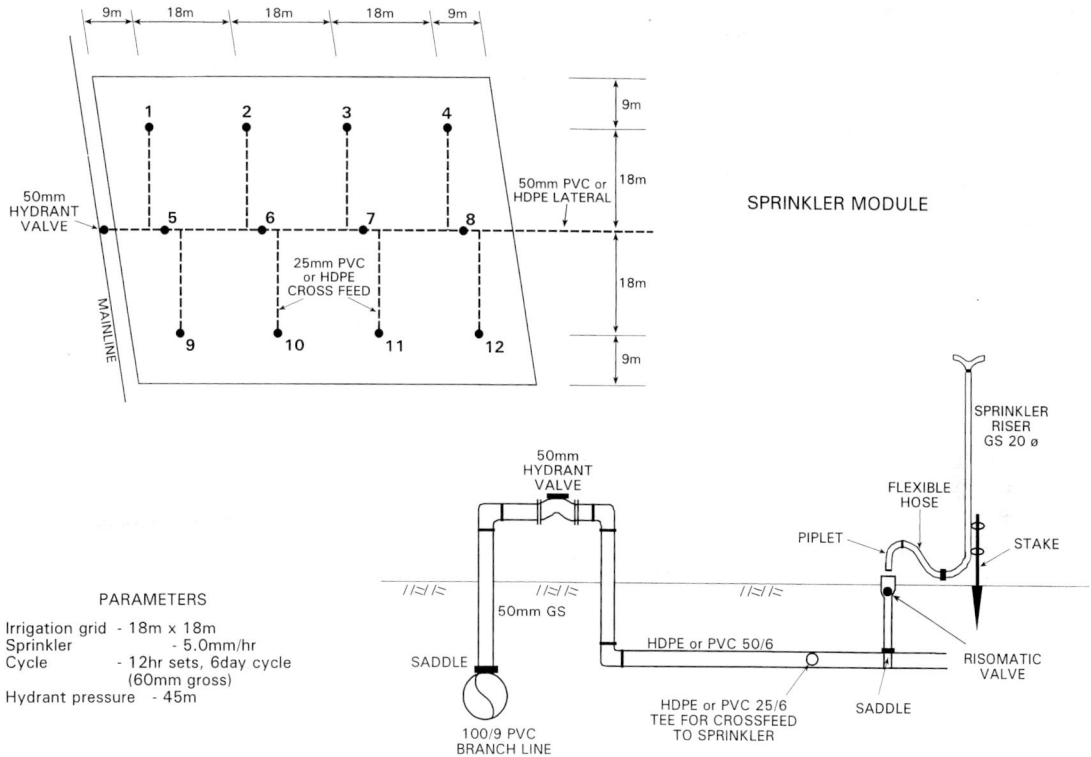

PARAMETERS

Irrigation grid - 18m x 18m
Sprinkler - 5.0mm/hr
Cycle - 12hr sets, 6day cycle
(60mm gross)
Hydrant pressure - 45m

Fig. 5.11 Typical dragline sprinkler layout.

Fig. 5.12 Floppy sprinkler system. Source: Floppy Sprinkler (Pty) Ltd, South Africa.

Table 5.4 Typical operating ranges for rain-guns.

Category	Flow (l/s)	Gun pressure (bar)	Diameter (mm)	Lane spacing (m)	Run (m)	Area covered (ha)	Run time (h) for 25 mm application	Average precipitation (mm/h)
Small	8	4.8	63	52	200	1.04	9.0	6
Medium	16	5.5	90	66	400	2.64	11.5	8
Large	35	6.2	125	85	500	4.25	8.4	10

In sugarcane, Floppy sprinklers are installed as a solid set system typically on a 12×14 m triangular grid, which will give a gross precipitation rate of 4.2 mm/h. In 2002 the cost of infield equipment is approximately US$1800/ha including trenching and installation of lateral pipes.

The Floppy system has been tested extensively by the Water Research Commission[17] in South Africa and is a commercially proven system. It has found great favour on new smallholder sugarcane schemes owing to its simple method of operation, low energy cost and low maintenance requirement.

Rain-gun travelling irrigators

Rain-gun sprinklers are best run along dedicated towpaths running parallel to the cane rows and spaced at 65% of the wetted diameter (typically 70–90 m). The gun rotation is usually set at 270–330° so that the area behind the sprinkler in the direction of travel is kept dry to avoid tracking problems. Typical irrigation runs are 200–400 m long, but machines are available that will run up to 700 m. Table 5.4 shows the typical operating range for rain-guns.

The rain-gun system is versatile and operates well on undulating ground and for irregular shaped fields. Application rates can be adjusted by varying the speed of travel, operating pressure and nozzle size. The large throw of the sprinkler, however, means that wind-drift can severely affect water distribution, so that irrigating only at night may have to be considered. The travelling irrigator takes around 1 h to move and set up on each lane and to tow the gun trolley into position, but once started the machine can be left unattended until the run is completed. This method of irrigation is most common in Queensland on the lighter hilly

soils where irrigation is supplemental to rainfall. The cost of a large category hard hose irrigator in 2002 is approximately US$22 000 (excluding pump and main line), and it should be able to command an area of 25 ha at peak demand giving an approximate unit cost of US$880/ha. This fairly low cost is offset by higher operating (i.e. energy) costs compared with other overhead systems. Figure 5.13 illustrates a typical rain-gun irrigator in Australia.

Fig. 5.13 Typical rain-gun system.

Drip irrigation

In comparison to furrow and overhead systems, drip is a relative newcomer to the irrigation industry. Drip systems first came into prominence in the 1970s, and have been widely accepted by growers of high-value horticulture and orchard crops but in row crops such as sugarcane accounts for < 1% of the irrigated area on a worldwide basis.

Drip manufacturers, however, have invested heavily in research and development to improve component design and reduce cost, and this is resulting in an ever-increasing rate of conversion of irrigation systems to drip. Two notable examples are Hawaii where a massive conversion programme from furrow to drip was implemented in the 1980s, and in India[18] where the Government has been subsidising farmers on a drip conversion programme since the 1990s.

Drip is a high-tech system requiring a good knowledge of soil–water relationships, water treatment and irrigation scheduling to achieve the best results. The potential benefits are:

- a higher cane yield;
- a saving on gross water use;
- a saving on operating costs;
- a reduction in weed germination and growth;
- a flexible yet precise method of water application;
- the option of applying fertiliser through the drip system; and
- the option of a manual, semi-automated or fully automated control system.

Drip irrigation is not tolerant of poor management, and the aspects that need to be understood and appreciated are:

- It is not a visual form of irrigation; consequently it requires careful monitoring and crosschecking of flow meters, pressure gauges and application times to assess whether the system performance is correct or if a fault has occurred.
- It requires a clean water supply and meticulous attention to filtration, backwashing, flushing and chemigation to keep emitters free of blockage, which in the worst instances could cause irretrievable system damage.

- It requires skilled operators that are well trained in the mechanics of operating and maintaining the drip system, and are aware of the consequences if the maintenance tasks are not undertaken correctly.

A poorly maintained furrow or overhead system will continue to work tolerably; but a drip system will fail disastrously. This has to be remembered always.

Operating principle

The operating principle of drip systems is to apply a precise and uniform flow of water via an emitter direct to the root zone of the crop. Flow rates at each emitter are very small, typically 0.8–4.0 L/h, hence the term 'drip' or 'trickle' irrigation. The wetting pattern (or moisture bulb) formed under each emitter varies with the soil type. In a sandy loam there would be more of an elongated vertical pattern, i.e. carrot shaped, whereas in clay soil the pattern would be more lateral, i.e. onion shaped.

The emitters in sugarcane need to be spaced so that the moisture bulbs join up to give adequate moisture all along the cane row. It is important that the selection and spacing of emitters are matched to the soil type and crop; consequently, it is worthwhile undertaking field tests with various emitter spacings on sample soil types when designing a system. In sandy loam the emitter spacing is typically 300–500 mm, while in clay it will be 600 to 1000 mm. Emitter flow rates are generally chosen to provide an application of 1.0 mm/h.

The water supplied to the roots is conveyed in tubes and pipes to the emitter from the primary source, which gives a very efficient irrigation system of 90–95%. Evaporation and other losses are minimal in a drip system.

System components

Starting from the downstream end the main components of a drip system include:

- *Emitters*. Manufacturers have developed many different forms; but all contain labyrinths for agitating the water flow to prevent clogging, and some emitters contain a diaphragm for pressure control. Emitters are either mounted in-line or are formed integrally with the dripper line tubing.

- *Dripper line (or laterals).* These flexible plastic tubes range in size from 16 to 25 mm diameter and 0.25 to 1.2 mm wall thickness. Design operating pressure is typically 1.0 bar, and lateral lengths above 500 m are feasible with the right topography and emitter specification. Thin walled dripper line is often called drip tape. Laterals are either positioned on the surface or are buried some 100 to 200 mm below the surface. The dripper line is placed with the emitter orifice facing upwards so that any sediment is pushed along the bottom of the tube to the flushing point. There is then less risk of sediment entering the emitter labyrinth.

- *Mains and sub mains.* This is the pipe network for connecting the dripper lines to the filter station. Irrigated areas are divided into panels or blocks with a manual or automated valve controlling water flow to the panel. On large schemes, secondary filters should be installed at the valve stations to protect dripper lines against dirt ingress that may result from a mains pipe failure.

- *Flushing mains.* This is an optional but recommended pipeline and valve system connecting the downstream ends of the subsurface drip laterals for providing a quick and simple method of flushing dripper lines.

- *Filter station.* A sand-media, screen or disk-type filter system is required to remove dirt and organic matter from the raw water supply. Filtration down to 125 micron is normally required to meet the operating specification of emitters. Filter backwash systems are usually automated and are actuated either by a timer or a maximum pressure differential across the filter. A 1.2 m diameter sand-media filter will have a throughput of 50–90 m^3/h, whereas an 18-unit disk filter (see Fig. 5.14) on a similar footprint would have a throughput of 500 m^3/h. The filter station is probably the most vital component of a drip system.

- *Fertigation system.* Drip irrigation is an ideal carrier for fertiliser (nitrogen, phosphorus, potassium and micro-nutrients), but it must be

Fig. 5.14 Eighteen-unit disk filter system. Source: Arkal Filtration Systems, Israel.

water-soluble or in liquid form. Since fertiliser often contains impurities, the injection point should be upstream of the filters or the fertigation line should have an in-line filter. Fertigation equipment can be fixed or mobile, and the simplest system will comprise a mixing tank, flow meter and Venturi injector located on a bypass line. The drip system needs to be well flushed before and after fertiliser injections.

- *Chemigation system.* Depending on site conditions, the chemicals that may have to be periodically injected to prevent clogging of emitters include trifluralin (root inhibitor), chlorine (control of algae and bacteria) and sulphuric/hydrochloric/nitric/phosphoric acids (control of pH and mineral scale). The drip system needs to be well flushed before and after chemical injections. Chemical injection equipment tends to be portable, self-contained pump units, since injection frequency and volumes are low.

- *Pressure source.* The operating pressure of a compact drip system on level terrain will be around 4.0 bar at the inlet to the filter station, and this is normally provided by a pump. This pressure can sometimes be provided by gravity in countries with upland reservoirs, e.g. Mauritius.
- *Control system.* The options can range from manual, semi-automatic to fully automatic control systems. Filter backwash operation should always be automatically controlled, but all other valve and fertigation/chemigation operations can be manual if this is the grower's preference. Early generation control systems used hydraulic tubes or electric wires laid in trenches to link field valves to irrigation controllers; but these were prone to damage and other problems on large spread out systems. However, the latest generation of radio-operated controllers offer a much neater control solution for fully automating large-scale drip schemes.
- *Ancillary components.* Other equipment commonly found on drip systems include flow meters (flow monitoring), hydrometers (volumetric metering of irrigation applications), vacuum breakers (prevents suck-back of soil into emitters when irrigation is stopped), air valves (to vent air trapped in mains/sub mains), pressure gauges (for system monitoring) and pressure control valves (for system operation).

Useful information sources for drip system design, operation and maintenance are available from several publications[19,20] and from literature supplied by the manufacturers (e.g. Netafim, T-Tape, Hardie, Jain, etc.).

Surface drip

With surface drip the laterals tend to be placed in the inter-row rather than alongside the cane row. Two cane rows are therefore supplied from one lateral, and with a row spacing of 1.5 m the laterals would be spaced at 3.0 m. A clayey soil with a good lateral wetting pattern is required for this layout. A robust, thick-walled dripper line needs to be chosen to withstand exposure to the environment, and also the handling before and following harvest, i.e. reeling in and reeling out. Flushing mains are not so essential since the ends of the dripper line can be easily opened on an individual basis.

Surface drip was the favoured system initially amongst cane farmers, but in recent years it has been largely replaced by subsurface drip. Surface drip is simpler to install and easier to check and repair for blockage or leakage; but it is no cheaper than subsurface drip and has a number of disadvantages:

- requires additional labour to reel dripper lines in and out at harvest;
- dripper lines can get tangled and damaged in high yielding and lodged cane;
- the lines are at risk of rat damage;
- they would be totally damaged if a runaway fire occurred;
- there is a higher incidence of smut disease on the 'dry', non-irrigated side of the cane rows; and
- there is a higher weed population in the irrigated inter-row.

Subsurface drip irrigation (SDI)

Placing the dripper line underground removes the disadvantages of surface drip and enables a cheaper thin-walled drip tape to be used. This tape is sacrificed at replant; so for economy an SDI system needs a plant and ratoon cycle of at least 8 years. Some growers have suggested that the tape could be retained at replant by adopting minimum tillage practices, but this remains to be proven in practice.

For optimum yields, a dripper line should be buried adjacent to each row; but with a conventional row spacing of 1.5 m, this entails 6667 m of dripper line per hectare. Various tramline systems have been used in Venezuela[21] to extend this spacing to 3 m; but it must always remain compatible with the harvesting system. In a redevelopment project[22] in Swaziland, a dual row, 1.8 m spacing system was adopted being more suited to the track widths of the harvesters and cultivation equipment. As the drip tape is placed between the dual cane row (commonly planted 40 cm apart), it is protected from wheel loading and is an optimum distance

Fig. 5.15 Dual row subsurface drip system.

from the cane rows for wetting and for keeping clear of root growth. Figure 5.15 illustrates a dual row system with the end of the drip line exposed.

The average performance over the initial 4 years of operation in the Swaziland layout for water use efficiency has been 9.5 t cane/ML of water (i.e. applied irrigation + effective rainfall). This exceeds the performance of other furrow and overhead systems by over 20%.

Soil moisture instrumentation

The objective in irrigation is to deliver water to the cane roots at the right quantity and the right frequency to maximise growth. In an ideal world, there would be an instrument that would advise the soil moisture condition to the farmer so that he can schedule the next cycle of irrigation. Unfortunately, soil strata can be very variable, and single point moisture sensors are not necessarily representative of the soil moisture in the whole cane block. Furthermore, multi-point moisture sensors can be prohibitively expensive to install and monitor. Also, the early generation of moisture sensor (e.g. the gypsum block and tensiometer) only worked well on light sandy soils, whereas a high proportion of irrigated cane is grown on clayey soils.

As a result of this limitation on instrumentation, irrigation has tended to be scheduled on the basis of climate monitoring (i.e. evaporation and rainfall), which can be recorded accurately to provide a 'profit and loss' soil moisture balance. However, an empirical crop coefficient has still to be applied to the evaporation measurement to determine the net irrigation water requirement of the crop. Effective climate-based scheduling methods have been developed with the aid of advanced evapotranspiration formulae (e.g. Penman-Monteith) or mini-evaporation pans; but the farmer is always recommended to visually crosscheck the soil moisture using hand-augering.

In recent years a new generation of soil moisture instrument has emerged that when used in conjunction with automatic weather stations, data loggers, modems, PCs and specialist software can provide researchers and cane farmers with a first class set of tools for testing and optimising different irrigation scenarios. Some of the more common devices are described below.

- *Neutron probe.* This sensor is lowered into a vertical tube that is permanently embedded in the soil. It operates by emitting a radioactive signal and counts the rate at which the neutrons return. Specialist software then converts the count rate to soil moisture. Several readings at different levels are taken at each tube location to provide a moisture profile down the root zone. Each tube installation needs to be calibrated for accurate results.
- *Time Domain Reflectometry (TDR).* The operating principle is that an electromagnetic pulse

is sent into the soil via stainless steel rods called wave-guides. Soil moisture measurements are based on the change in the dielectric constant of the soil with variations in water content. Several manufacturers market TDR equipment, but it tends to be expensive and limited to scientific research.

- *EnviroSCAN.* This is a permanently installed vertical probe comprising typically of four to eight sensors positioned at different levels. The individual sensors create a high frequency electrical field that is measured and is a function of the soil-water content. The probe is linked to a data-logger for the continuous recording of readings at around 15 min intervals. The data are downloaded to a notebook computer or transferred by modem to a PC for processing by specialist software. The output is graphical, and over several wetting and drying cycles the full and refill points can be defined without special calibration. The access tube for the probe must be carefully installed because only a small volume of soil around the tube is measured, and if this is overly disturbed then the sensor readings will be erroneous. Figure 5.16 is an exposed view of the sensors, tube and data logger.
- *Diviner 2000.* This is a portable version of the EnviroSCAN probe with an instantaneous direct readout of the soil moisture profile from a single sensor that slides down the access tube. Multiple readings can be taken in just a few seconds. This portable device is more suited to routine irrigation scheduling, while the EnviroSCAN probe is mostly used for irrigation research and for testing different scheduling scenarios.
- *Aquaflex.* The operating principle of this device is similar to the TDR appliance, but the sensor is a 3 m long cable buried horizontally within the root zone. It is a fixed installation, soil-moisture monitoring device that is particularly suited to irrigation research on turf grass and shallow-rooted crops, where a single-depth moisture reading is adequate.
- *Delta-T profile probe.* This probe has evolved from the ThetaProbe soil-moisture sensor, and can be used as a portable measuring device or a fixed installation that can continuously monitor

Fig. 5.16 EnviroSCAN soil moisture probe and data logger. Source: Sentek Pty Ltd, Australia.

soil-moisture. The 0.5–1.0 m long probes are housed in a 28 mm diameter plastic tube and take readings at 4 to 6 levels. The probe measures volumetric soil-water content, and the operating principle is similar to the TDR device. Its application to sugarcane has not been recorded.

Drainage

The 'failures' that occur in irrigation schemes are often the cause of too much water rather than insufficient water. In new schemes irrigation designs must go hand-in-hand with the drainage policies to provide a complete and integrated farm plan. In many sugarcane areas soil salinity and sodicity are often common areas of concern, and good drainage is a prerequisite for keeping these under control.

A good knowledge of regional hydrology and soil substrata is needed by cane growers to assess the drainage requirements and to manage soil salinity/sodicity. Specialist assistance is usually available to farmers from Government Agricultural Organisations and sugarcane Extension Services.

The sugarcane plant itself is a very effective 'drainage tool', being able to transpire at peak canopy some 7 mm depth of water per day, which in a soil medium translates to a much greater depth of soil moisture. However, cane roots will not tolerate anaerobic root conditions under ponded water for more than a couple of days before there is a significant reduction in yield. Consequently, detailed attention to drainage is essential.

Drainage categories

There are two main drainage forms:

- *surface drainage* to control and dispose of surface run-off following rainfall; and
- *subsurface drainage* to maintain the water table below the root zone.

In sugarcane the water table needs to be maintained at least 0.6 m below the soil surface to provide optimum soil moisture conditions within the root zone.

Surface drainage

On free-draining soils, surface drainage is less critical since the rainfall can infiltrate vertically through the soil; however, on soils with a steep topography, conservation measures are necessary to prevent erosive run-off during intense storms. In contrast, good surface drainage is essential on heavy clay land with flat gradients, and in such conditions cane is best grown on ridges to promote soil aeration. Laser-controlled land grading is highly recommended for such terrain to provide an even down-slope to the tail-drain without flat spots that would cause ponding. Laser levelling is usually carried out with the joint purpose of providing efficient furrow irrigation as well as good surface drainage.

Surface run-off from cane rows is collected in tail drains that should be wide and shallow to allow easy crossing by machinery. It is good practice to line tail drains with a non-invasive grass to slow water velocity and filter out soil particles, e.g. *Vetiver zizanioides*, *Cynodon dactylon* and *Stenotaphrum secondatum* – as long as these latter two species are kept under control by cutting as they are invasive. Figure 5.17 illustrates a well-constructed tail drain in Swaziland with concrete bolsters to reduce soil erosion during grass establishment.

From the tail drains surface run-off will be fed into the secondary and main drains before it exits to the regional drainage system or to a tail water recycling system. The required dimensions for drain profiles, culverts and bridge crossings are a function of catchment size, storm intensity, topography and roughness coefficient. Design handbooks are

Fig. 5.17 Grass lined tail drain with wide and shallow profile.

often available to farmers from Extension Services, but in the absence of these a drainage engineer should be consulted.

The drainage coefficient is a useful parameter for calculating the design flows at culvert crossings, drop structures and in drain reaches, and appropriate design manuals will advise the correct sizing of structures. The coefficient is often expressed as litres per second per hectare, and is specific to local conditions. Criteria of 48 h or 72 h storm events with a 1 in 2 year or 1 in 5 year return period are often used to calculate the drainage coefficient. It is usually only valid for small farm catchments, and in subtropical cane growing districts the coefficient would typically range between 4.4 and 7.7 L/s/ha. The adoption of a higher and more conservative drainage coefficient would mean larger and more expensive drainage structures while the adoption of a lower coefficient could result in prolonged flooding as run-off is held back in the cane fields by an undersized drainage network. A pragmatic approach is needed whereby some element of flooding is tolerated, say 48 h, before the storm water has been dissipated.

Subsurface drainage

Good vertical drainage is required in the root zone to wash out salts and maintain beneficial water/air ratios. Controlling the water table to below 0.6 m (or much deeper in saline conditions) is a vital part of cane management and can be achieved in several ways:

- deep, closely spaced open drains;
- subsurface drains (e.g. tile drains, slotted pipe, mole drains, etc.); and
- well pumps.

Water table levels tend to fluctuate due to the wetting and drying effect of irrigation cycles and rainy/dry seasons. Where high water tables are a known problem, it is important to install and monitor a network of observation wells or piezometers. These can be simple, inexpensive 25 mm diameter tubes (usually plastic), which are slotted at the bottom and inserted into a 75 mm diameter hole that has been augured 2 m deep and have a gravel/sand

envelope to prevent soil ingress. The top of the tube should be capped to prevent insects and debris entering. A clay seal should be placed around the tube at the ground surface to prevent run-off water entering the hole. Water levels can be manually monitored using a 'plopper' tape, a well dipper or a float and flag. Alternatively, the new DIVER instrument manufactured by Van Essen Instruments and containing integral water level sensor, data logger and battery can be used for automatic monitoring.

An open drain network is not an ideal system for controlling water tables because of land loss, field access difficulties and high maintenance requirement. If the soil hydraulic conductivity is very high, then widely spaced open drains are effective for water table control (e.g. in the Florida 'muck' soils), but this occurs in very few cases. However, deep open drains are effective at intercepting spring lines where an upland free-draining soil meets a lowland clay soil.

The camber bed system was developed for draining and cultivating flat, heavy clay land, and is the predominant system in the Guyana cane industry. The camber beds are separated by open drains 900 to 1200 mm deep formed by a tractor-powered ditcher implement. The drain spacing is typically 6 to 15 m depending on the style of camber bed. The inter-bed drains are primarily for collecting surface water run-off, but they also help to control the water table within the bed. This groundwater control only works if there is a reasonable tilth within the camber bed to allow the horizontal movement of water. In its normal physical state there is negligible movement of water through the heavy clay.

A blanket subsurface pipe drainage network is generally not an economic system for the sugarcane industry. However, it can sometimes be forced on a scheme when irrigation accessions to groundwater have raised the water table to within the root zone. This has occurred in Swaziland on a particularly difficult duplex soil where subsurface drains have had to be installed at 20–40 m spacing. As an established estate, it could finance the work over a number of years. Subsurface drains are more commonly used in isolated areas to treat trouble spots arising from springs, rock bars and low spots.

A subsurface drain is typically constructed from proprietary slotted drain pipes (50, 75, or 100 mm diameter) placed in a narrow trench typically 1.2–1.5 m deep, at a slope of at least 0.5%, and with a gravel surround to filter out soil particles. Manholes, constructed from polythene or precast concrete, need to be positioned at bends or at junctions with collector pipes. The drainage pipe will have to be flushed clean periodically using a proprietary high-pressure drain cleaner with jetting nozzle.

Groundwater can be a valuable source of irrigation water if it is of good quality. Well pumps can therefore provide the dual benefit of reducing water table levels in conjunction with supplying supplementary irrigation water. In the Peruvian cane industry a high proportion of the surface water infiltrates into groundwater that is later pumped out and recycled into the irrigation system. Some 30% of the total irrigation supply is derived in this manner from groundwater pumping.

Well pumping for water table control requires the right hydrogeological conditions of permeable strata and the free vertical movement of water. Wells are sited on a triangular or rectangular grid so that their cones of depression overlap to effect a blanket reduction in the water table level. Well pumping should only be contemplated if the discharge can be re-used for irrigation or the substrata hydraulic conductivity is sufficiently high that a single well can influence a large area.

Environmental considerations

The storm and irrigation run-off from cane farms entering natural water courses and coastal zones can cause concern amongst environmental groups in that the farm chemicals (i.e. herbicides and pesticides), fertilisers and soil particles carried by the run-off will upset the natural habitat. Environmental legislation and controls are increasingly being introduced to monitor and improve the water quality of farm run-off. Cane farmers are mostly supportive of these new controls and will often provide the initiative for the mitigation measures.

Some of the techniques that cane farmers can adopt to improve run-off water quality and soil retention are as follows:

- *Mulching.* Retaining a cane trash blanket or incorporating it in the soil profile to reduce surface-flow velocities and filter out soil particles.
- *Ground cover.* Maintaining appropriate vegetative cover to non-cropped areas (e.g. tail drains, headlands, verges, etc.) instead of leaving bare soil.
- *Fallow cover crop.* Growing a legume cover crop in fallow fields to improve soil stability against run-off as well as acting as an ameliorant.
- *Minimum tillage.* Minimising the inter-row tillage operations at plough out and replant to reduce soil disturbance and transport.
- *Channel velocity.* Designing the drainage channel network so that the water velocity is less than 0.6 m/s in silty loam soils and 1.2 m/s in clay soil to minimise bank erosion and sediment transport.
- *Tail water recycling.* Installing tail water ponds and recirculation pumps to collect and recycle irrigation run-off and 'first flush' rainfall run-off. Tail water storage capacity should be capable of collecting at least 12 mm of 'first flush' run-off from the gross farm area.
- *Chemical mechanisms.* Applying soil-binding polymers (e.g. polyacrylamide) with irrigation supply water on surface irrigation systems to coagulate and settle out suspended soil particles.
- *Drop structures and silt traps.* Installing structures at sudden elevation changes in drains to dissipate water energy and trap silt. Planting vetiver hedges on steep channel reaches and outfalls to slow water velocity and filter out soil particles.
- *Environmental buffers.* Providing a wetland reserve with minimal flow velocity between the farm main drain outlet and the natural river/coastal system to filter out contaminants and sediments. This reserve acts as a remedial management option and needs to be used in conjunction with other solutions for preventing the initial loss of sediment and contaminants.

World Wide Web information sources

The Internet has become a colossal information source for publications, research papers,

Table 5.5 Useful irrigation and drainage web sites.

Web address	Institution	Web site content
www.asae.frymulti.com	American Society of Agricultural Engineers	Technical library (search under Irrigation & Drainage)
www.cati.csufresno.edu/cit	Center for Irrigation Technology – California	Research reports, training courses, links, testing
www.cgiar.org/iwmi	International Water Management Institute	Research papers and reports
www.engineering.usu.edu/iic	International Irrigation Center – Utah State University	General information, software, training courses
www.fao.org/ag/agl/aglw	FAO – Water Resources, Development and Management Service	Irrigation equipment database, statistics, software, training material, on-line documents
www.fao.org/iptrid	FAO – International Programme for Technology and Research on Irrigation & Drainage	Research reports, links, publications, contacts, projects
www.fao.org/landandwater/aglw/cropwat.htm	Food and Agriculture Organisation	Software download site for CLIMWAT and CROPWAT
www.icid.org	International Commission on Irrigation and Drainage	Research papers, meetings, databases, publications, links
www.ilri.nl	International Institute for Land Reclamation and Improvement	Drainage and water management publications
www.irrigation.org	US Irrigation Association	News, events, links, directory
www.irrigation.org.au	Australian Irrigation Association	News, events, links, directory
www.irrigationworld2000.com	Trade information database	Company information
www.itrc.org	Irrigation Training and Research Center – California	Research reports and training courses
www.sabi.co.za	South African Irrigation Institute	News, events, links, directory
www.waterengr.com/hec.htm	US Army Corps of Engineers	Hydrology and hydraulics software (some are free)
www.watertight.org	Center for Irrigation Technology	On-line irrigation scheduling
www.wcc.nrcs.usda.gov/nrcsirrig	USDA National Resources Conservation Service	Irrigation page – general info, links, software, manuals
www.wiz.uni-kassel.de/kww/projekte/irrig/irrig_i.html	Virtual Library Irrigation	IRRISOFT software; Irrigation-L discussion group; directory of irrigation companies and equipment
www1.silsoe.cranfield.ac.uk/ukia	UK Irrigation Association	News, events, links, directory

equipment and suppliers, manuals, software and training courses, and the topics of irrigation and drainage are no exception. There are many web sites from which useful software and literature can be downloaded, as well as providing links to other related web sites. In addition, problems and advice brought up in specialist discussion groups or through search engines can provide pointers. The Internet is a fast-changing environment and web addresses can sometimes alter, but some of the institutional web sites that provide excellent directories and links are given in Table 5.5.

REFERENCES

1 South African Sugar Association Experiment Station (1996) *Senior Sugarcane Certificate Course Notes.* SASEX, Mount Edgecombe, Durban.
2 Sweet, C.P.M. & Patel, R. (1985) Changes in cane yield of irrigated variety NCo376 due to season and their implications when evaluating field performance. *Proceedings of the South African Sugar Technologists Association,* **59**, 210–214.
3 McGlinchey, M.G. & King A.G. (1998) A computer crop model as a management tool: developments in Swaziland South African Sugar Industry Agronomists Association Annual Meeting.

4 Ellis, R.D., Wilson, J.H. & Spies, P.M. (1985) Development of an irrigation policy to optimise sugar production during seasons of water shortage. *Proceedings of the South African Sugar Technologists Association*, **59**, 142–147.

5 Clowes, M.StJ. & Breakwell, W.L. (1998) *Zimbabwe Sugarcane Production Manual*. Zimbabwe Sugar Association Experiment Station, Chiredzi, Zimbabwe.

6 South African Sugar Association Experiment Station (1997) *Herbicide Guide*. SASEX, Mount Edgecombe, Durban.

7 Thompson, G.D. (1976) Water use by sugarcane. *South African Sugar Journal*, **60**, 593–600, 627–635.

8 Holden, J.R. (ed.) (1998) *Irrigation of Sugarcane*. Bureau of Sugar Experiment Stations, Queensland, Australia.

9 Withers, B. & Vipond, S. (1974) *Irrigation Design and Practice*. B. T. Batsford, London.

10 Smedema, L.K. & Rycroft, D.W. (1983) *Land Drainage – Planning and Design of Agricultural Drainage Systems*. B. T. Batsford, London.

11 Walker, W.R. (1989) Guidelines for designing and evaluating surface irrigation systems. *FAO Irrigation and Drainage Paper 45*. FAO, Rome.

12 Ross, C.W., Sojka, R.E. & Lentz, R.D. (1994) Polyacrylamide as a tool for controlling sediment run-off and improving infiltration under furrow irrigation. In: *Proceedings Australian and New Zealand National Soils Conference 1996*, Vol. 2, *Oral Papers*, pp. 229–230. University of Melbourne, Australia, 1–4 July 1994. US Department of Agriculture, Agricultural Research Service. Kimberly, ID.

13 Sojka, R.E. & Lentz, R.D. (1994) Polyacrylamide (PAM): a new weapon in the fight against irrigation-induced erosion. *USDA-ARS Soil and Water Management Research Unit, Station Note No.01–94*, Kimberly, ID.

14 Sojka, R.E. & Lentz, R.D. (1997) Reducing furrow irrigation erosion with polyacrylamide (PAM). *Journal of Production Agriculture*, **10**, 1–2, 47–52.

15 MacKinnon, A. (compilator) (1998) *Ord River Irrigation Area – Best Practice in Irrigation Farming*. Kimberley Development Commission, Western Australia.

16 Kay, M. (1983) *Sprinkler Irrigation – Equipment and Practice*. B. T. Batsford, London.

17 Simpson, G.B. & Reinders, F.B. (1999) Evaluations of the performance of two types of sprinkler irrigation emitters installed on permanent and dragline systems. *WRC Report No KV 119/99*. Water Research Commission, Republic of South Africa.

18 Magar, S.S. (1995) The status of microirrigation in Maharashtra, India. In: *Fifth Microirrigation Congress, Florida, April 2–6*, pp. 452–456. American Society of Agricultural Engineers, St Joseph, MI.

19 Soopramanien, G.C. & Batchelor, C.H. (1991) *Drip Irrigation of Sugarcane – Operation and Maintenance Manual*. Mauritius Sugar Industry Research Institute, Reduit, Mauritius and Institute of Hydrology, Wallingford, UK.

20 BSES (1995) *Reference Manual on Drip Irrigation for Sugarcane*. Bureau of Sugar Experiment Stations, Queensland, Australia.

21 Leiva, E. & Barrantes, A. (1998) Increasing sugarcane yield with subsurface drip. *Sugar y Azucar Journal*, August, 26–31.

22 Merry, R.E. (2001) Dripping with success. The challenges of an irrigation redevelopment project. *British Society of Sugar Cane Technologists, Autumn Technical Meeting, Westonbirt, UK*, 11 October 2001.

Chapter 6
Sugarcane Agronomy

James E. Irvine

Sugarcane agronomy involves the integration of soil sciences and crop production. Because the world's sugarcane soils and methods of production are so diverse, it is impossible to treat them all in a thorough manner in less than many volumes. For this reason the present chapter is presented as an outline of soil and production management problems that can be expected from the initiation of a plantation to the end of a multiple-year crop cycle.

SOIL MANAGEMENT PROBLEMS

Soil erosion is probably the most costly of soil management problems, since once it has occurred the loss cannot be recovered. Gully erosion can result in entire fields becoming a total loss. Sheet erosion can remove topsoil from fields with only gentle slopes without management being aware of the loss. Wind erosion may be even more subtle since it may be written off as only a dust storm. The National Resource Conservation Service, United States Department of Agriculture (USDA), has publications for erosion control and these are available on the internet (www.nrcs.usda.gov). In the US, erosion of soil weighing > 11 t/ha per year may subject a farmer to penalties for violation of the farm's conservation plan.

Soil types may vary in susceptibility to erosion, with montmorilinitic clay being more susceptible to gully or sheet erosion than kaolinitic clay. Clays are more susceptible to wind erosion than sand. While sand forms dunes due to strong winds, clays may form dunes. These are frequently downwind from primary and secondary sand dunes and have a very different vegetation. Loess soils are broad areas of wind-born silt and clays that are fertile if

stabilised. In some cases, erosion may be viewed as beneficial in that the soils removed by gully and sheet erosion may end up as the delta soils that support a rich agriculture.

As can be expected, the worst gully and sheet erosion occurs in areas of potential torrential rainfall, especially where slopes are steep. The conventional erosion control method is to plant sugarcane along contour lines after these have been determined by level, tripod and a marked stake. Terracing works well where labour and slope permit. For steeper slopes, a row on the contour line is paralleled by the next row below, and so on until the maximum safe number is reached. The last row is then followed on the downhill side by a swale and a burm followed below by more rows, another swale and burm, and so on. The swale and burm store runoff water for soil moisture and break the force of the runoff in its downhill rush. The latter is more important because water running down the slope gains speed and force the further it goes, so the probability for erosion is greater on the lower slopes. Vetiver grass *Vetiveria zizanioides*, a native of India, is used in a number of countries as a barrier on the downslope side of sugarcane fields. Vetiver is a perennial that does not flower or form objectionable rhizomes but makes a thick, low, grassy hedge which is effective in slowing soil movement. Growers in Barbados, however, find that mechanical harvesting damages the vetiver sand.

Leaching occurs in coral, gravel and sandy soils unless they are underlaid by a bed of clay. Fertiliser elements applied to the crop can easily be removed by leaching with excessive rain or irrigation water. Nitrogen and potassium are the elements more susceptible to leaching and usually end up in the ground water. Some herbicides applied to cane have been detected in ground water and cause concern. After years of cultivation and irrigation, water tables tend to rise above those in nearby non-cultivated areas and the leachates become a reservoir of nutrients for following crops. For this reason, it may not be necessary to apply nitrogen and perhaps potassium to plant cane crops.

Waterlogging occurs when soil becomes saturated with water and the redox potential reaches unsatisfactory levels[1]. Roots then die and both water and nutrients fail to be absorbed by the plant. While waterlogging may occur naturally, it can usually be corrected by removing surface water with drains or pumps and little damage occurs if action is prompt and the leaves are above the water level. Damage is more severe when irrigation is short and farmers impound excess water in the fields fearing the next round of water may not come. When the next round comes, the fear remains and the field is again subjected to an excess. If the water has a high salt content, then more damage is added to the previous injury and losses are exacerbated.

Salt accumulation is common where irrigation water has a relatively high salt content (800 p.p.m. or more). It is usually suspected when a thin white crust appears on the highest part of the row. If a field irrigated with water having 800 p.p.m. salt is provided with adequate subsurface drainage, the yield of cane will not be damaged by the salt content. However, if subsurface drainage is not provided, salt will accumulate throughout the crop cycle and yields will be reduced. When the crop is ploughed out and replanted, the following crop will suffer unless a fallow period before replanting received showers that leached the salt. Replicated lysimeter tests[2] with carefully controlled salt contents showed that, when compared to cane irrigated with rainwater (6 p.p.m. salts), cane irrigated with water having 800 p.p.m. salts was not different in yield. With a 1875 p.p.m. salt content yields were 62% of rainwater controls and at 3008 p.p.m. the yields were 51% of rainwater controls. These tests were run with no accumulation of salts as would be expected in undrained fields.

Soil pH can be made less acid with the addition of lime or more acid by the addition of sulphur. How-ever, the pH correction requires large amounts of chemical and years to take effect. The reason for the large chemical requirement is explained by the large differences that occur. A pH number is the reciprocal of the logarithm of the hydrogen ion concentration. Numbers below 7 are acid and above are alkaline, and the difference between pH 7 and pH 6 is ten times the H ion concentration. Modern breeding and selection have done a lot to solve the problem and we now have varieties that are productive on soils that have a pH of 8.6 and suffer only occasional iron chlorosis. Other varieties grow on soils that are pH 4.2 and are tolerant of both the low pH and aluminium toxicity.

SITE SELECTION

Unless sugarcane is to be grown for forage, a new plantation site would, of necessity, be located near a mill and boiling house. Most of the suitable locations are between 30° north or south of the equator, and even beyond these extremes there are locations where sugarcane is grown successfully. However, there are also many places within these limits that are unsuitable because of lack of water, drainage, suitable soil, gentle to moderate slopes, market roads, and markets themselves. Large plantations will occupy the best land, consume the most irrigation water and provide the enterprises with the best infrastructure and technology. The smaller plantations will occupy the more marginal soils and slopes, perhaps depend more on rainfall and will develop their own technology based on what works for them and their neighbours. Given access to a market, water becomes the most important factor in site selection. Rainfall records and surveys for quality of potential irrigation water are essential in preparing a plan for development, whether the project is large or modest. Distance from the mill site is also important, since large tonnages of the crop must be transported economically. Transport by truck more than 100 km on good roads is probably uneconomical. Soil, slope and drainage are factors that are important to site selection, but can be improved with effort and expense.

CLEARING

Both large and small sites may have to be cleared if they are not in cultivation. In both cases, the land intended for sugarcane must have the vegetation cover removed. Before removal, all trees with value as timber should be marked and sold. Plants having no value as timber or other uses can be removed by chainsaws and bulldozers, and burned on site. Stumps should be pulled and burned as well. Ash from these fires will have useful nutrients when turned back into the soil. Following clearing, a rough grading with bulldozers and land planes will make subsequent activities easier.

PLANNING AND LAYOUT

Careful planning, with provisions for future expansion or changes, is essential for maximum efficiency in land use. The optimum balance of area is to have 90% in fields and 10% in roads, ditches and canals. This ideal may not be practical in many areas, especially those with irregular terrain. Flat land lends itself to large fields, minimising the need for roads, ditches and canals. Heavy clay soils may require fields that are narrower because low internal hydraulic flux means ditches should be closer together. Irrigation needs and type may require open canals for water transport. Fields with steep slopes will have rows following contours rather than the slope, and swales and burms to control runoff. Roads may need to be re-laid to allow the most efficient field design and the shortest distance to transloading sites and to the mill. Roads should never attack a steep slope directly but should cross contours at a narrow angle with switchbacks when needed. Trees bordering roads are pleasant, providing people shade and peace, but trees will also shade the crop, compete for water and provide a hazard for the aerial application of chemicals.

FIELD DESIGN

Field design, independent of slope, contours, ditches and roads, should maximise the utilisation of the surface area and access to the field[3-7]. Field shapes are generally rectangular except where property boundaries or geology force shape restrictions. In these cases, the geometry of the field may force the incorporation of rows that are shorter than the maximum length of the field (point rows) which, with ever shortening length, reduce the efficiency of mechanical operations in the field.

Row length may vary from 10 m in experimental plots to 400 m or more in commercial fields. A general rule is that the longer the row, the more efficient are the machines that cultivate and harvest the crop. Longer rows and wider fields increase the area in cultivation and decrease the area in roads. Exceptions to the general rule might be made for smaller farms or farms where hand labour and animal power predominate or boundaries and terrain limit field design. With ever increasing mechanisation, longer rows will prevail, and the main consideration in determining the maximum length may be access to machines that require repair or maintenance when stopped in the middle of the row. Machines that carry seedcane or fertiliser must have capacity to reach the end of the row before depleting the cargo.

The number of rows per field will vary depending on irrigation requirements, drainage and water table levels (see Chapter 5). In Florida, where fields are flat and internal drainage is good, fields may have 120 rows of 425 m. In Louisiana, fields are also relatively flat but irregular, and the area has a high water table and high rainfall, making fields 30 m wide with raised beds and high rows necessary for surface drainage.

Row widths vary from 60 to 200 cm, with the average probably at 150 cm. With no other limiting factors, the more narrow the row, the higher the yield of cane and sugar. As interplant distance decreases arithmetically, plant population increases exponentially[8]. Although plant weight and the number of stalks per plant decrease with decreased spacing, the increase in population is such that yield also increases exponentially with closer spacing. This effect may be more pronounced in subtropical cane areas where the growing season may be shorter.

If closer spacing of rows means higher yields, then the question immediately arises as to why all cane is not grown on more closely spaced rows.

There are several good answers to the question. Historically, row spacing has evolved from narrow to wide in response to mechanisation. A 1 m row spacing is comfortable for a man and a mule, while a 1.5 m row width is a comfortable width for operators of tractors, wagons and harvesters (1.8 m axles may provide more stability for harvesters, hence the wider rows in Louisiana). Seedcane supply is another factor since, if the same amount of cane is placed in the furrow, a field with 0.9 m rows would require twice the tonnage of seedcane as would a field with 1.8 m rows. If wider row spacings were an adaptation to machines, closer row spacing would require the modification of machines to achieve the goal of higher yields.

In the 1980s and 1990s, attempts to mechanise close-row spacings of sugarcane began with the modification of a tiller-shaper that could prepare soil and shape rows from 60 to 180 cm apart. In carefully controlled small plot tests, the plant cane yields of cane on 60 cm rows and 180 cm rows with three drills greatly out-yielded the standard 180 cm spacing. However, after these plots were harvested, the yields of the ratoon crop were poor in the 60 cm rows, mediocre in the triple drills and normal in the 180 cm rows. Similar trials[9] conducted on a commercial scale were even worse because plantation operators were not experienced enough to properly cover the 60 cm and triple drill rows. As a consequence long stretches of stand were lost. A compromise using a 125 cm row spacing proved easier to manage on a plantation scale, and a whole-stalk harvester modified for a 125 cm row was able to harvest the cane with commendable results. However, the increase in yield was not enough to stimulate the industry to change all of the machine fleet to accommodate 125 cm rows either singly or two at a time.

LAND PREPARATION

Once the site has been cleared and previous vegetation removed, it must be prepared for planting. If erosion control requires them, contours and burms should be established to impede the downhill flow of water. Intervening rows should parallel the contour lines. If not recently in cultivation and if irrigation is intended, the field must be shaped to conform to the type of irrigation to be used (see Chapter 5), and both irrigation and drainage are components to be considered. A level field can be used for pan irrigation, a field with a 1–2% slope is needed for furrow irrigation, while slope is less critical with cannon, sprinkler or drip irrigation. Drainage is important whether irrigated or not, since flooding rains can occur in deserts. Surface drainage should be addressed in all fields[4], and subsurface drainage should be considered in areas where salt accumulation is a potential problem or in areas where saturated soils may cause stubble decline. Once slopes, ditches and canals are placed and the necessary land-planning is completed, fields should be subsoiled. The subsoiler should have a curvilinear shank fitted with a hardened steel shoe and set so that the sole of the shoe runs under the hard-pan, lifting and shattering it. This has less power requirement than a tool with a vertical shank. Subsoiling can be done in two directions, one with the row and the other at a wide angle to the row. Subsoiling is more effective in sandy loams and less effective in clays. It is most effective in dry soils and is even damaging in wet soils.

Following the subsoiling, the field must be disced to break clods and improve tilth. Discing should be done with gangs of 45 cm cut-out discs. The cut-out discs are used for increased cutting efficiency, while smooth discs are used for increased efficiency in moving soil. Care must be taken to avoid discing when the soil is too damp. The repeated use of heavy discs will rebuild the hard-pan previously destroyed by subsoiling. Pre-plant application of phosphorus can be broadcast and disced in, as can insecticides for wire-worm and grub control, especially in areas that were grassy. Field roads must be established or rebuilt, giving attention to their drainage. Most are elevated slightly above the field level to keep the road bed dry and firm. In high rainfall areas a shallow V-ditch is placed between road and the row ends to provide drainage for field and road and to provide gentle access for tractors, wagons and harvesters.

NURSERIES

When a field has been prepared for planting, the manager must then transport tonnes of seedcane to the site for distribution down the rows. The cane to be cut for seed is chosen by variety, and the field from which it is taken is usually the closest to the field which is to be planted. This choice minimises transport costs and time. However, the closest field may be the poorest in seedcane quality and the loss due to variety mixtures, diseases, pests and weeds may far outweigh the gain in transport economy. Managers must insure that their seedcane supply is of high quality, since it is the best way to start a new crop cycle. The way to insure quality is to have a nursery system with pure varieties free of disease, pests and weeds.

Varieties have proven to be one of the most enduring investments a grower can make. Once a good variety is obtained, he usually grows his own seedcane and never has to buy again, unless a better variety becomes available. Hybrid sugarcane varieties were first produced over 100 years ago, and the world's breeders have done very well in creating varieties that are richer in sugar, higher in tonnage and resistant to diseases and insects. However, because varieties are bred primarily for local industries and because sugarcane is easily selected for micro-environments, it is rare that a really good, generally adapted variety is found. Farmers travelling to cane areas other than their own are frequently tempted to carry a few stalks home to see if they are better than the local ones. Many times this results in the importation of diseases or insects that become problems, and the imported variety fails because it is not selected for local conditions. It is far better for growers and millers to support their local variety programmes and let their breeders know of their interests and needs. Growers and technologists talk about yield of cane and/or sugar, whilst pathologists and entomologists talk about losses due to diseases and pests; but at the end of harvest, both field and factory management want monetary yield per hectare and per tonne.

Variety integrity is easily compromised when plantings are made from commercial fields that have not been rogued for volunteers or mixtures. Mixtures occur when a different variety is introduced after field reformation and remnants of the old crop are still viable. These sprout and grow with the new crop and a mixed field is established. Mixtures can, of course, be started by careless cutting of seedcane from misidentified varieties. Management must make certain that all varieties cut for nurseries are properly identified and carefully separated and mapped when planted in the primary nursery. Fields which are to be the source of nursery varieties must be inspected for those diseases that can be propagated with the seedcane. These include mosaic, yellow leaf syndrome, leaf scald *Xanthomonas albilineans*, smut *Ustilago scitaminea*, Fiji and ratoon stunting disease or RSD *Leifsonia xyli* subsp. *xyli*. Responsible field technicians can be trained to recognise these diseases, and they should select fields or portions of fields with the minimum of disease incidence. Fields with a high incidence of yellow leaf syndrome or mosaic should not be used for nurseries. Visible symptoms can be used for identification of diseases known to be in the area but, where possible, assays with specific antibodies should be used for both confirmation and surveys.

After visual inspection and (preferably) antibody assays, all seedcane intended for nurseries should be either heat-treated or propagated through meristem culture. There are various hot water treatments (HWTs) for different systemic diseases (see Chapter 3). The short HWTs are not appropriate for treating recalcitrant diseases like RSD and leaf scald, but the Australian system (24 h in running cold water followed by 3 h at 50°C) will cure smut, RSD and leaf scald. Very stringent HWTs over a period of days will cure mosaic, but the frequency of survivors is too low for commercial use.

Meristem culture has been used in Brazil for years as a source of clean propagating material for nurseries, and its acceptance is slowly spreading to other areas. In Brazil, the isolation and early culture of the meristem are done at a central location, and subcultures are sent to the participating mills for additional propagation before being potted and later transferred to field nurseries. In other countries the whole process may be done at one mill or, conversely the whole process may be done by a single commercial company. If the meristem is isolated properly, the process should provide plantlets that are free of mosaic, yellow leaf syndrome, RSD and smut. The process will not eliminate the leaf scald bacterium, which seems to survive in culture. As an added insurance, some laboratories doing meristem culture take meristems from plants derived from the Australian cold soak and HWT. It should be remembered that once liberated from any of the above diseases, the plants can become reinfected under field conditions.

Post-cure culture varies depending on the treatment used. Stalks treated with hot water should be planted in the primary nursery not more than a day after treatment. Heat treatment destroys the inherent inhibition of germination and all buds tend to germinate rapidly. If planted in soil infested with smut spores, the germinating buds may become rapidly reinfected. Plantlets taken from sterile culture after roots have formed can be grown in greenhouses, shade houses, or in specially prepared beds until the leaf sheaths and collars are about 10–15 cm high and the stems relatively stiff. At that point they may be transferred to the primary nurseries.

Primary nurseries, where cured plants are planted, should be as isolated as is practicable from other sugarcane and, if possible, placed on soil that has not been in cane for several years. Planting nurseries in the midst of other cane fields is an invitation to re-infection. The primary nursery should be large enough to leave 60–90 cm between plants in the row and 150 cm between rows. This spacing will provide rapid tillering and easy inspection of the developing plants. All plants that develop symptoms of the systemic diseases should be immediately eliminated (rogued). Plants should also be examined for variety purity, and any plant suspected of being a mixture should also be rogued. Following normal cultivation, the primary nursery may be used for seed in the secondary nursery. The ratoons from the primary nursery should then be used for the next round of heat treatment or meristem culture.

Secondary nurseries may be adjacent to the primary nursery, or placed at other locations depending on hauling distance and disease pressure in commercial fields. Secondary nurseries should also be inspected and diseased plants and mixtures should be rogued and destroyed. The inspections in primary and secondary nurseries should be conducted by workers who can identify both diseases and varieties. Several inspections would be required since symptoms of different diseases appear at different times, and it is difficult to look for disease and variety characteristics at the same time. Nurseries should be free of insects and weeds, since both reduce yields and both travel with the seedcane to new fields. Nurseries should be managed to give maximum yields, and may even be given nitrogen later than would be done for cane intended for milling. While farmers may prefer to plant cane with closer internodes for thicker stands, a case can be made for longer internodes packed with nutrients to feed the new shoots.

In small plantations, one primary and a few secondary nurseries may be sufficient, with commercial fields planted directly from the secondary nursery and from its first ratoon crop if inspection proves it to be clean. Larger plantations may require more than one primary nursery and more secondary nurseries depending on the extent of planting intended. Planning several years ahead is necessary so that secondary nurseries may be located close to areas that will be reformed and replanted. In very large operations it may be necessary to use secondary nurseries to plant commercial fields that will be used for seedcane.

COMMERCIAL PLANTING

Most sugarcane is planted following a fallow period after the harvest of the last ratoon crop. In those areas where seedcane is available and soil moisture dependable following plough-out, cane is replanted with only an interval long enough to remove the old stubble and reform the rows. The advantages of a fallow period are several: good timing with availability of seed and moisture, an alternate crop, a leaching period for salt control or a longer period for growth and a higher yielding plant cane crop. The primary advantage of successive planting is the maximisation of the area in cane at all times. In addition to these factors, the choice of systems depends on the local conditions and the business strategy of management.

Propagation of all commercial sugarcane is done with cuttings, most with the above-ground stalk or sometimes with divisions of the stool following the harvest. Cuttings may consist of the whole stalk or portions of the whole stalk. Wholestalk cuttings (with the terminal bud and leaf roll removed) are probably the most hygienic seed source. Whole stalks have two disadvantages. One is that, if not properly covered with soil, the upper portion may rise out of the furrow, and the second is that germination along the stalk is predictably irregular. In the basal portion, sett roots germinate first and buds later, and in the upper portion buds germinate first and roots later. Buds at the underside of the planted seed-piece will be slower to appear than those on top. Seed-pieces of from three to six internodes will appear to germinate faster when compared to the whole stalks but if carefully controlled plantings are made, the advantage apparent at two months after planting may disappear after six months.

Heat treatment of seed-piece cuttings for 30 min at 50°C before planting stimulates germination. This is the practice used in Hawaii as a control for pineapple disease *Ceratocystis paradoxa*. The spread of other diseases at planting is well documented. Whether cutting wholestalk cane to better fit it in the furrow or to germinate more rapidly, each slice of the cane knife presents the opportunity to spread RSD, leaf scald and gumming disease *Xanthomonas axonopodis* pv. *vasculorum*. Knives should be sterilised as they are used in the field for seedcane by passing the blade through a gas-fired torch on the headland, or by dipping the blades in a solution of carbolic acid (Lysol) for 30 s or more. When planting with billets cut by machines, the blades, elevator chains and teeth should be sprayed with disinfectant (Lysol again) before the machine enters each field. This caveat is especially applicable for fields intended for planting and probably is not practical for fields intended for milling. As with nurseries, all fields intended for commercial planting should be inspected for variety purity and the presence of systemic disease and insect pests. Fields infested with Johnson grass *Sorghum halepense*, Raoul grass *Rottboelia cochinchinesis* or guinea grass *Panicum maximum* should be avoided, especially if the seedcane is to be cut by machines.

The old method of hand planting two running stalks with a 10% overlap, reversing the tops and bottoms of the pairs, is hard to beat for good stands and economy of seedcane. The disadvantage is that it requires a lot of hand labour. This can be reduced by using tractor-drawn planting wagons where the cane is loaded by grab-loaders with the cane-tops toward the front of the wagon and the stalks inclined at 45°. A movable partition at the front of the wagon is operated from the tractor hydraulic system and the seedcane is pushed to the back of the wagon as the workers pull and plant the stalks in the furrows. The tractor operator and two workers can plant three rows in one pass through the field. Further reduction in labour can be achieved where a tractor-drawn drum planter removes cane from the wagon, drops it in the furrow and, in some versions, covers the planted cane with soil all in one trip. Further modifications include the application of pre-plant fertiliser and even pre-emergence herbicide all in one trip. The price paid for less labour is a great increase in capital costs (up to US$259 000 for a multiple-function machine) and usually a greater amount of seedcane that is planted.

After the cane is planted, it may still be covered with a hoe in some areas, but perhaps most of the world's seedcane is covered with discs. These may be a single small pair, with the concave sides facing the furrow and moved with sufficient speed to place soil over the cane without pulling it out of the furrow. Larger (45 cm) disc gangs may be used on a tool bar for one, three or five rows, and adjusted to move the desired amount of soil over the seedcane. (Following one row equipment with multiple row equipment should be avoided; even skilful operators find this difficult to do without damaging the original work.) The adjustments may be made in angle and pitch and in tractor speed. For a flat row, discs almost parallel to the row and no pitch and slow speed are used. A row with a high crown can be built by turning the forward edge of the discs out, adjusting for more pitch with the bottom closer and the top further from the row, and running the tractor faster. A triangular sweep behind the discs can remove excess soil, and a roller can break clods, provide desired compaction and a smooth surface for pre-emergence herbicide application. Depth of cover varies in different areas for different reasons. Many growers like 5 cm of soil over the seedcane, feeling that amount conserves moisture and heat. Less than 5 cm may allow wholestalk cane that has not been heat treated to rise out of the seedbed. More soil may delay germination. Extremes include covering with 30 cm or more in parts of Central America where the fields lack irrigation. Growers then open furrows to that depth to insure that the seedcane has sufficient moisture to germinate before the rains come. Growers in other areas plant in very deep furrows but cover lightly, adding soil after germination and as the shoot grows. The reasoning is that the plant will have a deeper stool that will resist lodging and provide more buds for the ratoon crop. Pre-emergence herbicides should be applied as soon as practicable following covering since rains may make it impossible to get a tractor into the field before the new weed crop appears.

CROP MANAGEMENT

After planting, soil moisture may be low because of the repeated stirring of the soil during reformation and preparing new rows and furrows. Germination may be enhanced by irrigation, if it is possible. Rapid germination and emergence helps the new crop avoid the ravages of pineapple disease. Following germination and emergence, there is little reason to enter the field with a tractor again if weeds are under control and row shape is satisfactory. Before the advent of modern chemical control, from five to six cultivations with discs were required to keep weeds under control and maintain row shape. For the flat culture preferred by managers for combine harvesting, raising the row height is not desirable. When cane is laid in deep furrows, light cultivation is used to cover and later to bring soil to the developing shoot. In those areas where furrow irrigation is practised, furrow dressing or repair should be done so as not to disturb the pre-emergence herbicide. With the wide spectrum of herbicides available for the control of almost any weed, every precaution should be taken to increase their efficacy (see Chapter 5). Where herbicides are not used, as with organic sugarcane or inter-row alternative crops, increased hand labour or shorter crop cycles may be needed to keep weeds to a minimum.

Timing of cultural practices is critical for good management. In addition to the optimum timing of herbicides, timing of irrigation, fertiliser application and pest control are also important, and the reader should see the appropriate chapters. However, the timing of these four management activities must be integrated. Fertiliser at planting is usually limited to calcium (if needed) and phosphate applied before the seedcane is dropped. Nitrogen and potassium are usually applied just before stalk elongation begins so that these elements will be available during the grand growth phase. However, if soil moisture is low, irrigation or rain will be needed so that the plants can absorb the nutrients. Studies[10-12] have shown that fertiliser applied to drought-stressed sugarcane will not help yields. If nitrogen is applied as a split application, the second application may be made under drought conditions and the nitrogen uptake may be delayed until its absorption would be closer to harvest than desired, and the late growth stimulated by excess nitrogen would lower the quality of the cane delivered to the mill. Timing of pest control methods is perhaps the most difficult for the grower. Large plantations may have specialists that act as scouts to monitor the pest populations in the field. Lacking these, growers may hire professional scouts if such services are available. Extension agents are helpful, if available. Most scouts use a threshold population to determine whether control measures (parasites, predators or chemicals) are indicated. Using control measures, especially chemicals, on a rigid schedule is not a good practice.

The cultural practices employed in sugarcane have various effects on the crop and some of these are obvious. As mentioned, cultivation is used to control weeds and to move soil to its proper place. Once the desired row shape is achieved and good chemical weed control is practised, there should be no further need for mechanical cultivation. It would be an unnecessary expense and could cause soil compaction, disturb pre-emergent herbicides and damage shallow roots. Water is essential for nutrient uptake, cooling the plant through evapotranspiration, facilitating biochemical reactions and providing the force for cell elongation that is the basis of growth. An excess of water in the soil ruins the redox potential, brings about an oxygen deficiency and causes roots to die[1,13]. An excess of water over a newly planted or a newly harvested field could cause death of the plant material in a week. If leaves are above the water level, the field may survive longer.

The seventeen elements essential for plant growth production are, in order of quantity required, hydrogen, carbon, oxygen, nitrogen, potassium, calcium, magnesium, phosphorus, sulphur, chlorine, iron, boron, manganese, zinc, copper, nickel, and molybdenum[14]. Silicon is essential for grasses but not for broad-leaved plants[14,15]. The first three elements listed are available in air and water and the remainder are available in the soil or applied by the grower. Hydrogen, carbon and oxygen are the most abundant in sugarcane and are the building blocks of the cell walls and sugars. Nitrogen is most likely to be deficient in soils and is the most needed in fertiliser. Nitrogen is a building block of amino acid, proteins and other organic compounds in cane. An excess of nitrogen in many plants, including sugarcane, causes a high shoot to root ratio, low sugar content and retardation of flowering. Potassium, like nitrogen, is taken up in excess of need, and its abundance makes it a major contributor to the osmotic potential of cells. Phosphorus, like nitrogen and potassium, is frequently a limiting factor in growth and production. It is a key element of photosynthesis, respiration and other physiological processes and is present in deoxyribose nucleic acid and ribose nucleic acid. Sulphur is seldom added as a fertiliser because of its ubiquitous distribution. It is a part of the amino acids methionine and cysteine, and is in some vitamins and co-enzymes. Calcium, if applied, is usually a pre-plant application as lime (calcium oxide and calcium carbonate) in order to raise the pH of acid soils. While a great deal of Ca is needed to raise the pH, very little is needed for plant nutrition. Calcium is useful in activating certain enzymes when bound to some proteins. Magnesium, too, is seldom absent in soils but is sometimes applied with Ca as dolomitic limestone. Magnesium sulphate can be applied as a foliar spray for chlorosis. This element has a central position in the chlorophyll molecule, as iron does with haemoglobin. Magnesium is important in energy transport and as an enzyme activator. Chlorine is usually applied with potassium as KCl, and like nitrogen and phosphorus, the plants absorb far more potassium than is required. Chlorine is essential in the photosynthetic process and for cell division. Iron is important in electron transport, as are the minor elements zinc, copper and manganese. In spite of a great deal of research, the roles of boron, molybdenum and nickel are not clear, and this is especially true concerning sugarcane.

CROP CONTROL

The enhancement of natural ripening by the application of chemicals is very successful under certain management and environmental combinations. While a number of compounds have been used on an experimental basis, only a few have been successfully used on a commercial basis. Embark (mefluidide), Ethrel (ethephon), Fusillade Super (fluazifop-*p*-butyl) and Polado (sodium sesqui salt of glyphosate) are the most familiar and Roundup (glyphosate-isopropylammonium) has by far the largest market share. Virtually all cane harvested in Hawaii is treated with glyphosate, and it is also used extensively in Louisiana and Florida. Florida growers find that the recommended 630 g/ha application rate causes unacceptable foliar damage and prefer 351 to 430 g/ha applied 35 to 40 days before harvest. Some growers apply 245 to 279 g/ha hoping for a longer delay of up to 50 days before the intended harvest. The application of ripeners is most efficient when done by airplane, and care must be taken to avoid hypersensitive targets. Ripeners do not affect all varieties the same, so local tests are necessary before making commercial trials. Even after proving experimental efficacy, the product may fail if the crop is not harvested in the window of maximum advantage. This window will vary with locations and years, but may be determined by local experience. The maximum advantage occurs between the point where treated cane becomes higher in pol % cane than untreated cane and the point where the untreated cane catches up with the treated cane through natural ripening. Harvesting before or after these point is futile and wastes money and time. If the local system cannot adjust to this window, then ripening must be left to the breeders and nature.

Fields with 75–80% of the stalks flowering are reported[3] to suffer a 20% yield loss. Flowering can be controlled by withholding irrigation water for a month before the induction period. Flowering can also be controlled on a commercial basis[3,16] usually by the application of ethrel (ethephon or 2-chloroethanephosphonic acid) at rates from 0.56 to 0.84 kg/ha a.i. before floral induction takes place. Higher rates may cause shortened internodes and lateral bud germination. The chemical is effective if applied after the crop shows four to five internodes above ground and before mid-September in the northern hemisphere or mid-February in the southern hemisphere. Induction varies with varieties and with latitude [16], so local tests are required to determine the ideal application time. Because commercial sugarcane is a short-day plant, cane in near-equatorial areas may be under constant induction photoperiod, so the prevention of flowering may be impractical and application may only result in delay of flowering. In near-temperate areas or areas where nights are cool (18°C for five days at 30° N or 30° S of the equator) at induction time, flower induction may fail. If the four to six weeks before induction are dry, flowering may also fail. Conversely, if that period is wet and cloudy and temperatures are not limiting, flowering induction will be more probable and chemical control would be indicated. While flowering control by ethrel is cheap, its practicality depends on the ability to predict the probability of induction before application.

HARVEST

Preparation for harvest should include planning for the sequence of fields to be harvested. Harvesting by age is a common practice with older ratoon fields harvested first because they may be higher in sucrose content, and they may be indicated for replanting, thus giving more time for reformation. Plant cane is frequently left for later to allow more time for growth and sugar storage. Varieties that ripen early are harvested at the beginning of the harvest period and those that mature later are harvested later. Fields that are farthest from the mill may be harvested early and those closest may be harvested last. If the lower sucrose fields left for last are closer to the mill the cost of transport per ton of sugar recovered is less.

Other preparations for harvest include the removal of impediments, and the focus is primarily on conditions that impede the functions of machinery. Thus, all burms, irrigation pipes, temporary drains, etc. that would stop or slow traffic should be removed. Fields should be dried off so as to provided support for infield traffic, as well as slowing growth to improve sugar storage in the stalk. Field roads should be graded and mowed and culverts and bridges conditioned to support heavy traffic.

Harvesting green or burned sugarcane is sometimes a management decision, but is sometimes beyond management's control. If harvest is scheduled following wet weather, burning may be impractical and green harvest may be required. Where burning regulations are in force, burning may be prohibited near communities, schools or airports. Burning may also be impractical if smallholders are producing cane in small allotments or the entire field is not burned to protect quality. The quality of mature green cane that is hand cut, hand stripped and shipped immediately is a standard to which burned cane is compared. The former can be achieved only if hand labour is abundant and about two tonnes a day per cutter (tc per man-day) is an acceptable level of production. Paid by the task, most cane cutters prefer to cut standing, burned cane and five tc/man-day is an acceptable rate. Soldier-harvesters cut green cane, which is burned in the windrow (or heap-row), and combine harvesters cut green or burned cane. However, green cane harvest by combines requires more fuel and time and causes more extraneous matter (EM) to be shipped, so burned cane harvest is much preferred. Compared to hand harvesting, harvesting burned cane by machine requires higher capital costs for machinery, and results in a faster rate of deterioration than green cane, so rapid transport and short storage time in the mill yard are essential.

Transport of cane from field to factory varies over the world. Camels in India can carry a 225 kg load over the weigh scales while a truck with two trailers in Brazil may deliver 60 tonnes. Obviously, the transport used depends on the local society and the laws regarding weight limits. For management in increasingly labour-scarce areas, loading cane directly from combine harvesters or with grab-loaders from the windrows is by far more efficient than having workers carry less than a camel load up a ladder to be placed in a small truck for transport. The balancing of the cost of labour against the capital and operating costs of the machines is tilting more and more in favour of the machines. Further, trucks are designed to run on roads, not in fields, and using them for what they are not designed to do greatly increases maintenance costs. In addition, the low clearance of trucks causes damage to the potential ratoon crop when their operators run with their wheels in the furrows and again when they cross over rows to take a cargo from the field (i.e. stool trampling). This work is much better done with cane wagons with sufficient clearance that are drawn by a tractor and an operator who is familiar with farming. Transloading cane from wagons to the trucks is done at the roadside and preserves both trucks and fields. With whole-stalk cane, the above operations provide surge capacity with storage in the windrows, then at the transloading site and again at the mill yard. Surge capacity is good to have if stoppages occur in harvest, transport or milling. Many areas have a target of 24 hours between harvest (or burn) and crush because that is the average time for dextran to be detectable in burned cane that has been combine-harvested. Longer periods between cut and crush mean that dextran formation accelerates and sugar losses increase. More and more mills and refineries are imposing penalties on dextran content in the juice of delivered cane. The long lines of small trucks waiting to be unloaded at many mills is an economical anathema.

FIELD FACTORS AND CANE QUALITY

Factors affecting cane quality are many. Besides the effects of disease, insects, varieties and weeds, all discussed in other chapters, other factors in the field affect the quality of cane as it reaches the mill.

Management is partly responsible and causes the most frequent losses both in tropical and temperate areas. Foremost of management's errors is the application of fertiliser, either too much or too late. The main fertiliser element that affects cane quality is nitrogen, and when the nitrogen fertiliser is cheap, the farm manager frequently oversupplies his crop with nitrogen as insurance for a good crop. If cane is purchased on the basis of tonnage, then perhaps the practice isn't too costly for the grower, but if sugar content is a factor in payment, both excess nitrogen and late application are costly to both the farm and the mill. Given water and warm temperatures, the plant will absorb nitrogen promiscuously, the growth of cane will continue uninterrupted and little sucrose will be stored. The result is that more cane must be harvested, transported and milled for less sugar. Sugarcane will absorb potassium in abundance as well, and will contain far more than is required by the plant. When potassium-rich sugarcane is processed, the potassium ion is not precipitated in clarification but continues through the boiling house and becomes concentrated in the molasses. There potassium impedes crystallisation of sucrose, raises the molasses purity and causes losses in the amount of sugar made. Most of the other nutritional elements are not absorbed in excess or do not occur in large quantities in the soil.

Extraneous matter (EM) is another field factor that affects quality. Farm managers may not understand its importance, but millers certainly do. That is why the EM content is penalised even when cane is paid for by the tonne. Consisting mostly of young stem tissue and both young and old leaves, EM can be as much as 25–35% of the cane delivered. At those levels, while harvesting efficiency may increase (less energy spent to remove the EM), transport and milling efficiencies decrease. Every tonne of EM transported and milled replaces a tonne of cane that should have been present. This affects quality both directly and indirectly. Since the cane stalk is rich in sucrose and the EM is very poor in sucrose, the EM content dilutes the sucrose content per tonne of cane, and both miller and grower suffer directly. Indirectly, the EM decreases sucrose in the milling process. With clean stalks, milling produces juice and bagasse.

The bagasse has a moisture content of around 40–50% as a mixture of juice and imbibition water. With stalks rich and EM poor in sucrose, the milling process produces juice that is diluted with leaf water and juice from the younger stem sections, whose sugar is mostly invert. To this is added starch from green leaves and sheaths which becomes pasted in clarification and inhibits crystallisation. Finally, the EM that entered the mill with little or no sucrose, leaves as bagasse enriched with sucrose-rich stalk juice.

Field soil seldom accompanies hand-cut cane to the mill, but it is a common contaminant in cane that is machine-cut. With burned cane, soil frequently adheres to the sticky exudates that coats the burned stalk, as soil is gathered as the grab-loader rolls the cane along the ground or as the base cutter of the combine harvester blasts the chopped stalks with soil as the machine progresses along the row. Soil is also trapped behind leaf sheaths in both combine and soldier-type harvesters, and both harvesters types can gather the stools of lodged cane, roots and attached soil, and send these to the mill with the cane. Most dramatic are the amounts of soil that can be delivered with cane that has lodged in wet weather and then harvested with soldier-harvesters and loaded by grab loaders. Irrespective of the method of delivery, soil containing sand will wear mill rolls rapidly, suspended clay will give high brix values and lower purity, and will also add objectionable colour to the sugar. Because of the increased expense, lower yield and colour penalties, growers sending cane with high EM (cane tops, leaves, dead stalks, roots and soil) to the mill are hurting the mill and themselves if paid for sugar.

Burning cane before harvest is the most efficient way to reduce EM levels (including weeds) but, as indicated above, it may increase the soil content. Burning is done in various ways. Burning standing cane is the most widespread practice. Hand- or soldier-harvested cane is often burned as it lies across the cut field in windrows after harvest, and after hauling out the stalks, the remaining green trash is burned. Burning standing cane is usually practised when rows are 1.5 m or less apart and biomass yields are high. These burns, if properly done, pro-

duce hot fires, rapid combustion and tall plumes of smoke and steam. Although spectacular, these plumes are the most innocuous of cane fires and the primary pollutants are the carbonised leaf particles that rise with the plume, float for miles and sometimes fall in inappropriate places. As a result, these harmless but unsightly sooty particles are the major cause of complaints from communities in sugarcane areas. The fires from cane burned in the heap-rows are much less spectacular than those in standing cane, and the amount of trash reduction is less. These burns are practised where row spacing is wide and yields are too low to produce clean burns in standing cane. Burning green trash after hand-harvest or combine-harvest of green cane produces little fire but lots of heavy, unhealthy smoke, and has been banned in some areas.

Cane quality deteriorates faster in cane burned while standing for two reasons:

- the fire is hotter and burned stalks ooze juice, and
- the cane is usually cut by combine harvesters which cut the stalk into small billets.

Research in many countries has shown that bacterial dextran forms rapidly in burned and chopped cane, and processing problems follow after a single day's delay between harvesting and crushing. Green, whole-stalk cane in the windrow maintains quality from 4–8 days after cutting until it is burned. This technique gives the harvesting, transport and mill storage system some surge flexibility. While unburned harvested cane will slowly dehydrate and increase invert sugars, dextran formation usually follows burning. Chopping burned cane with combines, or crushing whole-stalk cane with chain slings during transport and storage, causes wounds where the dextran bacteria can enter and multiply. These bacteria (*Leuconostoc mesenteroides*) are ubiquitous in sugar mills, ditches, settling ponds and cane fields, and can be expected to enter any wounded cane and rapidly attack the sugar in moribund cells.

Weather has obvious effects on sugarcane quality. Most important is the combination of temperature, moisture and sunlight that drives the photosynthetic cycle. When these three factors are optimum for photosynthesis, the crop grows rapidly. If either temperature or moisture becomes limiting, growth stops and sugar is stored in the internodes, improving cane quality. If temperatures fall below about 10°C, photosynthesis slows to a stop and cane quality becomes static. Temperatures of 0 to –2°C may cause browning of leaves, and –4°C may kill terminal and lateral buds, and a few young internodes. Below this temperature more mature internodes are affected and at –11°C freeze cracks may be seen before the stalks thaw, and the stalk is usually frozen to ground level (these temperature and damage relationships will vary with variety and canopy cover). The effect of freezes on quality are well documented, with the most important effect being the production of dextran in the damaged internodes. The bacteria (*Leuconostoc mesenteroides*) enter the frozen buds, growth cracks and freeze cracks. They then multiply in the moribund and dead tissue, producing dextran in the course of their metabolism. Dextran is carried through the processing system at the mill and, like starch, causes a reduction in crystallisation of sucrose. Completely frozen cane may be unfit for sucrose crystallisation three weeks after a hard freeze and thereafter is used only for molasses. Less damaged fields may give reduced sucrose yields only several months or so after a freeze occurs. It is imperative that management surveys the damage to all affected fields (those with damaged canopies) to determine the effect of the freeze on stalk tissue. Fields with completely frozen stalks must be harvested and milled first, those with partially frozen stalks come next and those with no damage to stalks or leaves are left for last. With this or a similar strategy based on damage assessment, serious economic loss can be minimised. Should all fields be severely damaged, decisions become more difficult. A general rule is that cane with the highest sucrose content will produce crystal sucrose longer because the purity has further to fall than in low sucrose varieties. After the freeze-damaged fields are harvested, they can be cleaned and prepared for ratooning as is normally done since, if there is sufficient soil cover, the stools will not have been severely damaged.

RATOONING

Cultural practices for ratoon crops are very similar to those for the plant cane crop. Immediately after harvest, weather and harvest activities permitting, the inter-rows or middles should be re-defined with small subsoilers or small ploughs followed by discs or Lilliston gangs to destroy billets and weeds and to bury trash. This practice also cleans the middles, facilitating the flow of furrow irrigation water or excess rain. If seedcane and hand labour are available, replanting gaps should be done at this time. Gap planting is practical only if the gaps are one metre or more in length. Smaller gaps will probably be closed by tillering[3,13] from plants adjacent to the gap. Extensive gaps suggest that the field be reviewed with ploughing out and replanting (reformation) in mind.

Once the ratoon crop has germinated and any gap planting has been done, fertiliser should be knifed into the side of the drill without, of course, damaging the stools. In many areas, application of nitrogen is sufficient, although the amount applied may be twice that for plant cane. Phosphorus application as a side dressing is futile, since phosphorus is bound to the soil at the surface and little reaches the roots. Phosphorus applied before planting should therefore be sufficient for several ratoon crops. If tests indicate that additional phosphorus may be needed for older ratoons, it can be knifed into the inter-rows. Potassium application may also be included if tests suggest it would improve yields of older ratoons. Herbicide practices, irrigation (if any) and cultivation should follow the general practices for plant cane.

The yields of ratoons crops show a general decline year after year until the manager decides that it is time to replant. If plant crops are 18 months old before harvest, then the 12-month ratoon crop that follows will show a dramatic difference in yield. If the plant crop is only 12 months old, the following ratoon crop may give the same yield or even more, but successive ratoons will be weaker and weaker. Poor management will hasten the decline. Most varieties show this tendency, and one of the reasons is that breeders have not been very successful in selecting for ratooning ability. Since replanting is the most expensive phase of sugarcane culture, good ratooning varieties are especially valuable. A few have been identified and the trait has been exploited. One of the factors associated with these successful varieties is tolerance to diseases, especially RSD. For those varieties that are intolerant, heat-treatment or meristem culture is essential for ratoon longevity.

When to terminate a field's production is a difficult decision. Plantations in Hawaii grew plant cane only, but as a 2-year crop. Louisiana's tradition was plant cane and one, and sometimes two, ratoons, while Florida's was similar but with a higher percentage of second ratoon fields. Texas averaged five ratoons, and the average has improved with heat treatment. Tropical areas with benign climates and good weed control do even better.

The best decision making is done with long experience with the crop and a knowledge of variety performance under local conditions. Since this experience and knowledge are not always available, they can be supplemented by computer programs that integrate historical yields of fields, varieties, weather and economics in assisting management to reach a plough-out decision. Better growers have higher standards for their fields and many will not allow a field to continue if its last harvest declined to the regional average and the regrowth after the last harvest was not showing promise of improvement. The manager should review the past yields and the predicted yield (based on inspection) and if the field under question would not allow the attainment of the coming year's goals, it should be ploughed out and replanted.

Replanting can be carried out either by planting seedcane between existing cane rows, or by entirely removing the old crop and reforming the land and rows. Inter-row planting is not widely practised but has been used primarily in South Africa and Brazil where slopes and soil are such that total reformation results in unacceptable erosion. In these cases the old contour-planted crop is sprayed with systemic herbicide and left standing while the inter-row area is furrowed, planted with seedcane and covered. After the new rows have germinated, the old rows which formed a barrier to erosion, are removed, and the new crop is fertilised and cultivated.

Total reformation involves the removal of the stubble of the old crop following harvest and involves deep ploughing to lift and turn the stubble pieces so that they dry. In addition to facilitating the re-establishment of rows, this practice decreases disease incidence, borer infestation and variety mixtures. A special rotating tiller was developed by Copersucar in Brazil which hastened the death of stubble pieces by knocking off soil and tossing the pieces high in the air so as to fall on the soil surface where they dry rapidly. After the previous crop is destroyed, replanting may be immediate if seedcane and labour are available, or may be left for a fallow period for special weed or insect control, salt leaching, a cover crop or an alternate cash crop. Whatever use is made of the intercrop period, the new rows should be formed as soon as possible to avoid working wet soil at the last minute.

MONOCULTURE

Sugarcane tends towards monoculture since the building of a mill and the need for efficient transport from field to factory stimulates the concentration of continuous culture in the immediate vicinity. In areas very favourable to sugarcane culture, it may be grown almost to the exclusion of other crops. While agronomists may be taught that monoculture is deleterious to production, history offers examples to the contrary. When the Dutch occupied north-eastern Brazil, they recorded a field of Creole that was 50 years old before being replanted. In Louisiana, sugarcane has been grown continuously as a monoculture for over 200 years, with only fallow summers between ploughing out ratoon crops and replanting more cane. In most areas ratoon crops do get weaker with succeeding years and replanting is common. Only in some circumstances is a year or more allowed between the last ratoon crop and replanting. Weed control may be a factor, and 1 or 2 years of cotton between cane crops allows the use of herbicides that could not be used in cane. Insect control by fallow flooding or irrigated rice production may control weeds as well as soil insects. Perhaps the most intensive type of monoculture is the use of successive plantings of the same variety cycle after cycle. This practice may lead to the build-up of diseases that could be avoided simply by alternating varieties.

SPECIALTY CROPS

Organic sugarcane

While there is little if any market for organically grown sugarcane, there is a small but profitable market for organically produced sugar. To be sold as organic sugar, the product must be produced under strict control both in the field and the factory. Sugarcane must be grown on fields that have not received inorganic fertilisers for at least 3 years. The only fertiliser that can be applied must satisfy the definitions and rules of the market. Nitrogen may come only from guano, manure composts or nitrogen-fixing organisms. Mineral elements must be mined from natural deposits, e.g. phosphate as rock phosphate, and potassium sulphate from the Great Salt Lake or the Dead Sea. Minor elements may be used as sprays to correct deficiencies determined by tests. Weed control by application of manufactured chemicals is forbidden. Insect control must be by natural predators and parasites or by rotenone (if used with caution to avoid non-target organisms), natural pyrethrum (not containing piperonyl butoxide) or *Bacillus thuringiensis* preparations unaltered by genetic engineering. Chemical ripeners and floral control chemicals are not permitted. No transgenic varieties may be grown. Cane may be harvested by machine, but must not be burned. Mechanical processing for the extraction and purification of sugar is permitted as long as no chemicals are added. The most efficient producers of organic sugar are in Florida, where the histosol is rich in most fertiliser elements and the factory processing relies on micron filtration. The individual fertiliser compounds and other items permitted or prohibited are regulated by the Organic Crop Improvement Association (http://www.ocia.org).

Transgenic sugarcane

Genes taken from other organisms, even microbes and animals, have been engineered into sugarcane, and the plants wait for government approval and public acceptance before commercialisation. Foreign genes have been introduced into wheat, cotton, soybeans, maize and potatoes, and these transgenic crops are grown commercially. However there is still reluctance to accept them in Europe, and also parts of the western hemisphere but on a lesser scale. The first genes (herbicide and virus resistance) of commercial value in sugarcane were genes important to the farmer. Crosses have been made using parents with these genes and field tests show that the genes are expressed in a more or less simple Mendelian fashion. Genes for resistance to disease, insects, cold tolerance, and drought are available, and their introduction into sugarcane is in progress. Breeders using these and other transgenes will make formidable progress in breeding sugarcane in the near future. A far greater understanding of the genetics and improvement of sugarcane has evolved from molecular genetics.

Later genes introduced into cane were those that express proteins that would add significant value to the crop independent of the sugar content. The products of these genes are extracted by a micron filtration process with minimum heat and chemical additives[17]. The potential value of the bio-insecticides, food additives, pharmaceuticals and other high value proteins is high enough to think of sugarcane as a pharmaceutical crop, as it was centuries ago when sugar was sold in European pharmacies. The great advantages that sugarcane has over other crops for the production of specialty compounds are the enormous amount of biomass produced and the existing infrastructure and knowledge needed to handle and process an industrial crop.

INTER-CROPPING

Inter-cropping may refer to either growing alternative crops between crops of sugarcane, or between the rows of existing sugarcane fields.

Production of alternative crops between crops of cane is limited to the fallow period between plough-out of old and less productive ratoons and the reformation of fields for planting a new cane cycle. Where sugarcane is planted immediately following harvest and plough-out and the field is replanted to sugarcane (successive or succession planting) then there is insufficient time for an alternative crop. Where the interval between cane cycles is longer, short-term annual crops are indicated. These may be soybeans, rice, cotton, maize, etc., and each has its advantages besides that of being a cash crop. Rice that is flooded has the advantage of a weed control programme that reduces the population of gramineous weeds that are difficult to control with sugarcane weed control programmes. Maize and sorghum are alternative crops that benefit from weed control provided following sugarcane, and require less irrigation water than does the latter. Neither cotton nor maize and sorghum require the extensive land reformation needed for flooded rice. Planting maize and sorghum may, however, increase the incidence of mosaic and the frequency of stem borers. There may be times when the land is better left fallow than planted to an alternative crop. An example would be if the alternative crop were irrigated and the irrigation water would increase the salt content of the soil, while a field left fallow might improve with an occasional shower.

Planting alternative crops between the rows of sugarcane (usually plant cane) is widely practised where weeds are controlled and a quick cash crop is required. Tests have frequently shown that planting between rows reduces the yield of the sugarcane crop. However, economic circumstances may favour an early income over a small sacrifice later, especially when the plant cane crop endures for 18 months. Examples of inter-row crops that have been grown with sugarcane are black beans in Colombia, cucumbers and tomatillos in Mexico, sugar beets in the Northwestern Frontier Province (Pakistan), Irish potatoes in Louisiana and radishes in Java.

BASIC ECONOMICS

For centuries, the value of the sugarcane crop was measured in terms of the yield of cane per unit of land area. If the cane was used on site for forage, syrup or loaf sugar, the main operating expense was labour and the capital expense was a simple mill and animal power. With the centralisation of mills, the sale of cane came to be based on the stalks delivered free of leafy trash and soil. In more recent times, the value of the crop was determined by the yield of sugar from the cane. While all of these scenarios still exist in various parts of the world, the basic valuation of the crop is based on money received less operating costs. With that concept, the manager can calculate the value of each thing he does. The cost of each input (land preparation, seedcane, tilling, water, fertiliser, herbicide, ripener, etc.) is subtracted from the money received and the balance is operational (not capital) gain or loss.

Thus it becomes obvious that poor land preparation will detract from the yield of the first harvest and all ratoons, no matter how good the quality of each subsequent input. For this reason, more is spent on land preparation and planting than all other activities. When the most valuable product of the crop is sugar, as it usually is, then any practice that affects the yield or quality of sugar must be evaluated in terms of its economic return. Variety selection is an important management factor, as its sugar per tonne yield is most important for the mill, and sugar yield per unit area is most important for the field. However, grower and miller must recognise that the yield of money per tonne and per unit of land are the really important goals. So any field factor that increases return and reduces cost of production, harvest, transport, and milling will increase operational profit. Any operational decision or effort that limits yield or quality will increase losses. Over-application of nitrogen reduces sugar and may increase tonnage which increases harvest, transport and milling costs per tonne of sugar produced. Application of fertiliser when water is inadequate is a waste of money because the fertiliser will not be effective in a severe drought. A high tonnage variety with relatively low sugar yield should be planted near the mill to minimise transport costs of the sugar produced.

REFERENCES

1 Benath, L.L. & Monteith, M.H. (1966) Soil oxygen deficiency and sugarcane root growth. *Plant and Soil*, **25**, 143–149.
2 Lingle, S.E., Wiedenfield, R.P. & Irvine, J.E. (2000) Sugarcane response to saline irrigation water. *Journal of Plant*

Nutrition, **23**(4), 469–486.

3 Bakker, H. (1999) *Sugar Cane Culture and Management.* Kluwer Academic/Plenum, New York.

4 Barnes, A.C. (1974) *The Sugar Cane.* Wiley, New York.

5 Humbert, R.P. (1968) *The Growing of Sugar Cane*, 2nd edn. Elsevier, Amsterdam.

6 Hunsigi, G. (1993) *Production of Sugarcane.* Springer-Verlag, Berlin.

7 King, N.J., Mungomery, R.W. & Hughes, C.G. (1953) *Manual of Cane-growing.* Angus & Robertson, Sydney.

8 Irvine, J.E. & Benda, G.T.A. (1980) Sugarcane spacing. I: Historical and theoretical aspects. *Proceedings of the International Society of Sugar Cane Technologists*, **16**, 350–356.

9 Irvine, J.E., Richard, C.A., Garrison, D.D. *et al.* (1980) Sugarcane spacing. III: Development of production techniques for narrow rows. *Proceedings of the International Society of Sugar Cane Technologists*, **16**, 368–376.

10 Ingram, K.T. & Hilton, H.W. (1986) Nitrogen–potassium fertilization and soil moisture effects on growth and development of drip irrigated sugarcane. *Crop Science*, **26**, 1034–1039.

11 Wiedenfield, R.P. (1995) Effects of irrigation and N fertiliser application on sugarcane yield and quality. *Field Crops Research*, **36**, 101–108.

12 Yates, R.A. & Taylor, R.D. (1988) Water-use efficiencies in relation to sugar cane yields. *Sugarcane*, **1**, 6–10.

13 van Dillewijn, C. (1952) *Botany of Sugarcane.* Chronica Botanica, Waltham, MA.

14 Salisbury, F.B. & Ross, C.W. (1991) *Plant Physiology*, 4th edn. Wadsworth, Belmont, CA.

15 Anderson, D.L., Snyder, G.H. & Martin, F.G. (1991) Multiyear response of sugarcane to calcium silicate slag on Everglades histosols. *Agronomy Journal*, **83**, 870–874.

16 Moore, P.H. (1987) Physiology and control of flowering. In: *Copersucar International Breeding Workshop.* Cooperativa de Productores de Cana, Açucar, Álcool do Estado de São Paulo, Brasil.

17 Mirkov, T.E., Barrilleaux, A., Paterson, A.H., Yang, M. & Irvine, J.E. (2001) Processing of transgenic sugarcane for the recovery of high value proteins. *US patent application serial number 60/196,085.*

Chapter 7
Harvest Management

David Weekes

Good harvest management is crucial to the profitability of both the cane grower and the miller. The grower invests significant time and money to produce his crop but poor harvesting and transport operations can result in dramatic losses of recoverable sugar both from physical losses of cane infield and deterioration in cane quality before milling. Ongoing ratoon yields can also be depressed by poor harvesting practices. The harvesting and transport costs form a large proportion (normally 25–35%) of the overall cost of cane production and must be minimised. Very careful consideration must therefore be given to both the selection and the management of the harvesting system.

PRE-HARVEST BURNING

As the cane grows, the older leaves die and dry off. They may remain attached to the cane stem or gradually fall away. In either case the dead vegetation is generally known as trash, and varieties that readily shed dead leaves are described as self-trashing or self-shucking. The mass of leaf material remaining at harvest varies between varieties, and both the degree of self trashing and the leaf/stalk mass ratio can have significant effects on the cost of harvesting and follow-up work.

Burning the cane before harvesting removes most of the dead vegetation without causing significant damage to the interior of the cane stalk. In very intense fires there may be some charring or cracking of the outer rind. However, the cane stalks are killed during the burn and quality deterioration starts as micro-organisms (e.g. *Leuconostoc mesenteroides*) begin to convert sucrose into non-recoverable sugars (dextrans). Many of these

organisms enter the stalk through surface cracks generated by the burning process.

Burning for manual cutting has to be carried out sufficiently far ahead to allow the ash to cool, and, since manual cutting usually begins early in the morning, burning is normally carried out during the preceding day or evening. Depending on the timing of the burn, this adds at least 12–18 h to the time between cane death and processing (the 'kill to mill time' or KTM) when compared with unburned (green) cane. In dry conditions, burning is best carried out in the early evening when wind speeds are usually lowest and the risk of fires spreading out of control are minimised. Evening burning also reduces the time between burning and cutting. However, in wet weather, burning may have to be earlier in the day when the cane is driest, if a satisfactory burn is to be achieved.

Problems frequently exist with matching the quantity of cane burnt to the factory capacity and availability of cutters. Excessive burning, factory breakdowns or poor cutter availability lead to cane remaining uncut for additional time, sometimes several days, which increases the losses of recoverable sucrose. As a consequence, some sugar millers (e.g. Fiji) apply a price penalty to deter burning.

The productivity of manual cutters or mechanical harvesters in burnt cane may be more than double that in green cane, since the dead leaves do not have to be separated from the stalk as part of the cutting process. In very heavy crops, mechanical harvesters may be unable to economically handle the mass of vegetative matter in green cane. Improved labour productivity was the main reason for the adoption of burning in many countries as labour availability declined and costs increased in the 1950s and 1960s. Manual cutters or harvester operators have a clearer view of the base of burnt cane, and the length of stump left uncut is often less in burnt than in green cane. Long stumps represent a significant loss, especially since sucrose levels are higher in the base of the stalk than in the upper parts. Dropped and missed stalks are more easily detected in burnt cane, and the wastage rate is normally lower than in green-cut fields where significant quantities of cane may be concealed by the trash.

Long stumps, green tops, trash, and residual cane can harbour pests such as rats, borers and diseases, which may then survive to the following crop. In some countries burning is also required to remove dangerous animals (e.g. snakes and insects) or irritant weeds from the crop before manual cutters entering the fields. Weil's disease, transmitted by rat urine, can also be a danger to the labour force when unburned cane is cut manually.

In areas that still have abundant cheap labour, green cane can be cut manually and cleaned to standards that equal burnt cane. In most areas, however, green-cut cane has higher levels of extraneous matter (dirt, leaves and other material which does not contain sucrose) than cane that has been burned before harvesting.

Manually cut cane usually has extraneous matter (EM) levels varying from 3% to 5% in burnt cane and 7% to 10% in green cane. Mechanical chopper harvesters typically give EM levels varying from 5% to 7% in burnt cane and 8% to 12% or higher in green cane. Chopper harvesters can produce cleaner cane samples from green cane, but usually at the expense of greater infield cane losses through the extractor fans. As a result of the lower EM levels, the transport payloads for burnt cane are normally better than for green cane. In the absence of large quantities of trash, the loading process is generally easier and less cane is left behind in the field. In areas where furrow irrigation is practised, burning may be necessary to avoid infield water flows being impeded by cane trash post-harvest.

GREEN CANE HARVESTING

If the cane is cut green, then varieties that are both self trashing and have a low percentage of vegetative material in relation to their stalk mass are desirable for the harvesting operation. Many cane-breeding stations now take these aspects into consideration when selecting new varieties. When cut green, the stalks are not immediately killed, and sucrose is initially lost when the bud germination process is initiated. Sucrose reducing micro-organisms also enter the cane through the cut ends or through split and crushed areas caused by the harvesting process. The cutting to processing time in green cane, for which the expression 'kill to mill time' is also used, must therefore also be minimised in the same way as for burnt cane. Green cane harvesting avoids the time lost between burning and cutting, but the subsequent rate of deterioration may be similar to that of burnt cane.

A commonly used empirical figure is a loss of one percentage point of recoverable sucrose per day in wholestalk cane. The rate of deterioration is, however, strongly related to ambient temperatures, humidity, and the number of available entry points for micro-organisms into the cane stalks. Trials under the prevailing ambient conditions are normally the only way to provide a guide to the actual rate at which recoverable sucrose is being lost.

Until recently, the mechanised harvesting of green cane with chopper harvesters was not economically feasible, as they were unable to give adequate trash separation. Wholestalk harvesters often could only work with green cane, which then had to be burnt after cutting for trash removal. With additional engine power and extractor fan capacity, chopper harvesters are now able to give reasonable trash and top separation in green cane, and mechanised green harvesting has been adopted extensively in Australia and Brazil.

Empirical rules common within the sugar industry are that the KTM time for wholestalk cane should not exceed 72 h, and for chopped cane KTM should not exceed 24 h. These are not periods in which there is no loss of recoverable sucrose, but only the period during which the deterioration may be economically tolerable. In the conditions of high temperatures and humidity commonly found in cane growing areas that are close to sea level, the tolerable delay may be shorter, while in high altitude areas the lower ambient temperatures will reduce the rate of deterioration. As a general principle the KTM time must always be minimised, and cutting cane green can be an important part of this strategy.

The residual trash and cane tops left after green cutting are usually spread evenly over the field to produce a 'trash blanket'. This requires additional work following manual cutting, but take place automatically as a part of chopper harvesting. The trash cover can provide significant benefits of soil moisture conservation in areas with low rainfall since the trash shades the soil surface and limits evaporation. In areas with heavy rainfall or cool weather, however, a thick trash blanket can adversely affect ratoon regrowth. A trash blanket also protects the soil from erosion during the early stages of the ratoon regrowth, and this is a strong reason for green cane harvesting on land susceptible to erosion. The sugar industry in Barbados converted to burning during the 1960s, and suffered heavy yield declines as a result of increased moisture losses and soil erosion. A successful return to green cutting was finally achieved.

An even trash blanket provides effective weed suppression without impeding cane regrowth in the early stages of the ratoon cycle. This gives reduced requirements for herbicides and mechanical or manual weeding. The resultant cost savings have been one of the principal reasons for the increased popularity of green cane harvesting in Australia.

Trash ploughed into the soil during the replanting process can benefit the soil organic matter content. It can, however, impede primary cultivation and inter-row tillage in ratoons, although this has largely been overcome by the development of suitable implements.

Public and political pressure against the atmospheric pollution resulting from cane burning is becoming stronger in many countries, even where the sugar industry is a major employer. Legislation against deliberate cane burning is now common (e.g. Colombia). Cane burns in areas comprising many small farms can be dangerous, as it is not normally possible to establish a satisfactory network of firebreaks. Cane in such areas (e.g. the Mumias outgrower area, Kenya) must therefore be cut green. Fires in trash blankets, however, cause severe damage to cane stools and ratoon regrowth, and in areas where there is a high risk of malicious or accidental fires, trash conservation may be unacceptable.

Work has now begun into the economics of 'whole crop harvesting' where part or all of the trash is recovered for use in the generation of nonfossil fuel based electricity for export by the factory. This may be achieved by cutting and transporting the combined mass of cane and trash to the factory where it is sorted by a large fixed separation plant which incurs lower cane losses than in-harvester separation. Alternatively the trash can be picked up from the field post-harvest using conventional balers or silage machinery. Trials have indicated that the benefits of green harvesting can be achieved with only part of the trash left infield.

TO BURN OR NOT TO BURN

A decision on whether cane should be burnt or cut green therefore depends on many factors related to variety, climate, soil type, slope, irrigation systems, environmental pressures, pests and diseases, labour availability, delivery delays, and the economics of weed control. In general, cane grown under rain fed conditions on lighter textured or erodible soils should benefit from green cutting, while cane grown on heavy soils with little gradient, especially on furrow irrigated land or in areas of high rainfall and low ambient temperatures may be better burnt. There are, however, numerous exceptions to these examples, and trials under the prevailing local conditions would be essential to determine the optimum system.

Usually, manual cutters prefer cutting burnt cane owing to the greater productivity achievable, despite the increased levels of dirt and dust. However, once burning is established as common practice, it is generally difficult to revert to green cutting without large incentives being paid. The cane miller may hold differing views from the grower on the merits of green and burnt cane. Other than in areas where cutting standards are very tightly controlled, the superior freshness of green cane may be offset by a much higher percentage of EM that adversely affects sucrose extraction. Unless the local formula for cane purchasing deters excessive

EM deliveries, the miller might benefit from the superior cleanliness of burnt cane. Nevertheless, the miller's ideal raw material would be fresh *and* clean green cane.

UNPLANNED CANE FIRES

Cane fires can occur accidentally as a result of careless cane, trash or grass burning. Adequate precautions must always be taken to ensure that such burning is only carried out where there are adequate firebreaks and low wind speeds. Sudden wind gusts can easily spread fires into adjoining cane fields. Accidental fires are particularly common in fields adjacent to villages or housing areas where garbage and garden waste is burned. If the householders are also the owners of the surrounding cane, however, unplanned fires are far less common.

Mature cane is extremely vulnerable to arson by disaffected persons. Where the cane belongs to a parastatal organisation, fires may be started for political reasons. They may also be used to apply pressure to cane growers during labour disputes or to force higher harvesting wages. Small farmers hoping for early cutting and payment for their cane may also resort to deliberate burning, and simple vandalism can be a factor in some areas.

Heavy post-harvest trash blankets are susceptible to accidental or deliberate fires, and the resultant heat damage to the cane stools and ratoon regrowth will be much worse than a fire in standing cane of harvestable age.

Arson tends to be more serious than accidental fires as the perpetrators can pick susceptible upwind fields at times when few control personnel are available, such as at night or during public holidays. The extent to which an industry suffers from unplanned cane fires varies considerably according to the pattern of cane ownership, the political climate and the nature of the labour force, but adequate precautions and fire control measures *must* always be in place. The number of deliberate burns by cash-hungry small farmers can be limited by applying price penalties to burnt cane, or by withholding payments until the time when the cane would normally be scheduled for harvesting.

FIRE CONTROL MEASURES

On large cane estates it is normal practice to establish clean firebreaks, or to build roads to a greater width than necessary to provide a measure of fire control. Clean roads and tracks around housing areas can prevent many fires, and the early cutting of fields at risk from unplanned fires is a desirable precaution. A harvesting programme that produces bands of cut fields across the direction of the prevailing wind may also be necessary in industries with pronounced fire problems.

FIRE CONTROL EQUIPMENT

Self-propelled or trailed water tankers fitted with pumps and hoses should be kept full and readily available at night. There should also be easy access to reservoirs or points at which water tankers can be rapidly refilled. Heavy earthmoving plant, especially graders and wheeled tractors with heavy harrows can be invaluable for creating barriers to major fires and should be kept available and ready at times of high fire risks. Good prior organisation of personnel, transport, mobile equipment, fire beaters and communications equipment is essential. A system of radio-equipped watchtowers may be necessary on large estates for early fire detection.

In areas of medium sized private cane farms it is usual to arrange co-operative firefighting, but good telephone and radio links are essential to achieve a rapid reaction. Cane fires in their early stages are relatively easily controlled, and small water tank and pump units which can quickly be mounted into a pick-up truck are useful in this role. Larger backup units based on tractor-drawn or four-wheel drive truck-mounted tankers are also required, but in very dry or windy conditions a large fire is extremely difficult to stop, even by wide firebreaks, owing to the amount of burning trash blowing forwards from the main fire.

Areas with many small cane farms may have difficulty in acquiring adequate firefighting equipment. Large scale fires in such zones are however less common as the cane fields are not normally contiguous, the political and industrial reasons for arson are weak and the local people are far more conscious of the risks associated with careless burning of waste. The miller can, however, distribute or encourage the making of fire beaters and demonstrate controlled back-burning techniques.

REAPING AND TRANSPORT

General

The harvesting operation comprises burning (if applicable), cutting, loading, and transporting the cane to the mill. The manner in which the cane is stored and retrieved at the mill also affects cane quality but this is not normally regarded as an agricultural issue. Storage time should, however, be included in the KTM time. Optimising the harvesting (and factory storage) system is crucial to the economic success of the sugar production process. Bad harvesting practices can waste 15% or more of the sugar that has been expensively grown, and this percentage loss can represent the difference between a profitable and a loss-making operation. Incorrect harvesting practices can also seriously affect ongoing ratoon yields through carry-over of pests and diseases, soil compaction and damage to the cane stools.

Manual and mechanised harvesting

Manual cutting and loading of green cane into animal-drawn carts remain common in parts of Asia, Africa, the Caribbean, and South America where labour is both plentiful and cheap. For longer hauls, manually loaded trailers drawn by agricultural tractors, road or rail trucks are used. In exceptionally difficult conditions, the crop can be hand carried or loaded onto pack animals for removal from within the field to the transport unit. In areas with exceptionally low labour costs, cane is still cut and loaded manually for US$1.0 per tonne or less. No mechanised operation can compete with such low wage levels, but as labour becomes more expensive and scarce, it is normal practice to mechanise the loading operation while retaining manual cutting. Mechanised loading reduces harvesting labour requirements by up to 50%; but present day ma-

chinery costs, including depreciation and interest are likely to be in the range of US$0.5–1.0/t of cane loaded, depending on productivity, fuel and operator costs.

As the local economy strengthens, growers may be unable to pay enough to attract sufficient personnel to a hard and unattractive job. This difficulty may then be overcome by either importing transient labour from lower cost areas or by mechanising part or all of the harvesting process. Increased labour costs and shortages can also prompt a change from green to burnt cane cutting but this step must be taken carefully owing to the risk of soil moisture losses or erosion. In areas with high labour costs, such as Australia and the USA, fully mechanised cane cutting and loading have become the norm. Mechanical harvester costs are strongly linked to productivity, but the lowest costs for combined cutting and loading by chopper harvester are currently approximately US$3.0/t. It must be stressed, however, that adverse operating conditions, low productivity and high foreign exchange costs frequently result in this figure being multiplied several times over. There is no more effective way of wasting money than the inappropriate use of very expensive harvesting machines.

Fauconnier & Bassereau[1] evolved mechanisation guidelines based of the ratio of the daily wage to the local value of a tonne of cane. These stated that:

- if the ratio is below 0.25, cut and load manually;
- between 0.3 and 0.45, commence mechanised loading;
- between 0.5 and 1.0, fully mechanise loading; and
- over 1.0, mechanise both cutting and loading.

These guidelines remain valid, but the decision to mechanise is also influenced by local field conditions that may not favour machine operations (e.g. small fields, steep slopes, high rainfall, stones, etc.) and can justify manual operations being maintained well beyond the usual economic limits.

There have been many examples of premature mechanisation in lesser-developed countries. Unrealistic wage levels have at times created artificial labour shortages, particularly in state-owned industries where government limits may be applied to pay increases. The true costs or difficulties of operating complex machinery are also frequently understated by machinery salesmen and underestimated by local managers. Corruption has also been a factor in some attempts at premature mechanisation. Because of differing field conditions, staff skills and motivation, equipment productivity in lesser-developed countries is often much lower than for example Australia or the USA. At the same time, serviceability rates are often low owing to shortages of maintenance skills and difficulties with the supply of spare parts. Costs of spare parts and fuel prices are usually linked to hard currency exchange rates, and can therefore escalate far more rapidly than local wages. Mechanisation must therefore be approached with caution and only after thorough trials, training, and testing.

Manual cane cutting

Cutting cane by hand is hard physical work carried out under hot and unpleasant conditions. It is not normally regarded as an ideal job if alternative work is available, and in many countries cane cutters have low social status. Cane cutter shortages often occur as the national level of industrialisation develops and easier or better-paid work becomes available.

Manual cutter productivity varies significantly according to:

- whether the cane is burnt or green;
- the trash percentage and characteristics;
- stalk length and thickness;
- whether the cane is erect or recumbent;
- the presence or absence of weeds, especially vines;
- the crop yield (t/ha);
- the suitability of the tools used;
- the physique and age of the cutters; and
- the payment system and the financial aspirations of the cutters.

Productivity can vary from around 1 t/man-day for cutting and loading green cane (e.g. Vietnam, Indonesia) to over 10 t/man-day for cutting and windrowing burnt cane (e.g. Swaziland). Before mechanisation, Australian cutters claimed productivities in excess of 15 t/man-day when cutting burnt cane for mechanical loading.

The breeding of varieties that have low vegetative matter percentages, are self-trashing and erect is a long-term process. These characteristics can, however, make both manual and mechanical cutting much easier and quicker. It is normally possible to achieve higher productivity in heavier yielding crops providing that the stalks do not become recumbent and entangled.

The choice of cane knife can also have significant effects on the productivity of manual cutters. In many tropical areas, large edged knives (e.g. the panga, cutlass, machete, etc.) have evolved locally as general-purpose agricultural tools, but are not necessarily ideal for cane cutting. The long-handled cane knife, which originated in Australia and has been further developed in Southern Africa, tends to be heavier and have the weight concentrated at the blade end. This gives more efficient cutting when in the hands of a strong man, but is not popular among weaker male and female workers. The introduction of improved cane knives into a labour force must, however, be approached with care or opposition may develop. Cane knives with a bend in the blade can reduce stooping and give reduced stump length, but may prove unpopular as they are unsuited to other work, e.g. weed slashing, firewood cutting, etc. It is customary in most industries to supply cane knives and sharpening files to the cutters, either free or at a subsidised price, but there must be acceptance of the type of knife supplied.

Cane cutters are usually paid on a piecework (task) basis, either by the weight of cane or by the length of row cut. The length of row that constitutes a set task may be altered according to the cane yield (t/ha) to standardise the tonnage cut daily. Estimating the task is a common problem for cane cutting managers, and disputes concerning either weighing accuracy or tonnage estimation are common in a disaffected labour force.

Cane loading

Mechanisation of the loading process can reduce the labour requirement by up to 50% when compared with manual loading, but mechanically loaded cane normally contains increased levels of extraneous matter. Manual loaders can avoid picking up trash, and there is no movement of the cane over the soil surface to pick up mud and dirt. With manual loading, vehicular movement infield is minimised and soil compaction and stool damage is low, especially if the cane is hand carried to the field edge. As the cane is carefully aligned in the transport units, manual loading maximises the payloads, but the loading time is prolonged. Overall productivity per transport unit is therefore usually lower than mechanically loaded transport despite the higher payloads (Fig. 7.1).

The basic options for mechanical loading are:

- Manually or mechanically forming 3–6 t stacks which are winched as a bundle on to a trailer. For smaller stacks the winch may be built-in to the trailer, a system common in Southern Africa. For larger stacks the winch is mounted on a separate tractor unit (Western Kenya).
- Grab loading from windrows or small piles into high-sided trailers.
- Using continuous loaders which pick up the cane directly from the windrow, cut it into short billets and load it into an accompanying trailer via an elevator.

The optimum mechanised loading system varies with the topography and field conditions.

Winch loading

Manual stack-forming and winch-loaded trailers are used where labour costs are still relatively low, or where the fields are too small or steep for grab loaders to economically load from the windrow. A winch built into the trailer and driven by the tractor power take-off shaft can load smaller (3–4 t) stacks. Alternatively the winch can be mounted on a separate tractor. This allows larger stacks (6–8 t) to be loaded, fewer winches are required and the trailers are cheaper to construct (Fig. 7.2).

All the winch systems are mechanically simple, have a low capital cost and are suited to countries where the spare parts have to be locally made. A separate winch tractor may achieve 30–40 t/h depending on the stack sizes.

Although the bundle/winch system has the advantage of relative mechanical simplicity and low capital cost, these benefits are offset by the higher labour requirements to form the stacks and perform the loading operation. If the bundles are delivered direct to the factory, then hooking-on, lifting and discharging the bundles is also slow (*c.* 2 min/bundle). The stack system is well suited to operations in very small fields where grab loaders and trailers would have difficulty in operating on the short rows.

Fig. 7.1 Manual loading.

Fig. 7.2 Winch loading.

Grab loading

Grab loaders are more mechanically sophisticated than a winch unit, usually having hydraulically operated booms and transmissions that require specialised maintenance and spare parts. The formation and picking-up of grab-sized bundles by push piling can result in the incorporation of large quantities of extraneous matter into the cane. Piles formed by the cane cutters are preferable if labour costs permit the slight reduction in productivity.

Grab loaders are either the slewing type in which the loading boom pivots, or the tricycle type in which the entire machine must be turned towards the transport unit. Originally slewing loaders were mounted on standard wheeled or crawler tractors, but most are now built as self-propelled units fitted with a hydrostatic transmission. This permits rapid speed changes and from forward to reverse movement. The boom can pivot either through 90° to permit loading to one side only (e.g. the Cameco SP1800; Fig. 7.3) or, if conditions require the transport to be loaded from both sides, loaders are available which can slew through 180° (e.g. the Cameco SP 3000). Given high cane yields and an unrestricted supply of transport units, a well-

Fig. 7.3 Slew loaders.

Fig. 7.4 Tricycle loaders. Source: Bell Equipment Pty Ltd, South Africa.

operated grab loader can handle 40–60 t/h, but a more normal average would be 25–35 t/h.

Tricycle type loaders (e.g. the 'Bell loader' series; Fig. 7.4) are usually lighter and cheaper than the slewing loaders. They are extremely manoeuvrable through using individual hydrostatic drives to the front wheels and a single castoring wheel at the rear. Tricycle loaders can operate efficiently in fields without pronounced ridges, and are usually more stable on cross slopes than the slewing loaders. Because of their need to turn at right angles to the trailer, however, their operation is severely limited by high ridges. There is therefore a tendency for slewing loaders to be used in areas with relatively level fields with high ridges, while tricycle loaders are used in sloping areas and where cane is flat-planted.

Productivity of a tricycle loader is normally slightly less than a slewing loader, because a loader arm can slew through 90° more quickly than a complete machine can turn. Conversely, the tricycle loaders are mechanically simpler than slew loaders, and are generally easier to keep serviceable in adverse conditions. The most obvious drawback to tricycle loaders is their need to manoeuvre across the cane rows to approach the transport unit. In theory at least, this should cause in-row soil compaction and damage to the cane stools. In

practice, the actual effects seem to be economically insignificant, other than in particularly adverse conditions.

Push piling

For a grab loader to function satisfactorily, the cane must be formed into heaps that conform with the grab capacity, usually 400–800 kg. Ideally manual cutters should form these heaps, but this may not be possible owing to cost, traditional work patterns, or the use of wholestalk harvesters. If a continuous windrow of cut cane is formed, then it must be pushed into grab-sized heaps. Slew loaders are normally fitted with push pilers to form heaps, while tricycle loaders use one side of the grab to push up the heap. Both operations can bulldoze cane along the ground, which will result in mud from wet and sticky soils adhering to the cane when delivered to the factory. The piling action may also incorporate leaf trash into the cane heap unless the trash is kept well apart from the windrow by the cutters. Loader manufacturers have made attempts to reduce this problem, either by lifting the cane away from the ground or by inducing a rolling action to the heap to reduce dirt pick-up. However, few of these modifications have been successful, and the manual forming of heaps remains preferable to push piling.

Continuous loaders

Although once popular on large estates (e.g. in the Florida sugar industry), continuous loaders (e.g. the Cameco R6) are now tending to be replaced by chopper harvesters. Continuous loaders (Fig. 7.5) are large, complex machines that pick up the cane windrow by a system of chain elevators. There is therefore less pushing of the cane across the soil surface and less EM is picked up than with a push piler. The continuous loader cuts the cane into billets approximately 300 mm long that are elevated into an accompanying trailer. Options were usually available for air blast removal of light trash during the process. Continuous loaders have very high potential throughputs – in excess of 150 t/h – providing that sufficient transport is available. Their manoeuvrability is poor and they must be used in

Fig. 7.5 Continuous loader. Source: Cameco Industries Inc., USA.

large, level fields if their potential is to be maximised. Unless, however, the supply of cane and transport can be maintained, average productivity is likely to drop to 60–100 t/h, at which point it may be cheaper to use grab loaders.

Mechanised cutting

From the aspect of ongoing ratoon yields, manual cutting is always preferable to mechanical cutting since machines cause more damage to the cane stools and the risk of soil compaction is much higher. Machine basecutter knives are usually much less sharp than manual cane knives. This causes more splitting of the cane stumps and increases ingress for infective agents that will reduce ratoon growth. In extreme instances the machine may uproot complete cane stools.

The search for a satisfactory cane-harvesting machine started over 100 years ago and huge amounts of time and ingenuity have been devoted to the subject, particularly by generations of Australian cane farmers. Mechanised harvesting is currently dominated by combined cut/chop/load harvesters ('chopper harvesters') but the use of wholestalk harvesters and mechanical harvesting aids continues in some areas where topographic or other circumstances dictate.

Chopper harvesters

Most of the world's chopper harvesters are currently single-row wheeled machines such as the Case/New Holland 7000 (Fig. 7.6). Track-mounted machines such as the Cameco CH 3500 are available for use in wet conditions. Chopper harvesters top and then cut the standing cane before chopping it into sections (billets) approximately 250 mm long. These billets are passed through air blasts that remove the lighter trash before being elevated directly into an accompanying infield transport unit. Most chopper harvesters can now operate satisfactorily in green cane, although a balance has to be struck between productivity, cane cleanliness, and cane losses. With excessive air blast, lighter billets can be blown out with the trash[2]. Chopper harvesters are able to cope with a wide variety of crop, from low-yielding erect straight cane to heavy crops of tangled recumbent cane, although productivity is reduced in extreme conditions. Many wholestalk cutting machines are unable to handle recumbent (lodged) cane. Chopper harvester productivity is strongly linked to cane yield, whether the cane is burnt or green, row length, ridge form, stones and obstructions infield, and transport availability – as well as operator skill and motivation. Under the best conditions, single-row harvesters can maintain an average of 50–60 t/h (e.g. in burnt cane at

Fig. 7.6 Chopper harvester. Source: Cameco Industries Inc., USA.

the Ord River, Australia). The mean productivity under average conditions is, however, more likely to be in the range of 25–30 t/h after allowance is made for turning at row ends, delays while waiting for transport, etc.

Cane losses and EM levels rise significantly at high pour rates (the instantaneous throughput rate). EM levels in green wet tangled cane can exceed 15%, but average EM levels in green cane are likely to be 8–10%. The use of excessive extractor speeds can also result in cane losses of up to 15% of the crop without significant further reductions in trash levels[2]. As a result many mechanised operations in Brazil now deliberately restrict the pour rate.

Chopper harvesters produce large numbers of exposed stalk ends, and blunt or damaged basecutter and chopper blades also cause high percentages of split cane. Deterioration of the quality of chopper harvested cane is therefore faster than in wholestalk cane. Although there is an empirical rule that chopped cane must be milled within 24 h of harvesting, trials have indicated that significant economic benefits can be obtained by reducing the KTM time to < 12 h.

Wholestalk harvesters

The main alternatives to chopper harvesters are wholestalk harvesters. One type, the Louisiana or soldier harvester (e.g. the Cameco S30; Fig. 7.7)

grasps the standing cane between gathering chains before it is cut off at the base. The whole stalks are then passed through a topping device before being dropped back on the ground aligned across the rows in a windrow suitable for grab or continuous loading. Wholestalk harvesting in Louisiana has now mainly been replaced by chopper harvesting.

The alternative type of wholestalk harvester, largely developed in Brazil, effectively comprises the gathering, topping and cutting components of a chopper harvester, with the wholestalk cane being accumulated in a box at the rear of the machine. When full, the cane is dumped on the ground in a heap for grab loading.

Wholestalk harvesters have the advantage that they can directly replace manual cutters without any changes being required to a mechanised wholestalk loading and transport system. Wholestalk harvesters can therefore be used as a 'quick fix' if a labour shortage develops. There are, however, disadvantages. Most soldier type wholestalk harvesters can only cut green cane as the leaves and trash assist the gathering chains to grasp the stalks. There are no trash extraction devices and the cane must be burnt in the windrow after cutting. The chopper-based Brazilian machines can handle burnt cane, but wholestalk harvesters generally have problems handling recumbent, long or bent cane and are mainly used in lighter erect crops. As wholestalk harvesters operate independently of the transport

Fig. 7.7 Wholestalk harvester. Source: Cameco Industries Inc., USA.

units, their productivity especially by two-row machines can exceed 50–60 t/h, providing that the conditions are suitable.

Other than in the short term, the need to use separate machines for cutting and loading can be more complex and expensive than running a single chopper harvester which combines both tasks. For these reasons, the use of wholestalk harvesters worldwide is declining.

Harvesting aids

Development work continues with harvesting aids. These are simple machines, which do the hard work of cutting the cane but still require some manual input, usually to top and align the cane ready for mechanical loading (e.g. the Orbach cutter, the SASEX front-end cutter and the McConnell cutter). These types of cutter tend to work well in suitable conditions, usually light, erect cane, but are unable to cope with variations such as high yielding, recumbent or bent cane. It has also been found that the savings in labour achieved may not compensate for the increased complications and costs of operating the cutter.

Yield losses from mechanised harvesting

Mechanised harvesting operations accelerate the normal gradual yield decline in successive ratoons through soil compaction and stool crushing (chiefly by the transport units), the ripping out of complete stools by push pilers and basecutters, and higher levels of disease that can be encouraged by split stumps. Trucks, and trailers using old truck axles and wheels, can cause particularly severe infield damage as the high pressure tyres cause stool crushing and soil compaction in the cane root zone. This is made worse if the wheel track does not match the inter-row spacing.

Manually cut cane is normally formed into one windrow for every four or six rows, and the amount of infield movement by transport units and loaders is reduced accordingly. With single-row chopper harvesters, however, each inter-row is subject to two passes of the harvester wheels or tracks and two passes by the transport unit wheels. The effects of these repetitive passes is worsened if the units do not properly match the row spacing, and use high-pressure tyres. Poor basecutter blade maintenance and a crop poorly rooted in compacted soils will result in many complete stools being torn out of the ground, and ongoing ratoons will suffer rapidly diminishing stalk populations. As a result, mechanised harvesting can generate yield losses in the following crops of 5–20%. This has led to increasing interest in better matching of the row spacing to the machine's wheel track (e.g.

by dual row or 'pineapple' planting) and the use of low ground pressure infield transporters. Nevertheless, the possibility of yield depression should be taken into account by any grower considering a change from manual to mechanised cutting. This yield depression may lead to a reduction in the number of ratoons that can be taken before replanting.

Cane transport

The larger sugar factories process in excess of 2 million tonnes of cane per year, at milling rates which can exceed 20 000 t/day. As world cane sugar production is approximately 100 million t/year, then over 1000 million tonnes of cane must be transported annually. Cane is a low density, low value perishable product that must be moved from inside the field without causing serious damage to the soil structure and cane stools, and then transported at minimum cost to the factory. Deliveries must be sufficiently reliable to allow 24 h/day factory operation without excessive buffer stocks. Depending on the haul distance, the transport element can be more than half of the overall harvesting costs.

The density of grab-loaded wholestalk cane is usually between 200 and 300 kg/m^3 depending on the cane length, the amount of bent and twisted cane, the cane stalk weight, and the trash percentage. Grab-loaded wholestalk transport units must therefore have high volumetric capacity to ensure adequate payloads. Carefully arranged hand-loaded wholestalk cane can, however, reach 400 kg/m^3. The density of chopped cane is normally 300–400 kg/m^3, with short (under 250 mm) billets giving higher payloads at the expense of more exposed ends and a higher rate of deterioration.

In areas of intensive cane cultivation (i.e. a factory-operated estate), the average haulage distance to the factory may only be 5–10 km. When the cane is produced by private farmers, the haulage distance can exceed 50 km. The field-edge to mill haul may be flat or involve steep hills, the roads may be hard surfaced or dirt, private or public. In the latter case, transport units must comply with the local traffic regulations.

Infield transport

In some cane producing areas that have pronounced and reliable dry harvesting seasons, the soils may harden sufficiently to operate road trucks infield without causing economically significant damage to the soil structure or cane stools. This applies particularly to fully irrigated estates where irrigation can be withheld for some period before harvesting to induce cane ripening. Trucks, however, normally have axle loadings of at least 8 t and tyres inflated to 6–7 bar (100 p.s.i.), and their use infield risks soil compaction or crushing damage to the cane stools, either of which can severely reduce subsequent ratoon yields. Trucks are readily available for hire in many countries at low cost, principally because they can be used for alternative work outside the harvest season, and once on reasonable roads they can haul cane economically for long distances.

Where infield conditions are soft or the soils are easily compacted, reliable cane movement infield without excessive damage to the ratoon crops requires low ground pressure tyres, low axle loads and restricted payloads. These requirements conflict with the need to minimise costs on the field-edge to factory transport sector, which is usually best achieved with high payloads, high axle loads and high speeds. In these circumstances, agricultural tractors and trailers predominate for infield and short road hauls direct to the factory. In theory, tractors are able to carry out alternative work outside the harvesting season, but in most cane growing areas the land preparation season coincides with the harvest, and a separate tractor fleet must be used for tillage. Many tractors only do cane haulage work, and some manufacturers have responded by supplying specialised haulage tractors with uprated clutches, transmissions, and rear axles. In Southern Africa complete 'hauler' tractors are built (e.g. by Bell), which are very strongly constructed to withstand the stresses of constant haulage work (Fig. 7.8). Self-propelled, low ground pressure, infield haulage units are increasingly popular in Australia. With their larger diameter, lower pressure tyres, tractors have much better infield performance than trucks in soft conditions, and cause less damage to the soil and cane

Fig. 7.8 Tractor and weight transfer unit.

stools. In lesser-developed countries, however, the advantages of tractors are frequently negated by the use of trailers fitted with high-pressure truck tyres, which are used because of their low cost and easy availability. The ability to operate in soft conditions while minimising compaction can be optimised by the use of large diameter, low pressure trailer tyres, multiple tandem axles, weight transfer from the trailer to the tractor, and the use of four-wheel-drive tractors.

Once out of the field, tractors can haul trailers on the road singly or in trains, usually of between two and six units depending on whether the roads are public or private. The tonnage that can be hauled depends mainly on the topography, and braking ability downhill is often more crucial than traction uphill. In undulating areas a 60 kW (80 hp) tractor may haul 8–10 t in one or two trailers, while in flat areas a 110 kW (150 hp) tractor can haul five to six trailers with a total payload of > 50 t. At the field edge, trailer trains are usually separated and the trailers taken into the field individually. To facilitate this hitching and unhitching, they are usually of the turntable or 'free standing' type, which do not transfer any weight to the tractor. The hitching and unhitching process, however, incurs time and cost penalties which will only be worthwhile on longer hauls.

Transloading

The obvious answer to the problem of the incompatibility of the infield and road haul requirements would seem to be the operation of separate transport units, and to transfer the cane from one to the other at the nearest possible point to the field edge. However, transloading incurs costs usually in the range of US$0.50 to $1.00/t, and the saving in the long-haul transport must be at least enough to cover the cost of transloading. Obviously the case for transloading becomes stronger as the haul distance increases and an empirical rule is:

- haul distance < 12 km, use the infield transport direct to the factory;
- haul distance > 12 km but < 21 km, transloading may be cost-effective;
- haul distance > 21 km, transloading should be cost-effective.

As is usual with such empirical rules, there are many exceptions because of variations in the local haulage conditions and labour costs.

Transloading chopped cane is relatively simple as it can be poured from the infield trailer to the long-haul unit using high-tipping or elevator trailers (Fig. 7.9). This operation is almost universal in the Australian industry, where specialised infield haulage units transfer the cane either to rail trucks

Fig. 7.9 High-tipping unit.

(bins) or large (up to 50 t capacity) road trucks. An alternative is the use of relatively large (usually 10 t) containers which are transferred from an infield chassis to the long haul unit either by large forklift trucks (e.g. USA, Swaziland) or by container transfer equipment built into the trailers (e.g. Queensland). The container system can work with whole stalk as well as chopped cane (e.g. Swaziland).

Transloading wholestalk cane is more labour-intensive and expensive than chopped cane as pouring is more difficult, although not impossible (e.g. Colombia). Wholestalk cane is therefore most frequently lifted in bundles from the infield transport by cranes or gantry units for transfer to rail or road trucks. Cane bundles can also be dragged or rolled from single-bundle infield trailer to an adjacent multiple-bundle road haul unit (e.g. Western Kenya), which avoids the cost of a crane.

Many private farmers in South Africa use infield self-loading or grab-loaded trailers, which discharge the cane at transloading stations ('zones') where the cane is transferred to large road trucks by grab loaders or tractor-based cranes (Fig. 7.10).

In all instances, the cost of the labour required to connect and disconnect the sling chains and the operating costs of the crane or winch can be significant, and the personnel may be idle for long periods awaiting infield or long-haul units. The transloading system selected is therefore strongly affected by local labour costs.

Fig. 7.10 Cane transloading.

Field-edge to mill transport

Railway

Historically, animal-drawn carts supplied small plantation-based factories, while large centralised factories were fed by railway (Fig. 7.11). Many mills still operate narrow gauge (610 mm) railways (e.g. Australia, Fiji), while full-gauge railways are used in other areas (e.g. the Dominican Republic and Cuba).

Originally the rail trucks were taken infield on portable track for hand loading. However, moving portable track is laborious and has now largely been replaced, either by carrying the rail trucks infield on specialised trailers (e.g. Fiji), or by transloading cane from infield transport units to rail trucks at fixed sidings (e.g. Australia).

Cost analyses generally indicate that, while the operating costs of a railway system are low, the high capital costs preclude the construction of railway systems for new factories. Virtually all factories built in the last 50 years have been based on road transport. Conversely, the costs of converting a factory from a rail system to road haulage are also very high, and, as a result, some mills have continued to expand their rail networks despite the presence of high-quality government-maintained roads.

Satisfactory operation of a rail network requires high-quality management. Cane held in full rail trucks normally forms the factory buffer stock; but at the same time there must be sufficient empty trucks to be returned on time to the sidings for loading. In fully mechanised harvesting, any time lost waiting for empty trucks cannot be regained. Most cane railways have only a single track, and without very careful traffic control there is a risk of collisions and injury to staff. Radio linkage with the locomotives is now usual, and the Australian industry uses satellite navigation equipment to monitor the position of their trains.

Road transport

Since the 1950s, the improved capability of agricultural tractors and road trucks has resulted in a large increase in road transported cane (Fig. 7.12). Variations in climate, haul distances, infield, and road conditions, as well as operator costs, availability of haulage contractors, and government regulations, however, mean that there is no single 'best solution' for road haulage. The best transport option for each country, or even for each region within a country, must be individually assessed. Road trucks are not necessarily the ideal transport system. Where only gravel-surfaced or earth roads are available, their condition is often poor, especially during wet periods. In such conditions, trucks can lose their advantage of higher speeds and their limited traction and high axle loads can become a liability. Additionally, their higher speed potential can greatly increase the risk of accidents. Such road conditions tend to occur in lower wage areas,

Fig. 7.11 Railway transport.

Fig. 7.12   Truck transport.

and tractor haulage can then be economical over greater distances than might normally be expected. In Western Kenya, contractors transport over 3 million tonnes of cane per year with agricultural tractors and 15 t capacity weight transfer trailers on hauls of up to 35 km. Occasional trials with trucks confirm the ongoing validity of their choice.

Other transport systems

Local circumstances have dictated other unique solutions to the cane transport problem. In Guyana (3 million t/year), the heavy soils, flat terrain, and high rainfall combined with the absence of road building materials led to the evolution of a complex canal network within the estates. On this system, 35 kW (45 hp) tractors haul up to 150 t of cane in trains of small barges (punts). Repeated analyses of the present-day alternatives confirm that this is still the lowest-cost option for the circumstances.

It is therefore unwise to assume that any harvesting and transport technology can be successfully transferred to a different region. Careful analyses, particularly of the ability to maintain and operate large complex machines must be made, and practical trials should always be carried out before selecting a new transport system.

The ability of a transport system to handle both wholestalk and chopped cane is often essential at factories that are in transition from manual to fully mechanised harvesting. The cost of changing the transport fleet can be greater than purchasing the harvesters, and any factory considering a long-term change to chopped cane should consider an early start to replacing purely wholestalk transport units with ones which need little or no alteration to handle chopped cane.

Offloading systems

Offloading systems at the factory are linked to the loading and transport systems used, but some systems are far more efficient than others. Rail trucks are normally offloaded by placing them on a tipping platform (tippler), but the older platform designs which tip through only 45–60° can be ineffective when discharging heavily laden wholestalk trucks in which the cane is wedged (e.g. the Dominican Republic). The rail truck tippers in Queensland that rotate through 360°, discharging from the top of the truck, are far more effective (Fig. 7.13).

Road trucks carrying hand-loaded cane are usually of the flatbed type (e.g. Indonesia), and are most effectively offloaded using tipping platforms that elevate the front of the truck (Fig. 7.14). Hand-loaded cane can be loaded across the truck bed with a cable looped around the load which can then be discharged onto the ground by attaching the free end of the cable to a strong point and driving the truck slowly forward (e.g. Indonesia). In Fiji, the

Fig. 7.13 Bin tipper.

Fig. 7.14 A tipping platform.

same offloading cable cane is used for discharging into the cane carrier by provision of a winch to pull the load off the rear of the truck.

Lifting bundles of cane out of the transport using slings and a crane or gantry is frequently necessary where winch-loaded cane is delivered direct to the factory (e.g. Vietnam, Kenya); but the process is slow (approximately 2 min/load) and labour intensive.

High-sided vehicles must be used when loading wholestalk cane with grab loaders. Where trucks are used, they can be discharged to the rear using a tipping platform; but the most popular and ef- fective discharge method for grab loaded cane is the side rollout system (also known as the 'Hilo', chain net or basket system). This can also be used with tractors and multiple trailers. In this system the cane is loaded into a chain or cable net, one side of which is attached to one top edge of the vehicle body. A hydraulically operated hoist lifts the free side of the net to discharge it. This system is both quick (approximately 30 s/load) and has low labour requirements (Fig. 7.15).

Chopped cane must be carried in vehicles fitted with high sides or tipping containers (bins), the sides and bottoms of which must be constructed

Fig. 7.15 Rollout system.

from rigid steel mesh or solid sheet material. Single trucks can be end discharged using conventional truck tippers (e.g. Mexico, Indonesia) and self-tipping semi-trailers are also used (e.g. Barbados). Multiple trailer units must be discharged to the side, with the bin either pivoted at the top (e.g. Colombia, Papua New Guinea) or at the lower edge (e.g. Argentina, Florida). Detachable containers are used (e.g. the Cameco 'Portabox' system) where the complete container is removed from a specialist chassis and tipped by a separate mechanism (Fig. 7.16). Detachable containers have the advantage that they can be used to store the factory buffer

stock in a form where good stock rotation (first-in, first-out) is possible. Detachable containers must however have sufficient strength to withstand handling by forklift trucks and must be heavier than permanently mounted bins.

CANE STORAGE

Some cane stock must usually be kept at the factory to cover periods when the flow of transport units is interrupted. If the cane is loaded during daylight hours only, the factory may have to store 16–20 h

Fig. 7.16 Detachable bin.

milling capacity that may comprise several thousand tonnes. If mechanised loading or harvesting is being carried out 24 h/day, then stocks can be minimised or even total reliance placed on the continuous arrival of transport units (e.g. Ord River, Australia). Unless containerised, the factory cane stock is usually handled either by large wheeled loaders fitted with cane grabs or by a large gantry unit with travelling cranes/grabs. Wheeled loaders have lower initial capital costs and are more flexible, but gantries have lower long-term operating costs.

Good stock control is crucial to effective factory operations. To minimise cane deterioration, the stock must be kept to the minimum necessary for continuous factory operation. Where cane is stocked in a large heap, the oldest cane is at the bottom and is inevitably used last (first-in, last-out). Unless adequate steps are taken to divide the stock into several heaps, the first cane may remain at the bottom of the pile for several days, with adverse effects on its quality. Because of its faster rate of deterioration, storing chopped cane in loose heaps is particularly risky. Where a factory receives mixed chopped and wholestalk cane, the chopped cane should always be fed directly to the mill, and the buffer stock should comprise wholestalk cane. If a factory is operating with a large proportion of chopped cane, then storage in containers (which may comprise rail trucks) is the best system, providing that a first-in, first-out rotation system is practised. Minimising chopped cane stocks by 24 h/day harvesting is preferable.

CO-ORDINATION AND CONTROL OF HARVESTING

Adequate planning of the harvesting process is essential if sugar production is to be maximised. It is necessary for the cane to be cut at the correct age, as losses in growth potential and sucrose content will be incurred if the cane is cut too young or too old. Where ripening is induced, either by withdrawing irrigation or by chemical application, the necessary steps must be taken up to 6 weeks before the proposed harvesting date. Where natural ripening is relied upon, the cane quality normally increases slowly as growth slows in the dry season, and then declines again with the approach of the next wet season. The harvest must be then planned so that the maximum amount of cane is harvested at the peak of the maturity curve. Some varieties mature earlier and others later in the season, and the harvest programme must reflect this. All this planning must also accommodate the need for some fields to be harvested early to allow replanting operations.

The harvesting schedule drawn up before the start of the season must reflect all these requirements, and accurate estimates of the crop size and cane yields are essential to the planning process. On a day-to-day basis during the harvest, the manager or co-ordinator must, as far as possible, maintain a continuous supply of fresh cane to the factory, since factory stops are expensive in terms of idle labour, wasted fuel and lost juice quality in the process house. At the same time, an excessive stock of burned or cut cane, either infield or in the caneyard, must be avoided as it will deteriorate. Excessive burning for hand cutting will result in the surplus deteriorating for a further 24 h before it can be cut and milled; burning too little causes the factory to run out of cane and upsets the cane cutters. The harvesting manager must juggle all these considerations, possibly together with the risk of rain or an irregular turnout of labour.

Within an estate or small outgrower cane-production system it is essential to arrange cane loading and transport as soon as possible after cutting. Cane from partly cut fields can be loaded manually, but part-cut fields can present difficulties for mechanical loading. The correct co-ordination of burning, cutting, and transport is therefore vital to cane quality.

A factory without cane is very conspicuous, and it is tempting to maintain excess mill stocks to avoid factory time losses. However, this will lead to cane deterioration and loss of recoverable sucrose. A cane harvesting manager's principal task is to strike a balance between these extremes.

REFERENCES

1 Fauconnier, R. & Bassereau, D. (1970) Techniques agri-
   coles et productions tropicales. In: *La Canne à Sucre*, pp.
   248–250 Paris: Maisonneuve et Larose.

2 Whiteing, C., Norris, C.P. & Paton, D.C. (2001) Extraneous
   matter versus cane loss: finding a balance in chopper harvested green cane. *Proceedings of the International Society
   of Sugar Cane Technologists*, **24**, 276–282.

Chapter 8
Cane Payment Systems

Martin Todd, Gareth Forber and Philip Digges

INTRODUCTION

Cane payment systems define one of the most important relationships of any cane industry, since they determine how revenues are distributed between growers and millers. Cane payment systems also play a central role in determining the incentives that growers and millers face. Not only do they heavily influence the incentives to improve technical efficiency, they also have far-reaching implications for investment decisions.

In this chapter the different types of cane payment systems that can be found in cane industries around the world are analysed. This analysis includes a discussion of how cane payment systems allocate revenue between grower and miller, and how the incentives embodied in different cane payment systems can influence the technical performance of an industry, as well as the incentives to expand production. Specifically, the following issues are considered:

- different types of cane payment system;
- sampling and testing of cane;
- length of the campaign;
- average and individual cane quality;
- grower/miller incentives to expand;
- incentives to produce refined sugar; and
- incentives for different socio-economic groups.

DIFFERENT TYPES OF CANE PAYMENT SYSTEMS

The number and different types of cane payment systems in existence are an indication that there is no single formula for an efficient and effective cane payment system (Table 8.1). However, payment systems used to divide revenue between growers and millers can be separated into three broad groups:

- fixed cane price systems;
- fixed revenue sharing systems; and
- variable revenue sharing systems.

Fixed cane price

Although revenue sharing agreements tend to form the basis for payments to growers in many sugar industries, there are still some very large sugar industries where the price of cane is not linked directly to the value of the sugar produced. Instead, growers receive a fixed price per tonne of cane. Industries where a fixed cane price is still used include China, India, and Pakistan.

The exact nature of fixed cane price payment systems differ between countries, and even within countries. This is the case, for example, in India. In Uttar Pradesh, and other northern states, the price of cane is set at a fixed level with no premium or discounts paid for cane quality. Growers are therefore paid on the basis of weight alone. In Tamil Nadu and other southern states, the cane price is based on a standard 8.5% sugar recovery rate, with a premium if the recovery rate exceeds this level.

The key weakness of fixed cane price systems is the lack of a link with the actual sugar price. This creates a lopsided arrangement where growers and millers do not share price risk. Instead, the burden of price volatility is placed solely on the miller. This creates problems during times when the sugar price falls and millers' margins are eroded because of their obligation to pay a fixed price to

Table 8.1 Summary of revenue sharing arrangements.

Cane industries	Cane payment system	Revenue share based on revenue from:
Australia	Revenue share	Raw sugar (millers retain molasses)
Brazil	Revenue share	*Cristal** sugar and ethanol
Colombia	Revenue share	Raw/mill white sugar (millers retain molasses)
Fiji	Revenue share	Sugar, molasses and other by-products
Jamaica	Revenue share	Sugar, molasses and other by-products
India		
Maharashtra	Co-operative	Sugar and molasses
Uttar Pradesh	Fixed price (flat rate)	n.a.
Tamil Nadu	Fixed price†	n.a.
Mexico	Revenue share	*Estandar* sugar (millers retain molasses)
Philippines	Revenue share	Raw sugar and molasses
South Africa	Revenue share	Raw/refined‡ sugar and molasses
Thailand	Revenue share	Raw/white/refined§ sugar and molasses
USA		
Louisiana	Revenue share	Raw sugar and molasses
Hawaii	Integrated grower/miller	n.a.

* *Cristal* sugar is a very high polarity raw sugar produced by the Brazilian sugar industry.
† Growers receive quality premium based on sugar yield per tonne.
‡ Growers share in only part of the premium earned from exports of refined sugar.
§ Growers share in only part of the premium earned from exports of white and refined sugar.
Source: Industry Sources, LMC International Ltd.

growers. This situation is common in India, Pakistan and China. Some millers issue credit notes to their growers, rather than paying in cash, as a short-term solution to the problem.

Fixed revenue sharing

Under a fixed revenue sharing arrangement, revenues are shared on the basis of a *fixed* percentage distribution between grower and miller. Although these systems ensure that cane prices and millers margins are linked to sugar prices, the sharing of revenues on a fixed (percentage) basis can weaken the incentive to improve technical performance for both growers and millers.

Improving cane quality, so that mills are able to recover more sugar, may not be a priority for growers when they must share this additional sugar output with the mill. The disincentive is even more stark in the case of millers, for whom the investment in the equipment needed to improve sucrose recovery can run into several million dollars, yet mills receive a relatively small share of the value of any additional sugar produced. This is because, in the case of most revenue sharing agreements, growers are allocated a greater percentage share of revenues than millers. Under such systems, there is therefore a tendency for mills to concentrate on investments in which they retain the full benefit. These often focus on investments that lower costs rather than improve sucrose recovery rates.

Variable revenue sharing

The third and most sophisticated type of revenue sharing arrangements are those that use variable revenue sharing. Under these arrangements, the formulae ensure that, beyond a benchmark level of cane quality and factory efficiency, any incremental improvement in *cane* quality is entirely to the benefit of *growers*, while any improvement in sucrose recovery in the *factory* benefits only the *mill*.

While, from a technical and economic perspective, variable revenue sharing arrangements offer significant advantages over fixed price systems, variable revenue sharing agreements are much more complex and costly to administer.

PAYMENT FOR WHICH SUGAR PRODUCTS?

Some cane payments systems take account of payment for different types of sugar (raw, white, and refined). Others go further, including revenue from the sale of by-products such as molasses. For example, in Australia and Colombia, millers retain all of the revenue earned from the sale of the molasses, while the cane payment systems in South Africa and Thailand make provision for this revenue to be shared between growers and processors. In Brazil, growers and processors share revenue from the sale of ethanol as well as sugar (Table 8.1).

INCENTIVES TO IMPROVE TECHNICAL PERFORMANCE

Growers and processors can both influence the amount of sugar that can be produced from a given quantity of cane. Ideally, cane payment systems should provide an incentive for growers to maximise the sugar content of the cane and for millers to maximise the recovery of sucrose at the factory. However, achieving this ideal is not easy. The first hurdle to overcome is measuring the sucrose content of cane in a manner that is fair to both growers and millers. This requires testing procedures, which are often highly complex.

SAMPLING AND TESTING OF CANE

An integral part of cane payment systems is the sampling method employed to establish the cane's sucrose content. There are two main techniques used:

- the Direct Analysis of Cane (DAC) method, which usually involves drawing a sample of cane as it enters the mill yard;
- the (indirect) sampling of cane juice, after the cane has entered the factory; this is the so-called first expressed juice (FEJ).

From the growers' point of view, core sampling of cane seems to offer the fairest means of establishing the quality of cane delivered to a factory. The DAC method is capable of providing representative samples and operates completely independently of milling operations. Moreover, because growers' cane is sampled as it enters the cane yard, not as it enters the mill, growers are not penalised for any deterioration in the quality of their cane that might result from it being stored for long periods in the cane yard, and/or any sucrose losses sustained as a result of cane washing.

By contrast, if their cane is tested *indirectly*, i.e. by the FEJ, growers suffer the consequences of any deterioration in cane quality that occurs in the mill yard and/or sucrose losses during cane washing. Despite reservations in some quarters concerning the equity and accuracy of indirect analysis, this remains a common means of cane testing, not least because it is relatively inexpensive to establish and operate. Moreover, not every cane industry washes cane after delivery, and, in many industries, the efficient co-ordination of harvesting and cane transport means that post-harvest sucrose losses are minimal.

Clearly, the introduction of sophisticated testing techniques brings with it additional costs. While the advantages of testing for cane quality are likely to outweigh these costs for many countries, in countries where there are many small growers (which in some countries, like India, number in their thousands, if not tens of thousands, for each mill), each delivering small quantities of cane, the administrative burden and cost may prove too great. Instead, less technically efficient cane payment systems may be used that are much simpler and cheaper to administer. However, such systems often rely on payments based on the average quality of cane delivered by all growers. This raises important questions over the extent to which such cane payments systems can provide incentives to individual farmers to improve cane quality.

PAYMENT SYSTEMS

Systems based on the average quality of growers' cane

Cane payment systems that are based on the *average* quality of cane delivered to a mill by all growers provide little incentive for *individual* growers to improve cane quality. For example, a cane grower who produces cane with a high sucrose content will only receive payment based on the average quality of all the cane delivered to the mill. Consequently, the grower is unable to influence the price received for his cane, and has no financial incentive to produce cane of higher quality.

Such cane payment systems therefore shift much of the responsibility for improving cane

quality onto the mill. This is illustrated by the situation in Mexico, where the mills tend to devote a considerable amount of time to extension activities. Since the privatisation of the industry in the early 1990s, Mexican mills have succeeded in raising the average quality of cane through improved scheduling of harvesting, transport and milling. Figure 8.1 shows that sucrose content has increased by around 15% over the past decade, while fibre content has reduced as a percentage of total cane.

Systems based on the quality of the individual grower's cane

Systems that provide the *individual* grower with an opportunity to increase revenues through raising cane quality (such as those operating in Australia, Jamaica, Mauritius, Thailand and South Africa) are considered to be more effective in terms of targeting incentives at growers themselves.

However, the impact that such payment systems have on the price paid to individual cane farmers differs according to whether the industry operates

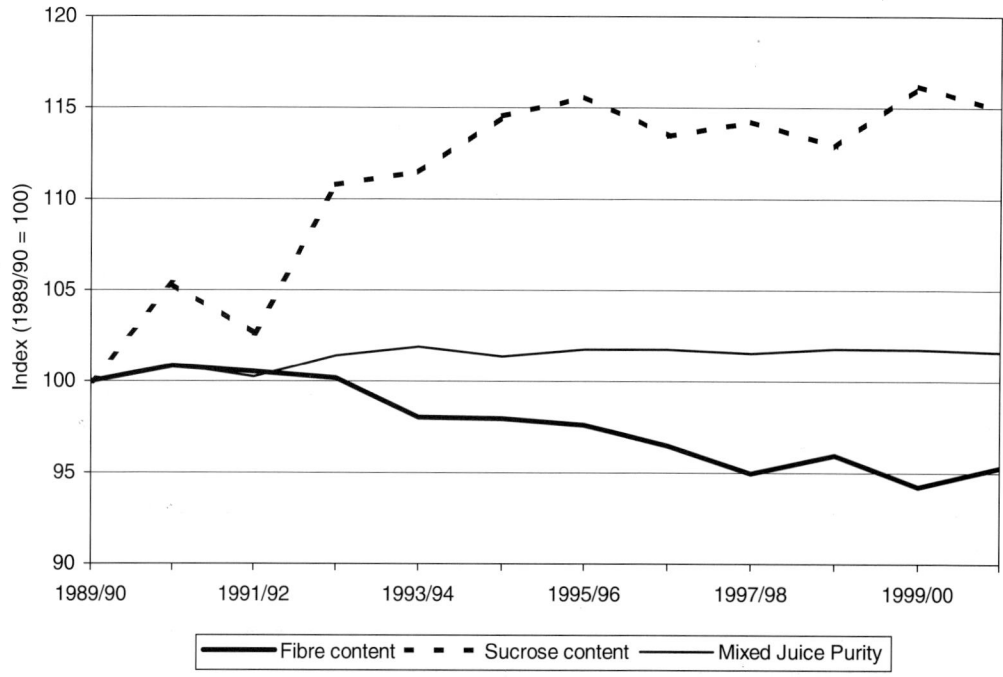

Fig. 8.1   Field performance in Mexico from 1989–90 to 2000–01. Source: Comité de la Agroindustria Azucarera, Dessarollo Operativo Campo-Fabrica de las Zafras.

a fixed revenue sharing system or a variable revenue sharing systems on the cane quality of individual grower basis.

- Under fixed revenue sharing arrangements, individual growers compete for a fixed proportion of the revenue from sugar production. For example, cane farmers in Thailand receive a more-or-less fixed share of revenue of approximately 70%. However, individual farmers are paid a cane price that is based on the quality of the cane that they deliver. This means that growers who deliver high quality cane receive a higher cane price than growers who deliver low quality cane. But, because growers are paid out of a fixed pool of revenue, they are effectively competing with each other for a greater share of the revenue pool that is allocated to growers as a whole.
- Under a variable revenue sharing system, while growers and processors compete among themselves on an individual basis, they also compete collectively against millers in an attempt to secure a greater proportion of total industry revenue. In theory, this type of payment system should provide the most effective means of encouraging growers and millers to enhance efficiency at both the field and factory levels.

However, in addition to the administrative difficulties already discussed, note that even with the most sophisticated payment systems, improvements in cane quality cannot be guaranteed. In Queensland, for example, cane quality actually *declined* between the 1960s and the late 1980s, despite the incentive for improving cane quality provided by the industry's Commercial Cane Sugar (CCS) formula (Fig. 8.2). This is because growers' investments were directed primarily at lowering *costs* (notably by extending the ratoon cycle and adopting mechanical harvesting), which was to the detriment of cane quality.

Incentives to improve the cane quality and sucrose recovery

Table 8.2 shows the extent to which each country's cane payments systems rewards growers and millers (as a group) for improvements in cane quality

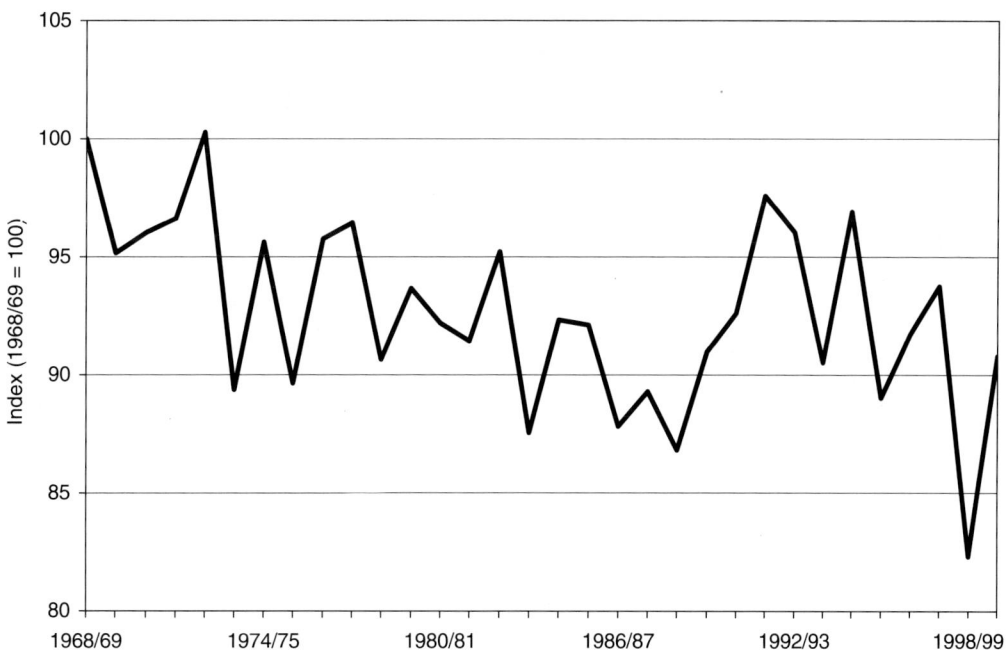

Fig. 8.2 Commercial cane sugar in Australia from 1968–69 to 1999–2000. Source: Australian Sugar Milling Council, LMC International Ltd.

Country	Increase in grower revenue per tonne of cane as a result of a 10% gain in cane sucrose content (%)	Increase in mill revenue per tonne of cane as a result of a 10% gain in factory recovery rate (%)
Australia	15	29
Brazil	15	20
Colombia	10	10
Fiji	10	10
India (Uttar Pradesh)	0	46
Jamaica	16	26
Mexico	10	24
Philippines	10	10
South Africa	10	10
Thailand	10	10
USA	10	
Louisiana	10	10
Florida	10	20

Table 8.2 Indicative incentives for growers and mills in different industry cane payment systems.

Source: Industry sources Ltd.

and factory efficiency. The table presents estimates of the percentage change in farmers' revenue that results from a 10% increase in cane sucrose content. The equivalent figure for millers, showing the effect of a 10% increase in factory sucrose recovery, is also presented.

The system operated in Uttar Pradesh (India) provides no incentive for the individual grower to raise cane sucrose content, because payments are made on a flat price basis, regardless of cane quality. Although Fiji and Mexico operate fixed revenue sharing agreements for the industry as a whole, at the *individual* level there is little incentive for growers to improve cane quality. This is because individual growers are either paid on the basis of weight (Fiji), or on the basis of the *average* quality of all cane delivered to a mill (Mexico).

On the factory side, those systems offering the greatest rewards to millers for improved sucrose recovery fall into two groups:

- Those systems, such as the Australian and Jamaican systems, where millers retain 100% of the incremental sugar revenues arising from improvements in factory efficiency.
- Those systems, such as the Mexican system, where the payment system stipulates a guaranteed recovery rate. In practice, the mill may, or may not, achieve this guaranteed recovery rate.

If the mill's actual recovery rate does exceed the guaranteed recovery rate (upon which grower payments are based), the miller is entitled to 100% of the incremental sugar revenue resulting from its achieving a recovery rate above the guaranteed level. However, if the actual recovery rate falls below the guaranteed rate, a proportion of all revenue must be shared with the grower.

In industries with a fixed revenue sharing split, millers share the financial benefits of greater efficiency with their growers. This is equally true for growers who share the rewards from an increasing sucrose content As a result, the rewards (and incentives) to improve field and factory performance are less than in those industries where growers and millers retain 100% of returns accruing from the efficiency improvements they achieve.

Limitations

Regardless of the incentives to growers enshrined in any particular cane payments formula to improve cane quality and sucrose recovery, factors *beyond* the control of the grower also have a profound effect on cane quality. For example, declining soil fertility and an increase in soil compaction and cane stool damage are cited as causes of the decline in CCS witnessed over the past 30 years in Queensland. In South Africa, cane yields and quality have been severely affected by the *eldana* borer. It is clear that, regardless of how effective the incentives of a given payment system may be, agroclimatic factors, pests and diseases are often important determinants of cane quality.

INCENTIVES TO EXPAND PRODUCTION

While there is a clear link between cane payment systems and the incentives to improve technical performance, payment systems also have far-reaching implications for other aspects of the sugar industry. Interestingly, the decision whether or not to expand the industry at both field and factory levels are heavily influenced by the way in which growers and millers are paid for their production. To understand the complexities surrounding this issue, however, it is important to bear in mind that there are two main ways in which industries can expand:

- by lengthening the campaign; and
- by increasing milling capacity.

Cane payment systems and the length of the campaign

Where growers are paid for the sucrose content of cane, the length of the milling season is an important issue. An extension of the milling season tends to reduce the season average sucrose content, because the additional milling takes place at the beginning and end of the season, when cane sucrose is low. In countries where the majority of cane is supplied by independent growers, this produces a conflict of interest between growers and millers.

- If millers were to accede to growers' wishes for a shorter milling season, they would be obliged to invest in additional milling capacity in order to process the year's cane production in a reduced number of days.
- Equally, if growers were to agree to a lengthening of the milling season (which may avoid the need for investment in additional milling capacity), growers would experience a reduction in the average sucrose content of their cane.

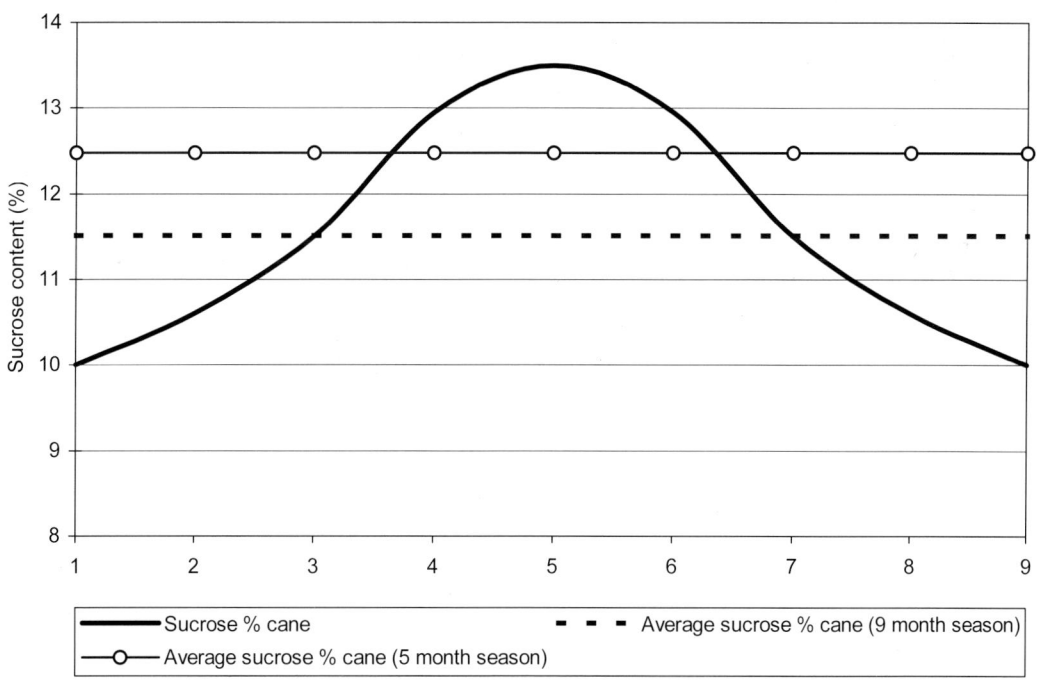

Fig. 8.3 Average cane sucrose content. Source: LMC International Ltd.

Even in industries where there is a relative payments system in place, the extension of the milling season is still a contentious issue. In Australia, for example, extension of the milling season has to be negotiated between individual mills and their growers. This is because the average sucrose content of cane is an integral determinant of the revenue accruing to growers and processors.

The trade-off between cane quality and season length is illustrated in Fig. 8.3, which presents a stylised version of the sucrose curve over the course of the harvest period. The high-point of the sucrose curve represents the mid-point of the harvest season. Owing to the shape of the sucrose curve, the 5-month average cane sucrose content is markedly higher than the 9-month average cane sucrose content. Factory recovery rates also follow a seasonal pattern. Recovery rates tend to be at their lowest at the beginning and end of the season; the incidence of wet weather during harvesting is much higher during these periods than in the middle of the harvest season. This results

in greater quantities of soil and sand entering the mill (interfering with sucrose extraction) and lower juice purity, which impedes recovery in the sugar house. Cane fibre content tends also to increase with the age of cane. Thus, the average recovery rate achieved in mills over the length of a season is also influenced by season length.

Figure 8.4 shows a stylised sucrose recovery curve, together with average recovery rates assuming a 5-month season and a 9-month season, respectively. Again, owing to the shape of the curve, the average recovery rate for the shorter season is significantly higher than the recovery rate for the 9-month season.

This analysis suggests that the benefits of longer season accrue mainly to millers due to greater utilisation of their milling capacity, despite a fall in sucrose recovery rates. While growers will benefit from greater use of their fixed assets (such as mechanised cane harvesters), the benefits are likely to be modest compared to those accruing to the miller as these fixed assets represent a relatively small proportion of total costs.

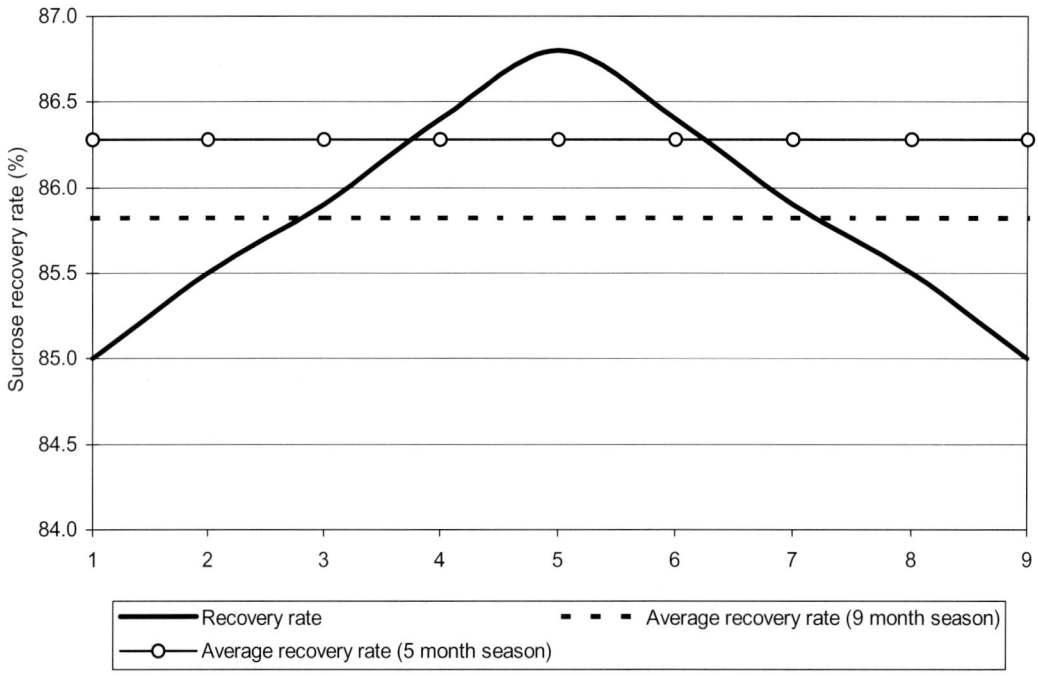

Fig. 8.4   Sucrose recovery rate. Source: LMC International Ltd.

Relative payment schemes

Relative payments schemes compensate cane growers for delivering cane at the beginning and end of the season. Such schemes are a feature of several cane payment systems, notably those in Australia, Jamaica, Mauritius and South Africa.

The purpose of relative payment schemes is to ensure that a grower supplying cane during week one of the crushing season will be paid relative to the average of all cane delivered to the factory during that particular week, *not* relative to the average for the season as a whole. Such schemes encourage growers to deliver cane regularly throughout the season, resulting in improved utilisation of milling capacity. Furthermore, these schemes permit group harvesting, a practice common in the Australian industry, to be organised without risking inequity between group members.

In other industries, mills have adopted less formal measures of encouraging growers to deliver cane at the beginning and end of the season. In Maharashtra in India, for example, growers are offered a bonus for cultivating early-maturing varieties.

Where relative payments systems are not operated *and* growers are paid on the basis of cane quality, growers are understandably reluctant to deliver cane off the 'peak' of the sucrose curve. A good example of this exists in Brazil, where independent growers concentrate their deliveries in those months where cane sucrose and juice purity are at their highest, leaving millers to process their own cane on either side of the sucrose peak.

Grower/miller incentives to expand

The decision to expand output is generally taken by individual growers and processors; it is rarely an industry decision. As a result, expansion tends to take place when there is an incentive for both growers and processors as individuals to do so. However, our analysis suggests that individual decisions to expand are often not in the best interests of the industry as a whole.

Why is there a discrepancy between the individual and industry viewpoints? The key to understanding this difference is how cane payment systems value the last unit of output, i.e. whether the marginal unit of output is valued at its *marginal* price or it is valued on the basis of the *average* price of sugar received from all sugar sales.

How the incremental unit of output is measured is important, since it will affect the production decision of grower and miller. It will also affect the allocation of resources within an industry. In many industries, the price that growers and processors are paid is based on the average of the prices earned by the industry in *each* of its markets. In this situation, additional production, which in many cases has to be exported onto the world market at the world market price, will receive a price above its marginal (world) price. This may encourage both growers and processors to expand output beyond an optimum level.

However, in sugar industries where markets are separated, and the marginal unit of output receives the price in the lowest price market, growers' and millers' production decisions will be determined by comparing marginal costs to marginal revenue.

The EU provides an example where, under a quota system, there is little incentive for individual producers to expand output, despite the industry's very high average sugar price. This is because the system of production quotas ensures that individual growers and processors earn the marginal value of any extra output, i.e. the world sugar price.

Interestingly, both Australia and South Africa altered their cane payment system precisely *because* it hindered expansion. In South Africa a 'pool' system where incremental production received the marginal price was replaced by a system based on average prices. As a result, incremental production is rewarded with the average price, rather than its true marginal value, providing an incentive for the industry to overproduce.

This is also true for Australia. However, since the average price in Australia is currently close to the world price, production should not expand far beyond an efficient level. This is because the average price paid to growers is not significantly different from the marginal price (i.e. the world price) that would have otherwise been received.

However, in other export-orientated industries, where this is not the case, offering the average price for sugar provides an incentive to overproduce, creating structural oversupply on the world market and, ultimately, depressing the world price of sugar.

Incentives to produce refined sugar

In many cane industries, the incentive to invest and/or expand the production of refined sugar may also be heavily influenced by the cane payment system. This is because in some industries, growers and millers share in the added value associated with upgrading raw sugar into refined sugar. In these instances, there is less incentive for millers to invest in new equipment to produce refined sugar. This is because millers will only receive a percentage of the premium earned from the sale of refined sugar. Industries in which part or all of the revenue from refined sugar is included in the revenue sharing arrangements include Thailand, Swaziland and South Africa.

The repercussions of this type of arrangement are evident in many industries. For example, the South African and Swazi sugar industries have been relatively slow to add value to their existing raw sugar output, since the miller will only receive a percentage of the added value from selling refined sugar.

The response of the Australian sugar industry to a change in the revenue sharing system in the late 1980s, when the industry ceased to share revenues from the production of white sugar and introduced revenue sharing based on the proceeds from the sale of raw sugar only, was a wave of new investment. As a consequence, refined sugar production and capacity increased dramatically. This trend is also mirrored in other cane industries, such as Guatemala, Colombia and Mexico, where millers retain the full value of the premium that they earn from the sale of refined sugar over standard quality white sugar.

IMPACT ON INCENTIVES FOR DIFFERENT SOCIO-ECONOMIC GROUPS

Even where cane payment systems succeed in providing positive incentives for both growers and processors to improve their technical performance, note that other factors may enhance or limit the ability of growers to respond to those incentives. Some of these issues are discussed below.

Integrated operations

In industries where cane production and milling are integrated, and particularly where they are under single ownership, as in the case of mill-owned plantations, sugar companies are able to optimise field productivity and factory efficiency to maximise profits across the entire operation. In this situation, a sugar company may opt to operate a mill for as long a season as possible – thus ensuring a high use of fixed assets and lower unit costs of capital – even though this may result in an overall lowering of average cane quality. Equally, the mill may decide that the costs of reducing the trash content of cane are more than outweighed by the gains within the factory, in terms of improved recovery and reduced wear and tear on machinery.

Co-operative mills

Mills that are owned co-operatively will tend, on average, to have more secure supplies of cane than those that rely upon independent growers. Although growers supplying co-operative mills will have an incentive to deliver high quality cane to the mill (because they share in the mills' earnings), it is still necessary to pay growers relative to one another to discourage 'free riding' by individuals. Unlike integrated plantation/milling operations, the trade-offs between cane quality and rewards to growers and millers are less clear cut. Furthermore, the co-operative sector in some countries (notably India) is highly political, with the result that optimising the efficiency of sugar production may not always be the first priority of the organisation.

Growers on leased or tribal land

For cane growers occupying leased or tribal land, the ability to respond to quality incentives may be blunted by concerns about the duration of growers' tenure. It is clear that farmers who have no security of tenure are unlikely to invest in relatively costly fixed assets (such as irrigation facilities or improved field drainage) that they cannot take with them. A further problem faced by growers occupying leased land is that raising credit tends to be difficult or impossible, a factor that has affected growers in both the Philippines and in South Africa.

In the Fijian industry problems of land tenure threaten to affect industry productivity, irrespective of payment incentives. In Fiji, most cane growers are ethnic Indians farming small (typically 4 ha) plots on leased land. Tensions between the Fijian and ethnic Indian populations prompted widespread concern that many cane growers would be forced to leave their plots if Fijian landlords refused to renegotiate leases. Although most leases are now being renewed, many growers are reluctant to invest time and money in the land, owing to fears that they will be displaced. This lack of investment is one of the causes of the alarming decline in cane quality and sugar recovery rates that has taken place over the past few years.

Smallholders

A fragmented field sector, where cane is supplied by a multitude of smallholder growers, presents particular barriers to the introduction of quality-based cane payments systems. The first problem is the difficulty of sampling cane from so many individual growers. This is particularly problematic where indirect analysis of cane is carried out, because it necessitates the identification of cane juice from numerous small batches of cane. Furthermore, the intensity of sampling is high if numerous individual growers are to be paid according to the quality of their own cane. This was a serious problem in Jamaica before the introduction of core sampling.

A more fundamental problem that arises for smallholders is whether the incentives provided by complex payment formulae (developed by mill technologists) are readily understood by growers. This suggests that the introduction of relatively complex systems may need to be accompanied by a specific extension programme so that incentives are fully appreciated.

The practical obstacles to the adoption of certain beneficial management practices limit the ability of the smallholder sector to respond to quality-based

incentives. In South Africa, it takes smallholders a long time to assemble a load of harvested cane, meaning that significant deterioration of cane takes place before the complete load reaches the mill. In Fiji, the industry has suffered major problems with the deterioration of cane since transportation of cane by trucks superseded the older system of rail transport. Lengthy queues of trucks at the mill gate result in long delays between burning and milling, with inevitable consequences for cane quality.

The Indian and Mexican industries have attempted to rationalise cane transport to avoid these difficulties. In northern India, growers deliver cane to outstations, from where the cane is transported by truck to the mill. In Mexico, groups of growers assemble joint loads of cane for delivery to the mill. However, under an outstation system, if the transport between outstation and mill is inefficient (as in northern India), cane quality still suffers.

Moreover, if cane quality is tested at the mill, it will be the grower who is penalised for cane deterioration as a result of transport inefficiencies. This potential inequity was resolved by Belize's industry by establishing cane testing at outstations. In theory, cane quality could also be tested in the field before harvesting.

RECENT PERFORMANCE OF CANE PAYMENT SYSTEMS

So, how have different cane payment systems performed in recent years? To conclude our analysis, we have compared the technical performance at the field level (sucrose content) of several countries, each operating different cane payment systems. The countries considered are India (Uttar Pradesh), which operates a fixed price system; Thailand and South Africa, which use a fixed revenue sharing system; and Australia, which uses a variable revenue sharing arrangement.

Note that changes in cane quality over time are affected by a host of factors, some of which are beyond the influence of growers. In reality, sucrose content is dependent on many factors including:

- weather patterns;
- ratoon cycles;

- changes in the regional distribution of production;
- the standard of transport infrastructure; and
- the length of time cane is stored before being crushed at the mill.

Although the fluctuations in cane quality from year to year cannot be attributed entirely to the grower, a number of interesting conclusions regarding grower behaviour under various cane payment systems can be drawn.

Figure 8.5 presents the sucrose content and cane yield, in index form, achieved in the Uttar Pradesh region of India over the past 15 years. In Uttar Pradesh, there is no incentive to improve cane quality as grower revenues are fixed per tonne of cane and decline per tonne of sugar with increasing sugar output. Since growers are not paid a premium for the quality of cane they produce, there is a tendency to concentrate efforts solely on the quantity produced. Indeed, Fig. 8.5 shows that this has been the case in practice. While the quantity of cane yield has increased dramatically over the past 15 years, the sucrose content of the cane produced

has consistently under-performed against the base year (1984/85).

Both Thailand and South Africa operate fixed revenue sharing arrangements. While cane produced in South Africa has consistently produced a higher sucrose content than that achieved in Thailand, the gap has narrowed in recent years. This is particularly true for the north and north-east regions of Thailand (Fig. 8.6).

In these regions of Thailand, the sucrose content of cane increased sharply during the early 1980s and again in the early 1990s. With the exception of the 1997/98 season, sucrose content was maintained at a level over 15% higher than that achieved during the base year (1979/80).

Importantly, the latter development coincided with the introduction of a new quality-based cane payment system. This trend suggests that the payment system may have had an important impact on the behaviour of Thai growers, providing an incentive to improve quality.

In Australia, the tremendous improvements in milling and processing technology that have taken place in recent years have led to a reduction in the

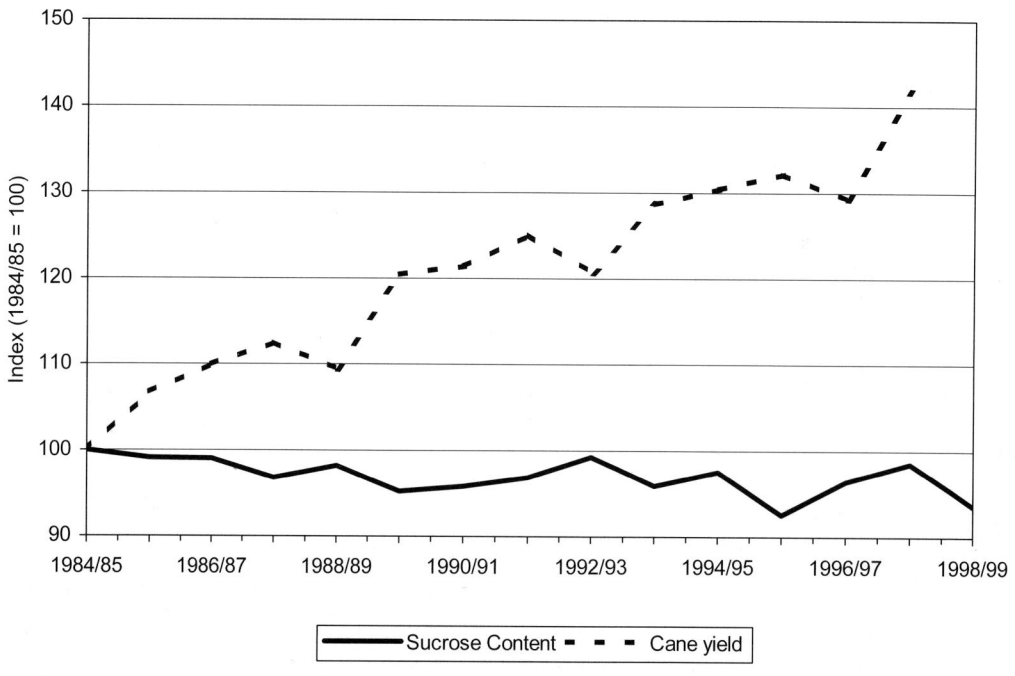

Fig. 8.5   Field performance in India (Uttar Pradesh). Source: India Sugar, LMC International Ltd.

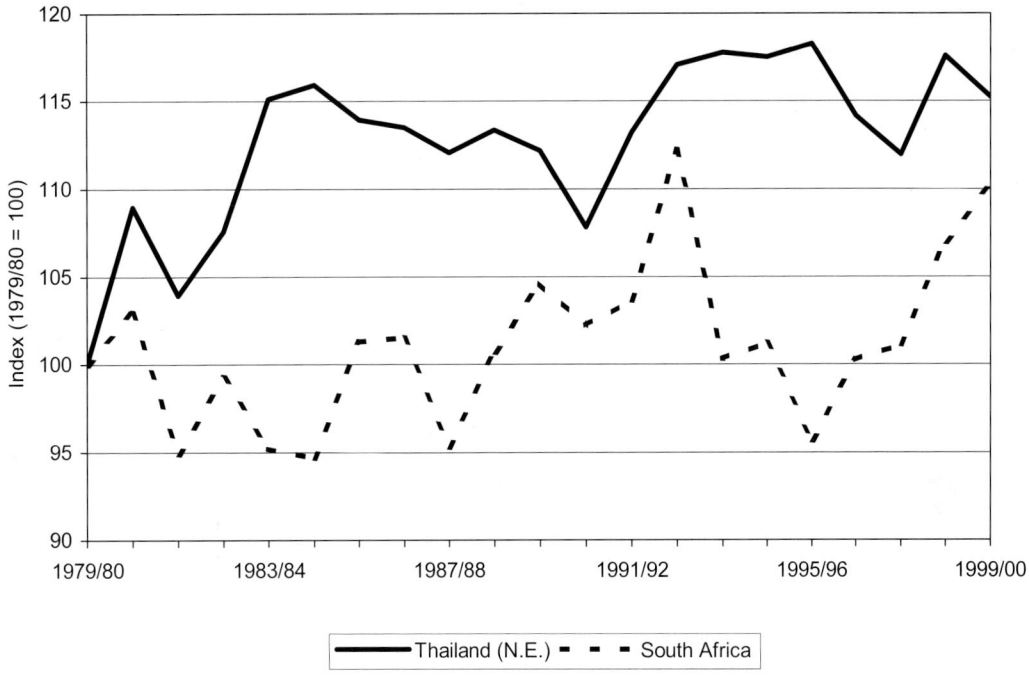

Fig. 8.6 Field performance in north/north-east Thailand and South Africa. Source: Sugar Milling Research Institute, LMC International Ltd.

growers' share of revenues. This trend has been re-inforced by the fact that cane quality, as measured by the CCS, has actually declined (Fig. 8.2). Using 1969–70 as a base year, the CCS index only reached around 90 in 1999–2000. The combination of these two trends has increased pressure on the margins earned by Australian growers.

The declining CCS experienced in Australia in the 1960s and 1970s is partly a result of changes in farming practices, which have included the intro-duction of mechanised harvesting and the decision to extend the ratoon cycle.

In all sugar industries, growers face a trade-off between maximising the quality of cane produced and minimising the costs associated with cane production. Despite the incentives that many cane payment systems provide to produce higher quality cane, growers may choose to employ cost-cutting measures, often at the expense of quality improvements, if they believe it offers the best way to improve returns. This seems to have been the case in Australia, where, from the growers' perspective, the benefits of mechanisation and a

longer ratoon cycle seem to have outweighed the returns offered by the cane payment system for higher quality cane.

CONCLUSION

Our analysis clearly shows that the importance of cane payment systems extends far beyond the establishment of rules under which revenue from sugar sales is divided between growers and proc-essors. Payment systems define the relationship between each group but, in doing so, exert a far-reaching influence throughout the sugar industry.

The division of proceeds between growers and processors will always lead to conflicts of interest between the two groups. As we have seen, in prac-tice it is difficult to guarantee that equity is always maintained while ensuring that incentives, for both growers and millers, are not distorted. It is not possi-ble to devise a perfect payment system, if for no other reason than industries evolve over time and payment systems often lag behind this development.

Arguably, the most appropriate structure is to integrate sugar operations so that the interests of both growers and producers are represented by a single sugar producer who is concerned by both factory and field performance. In this situation, sugar producers should decide to lengthen the season up until the point that the benefits gained from increased use of milling capacity equal the losses suffered from lower sucrose content. In other words, expansion decisions should be made, in theory, at or close to the margin, where economic efficiency is maximised.

Chapter 9
Project Planning

Ben Yates

INTRODUCTION

Earlier chapters of this book have dealt with the optimisation of the technical aspects of growing, harvesting and transporting sugarcane. These activities, however, do not take place in a vacuum. There needs to be an underlying economic reason for producing the sugarcane, and usually this will be a project requiring sugarcane as a raw material to produce sugar, ethanol or even energy derived from sugarcane biomass. Such projects do not just materialise – they require extensive planning followed by implementation of the plans. The purpose of this chapter is to provide an overview of processes involved in planning projects.

Project planning is concerned with the optimum use of resources and in a sugar project the relevant resources include land, water, people, and finance. The project cannot raise its finance unless an economic objective is being achieved, hence project planning will always involve an assessment of the markets for the proposed products and a comprehensive financial and/or economic analysis. The sustainability of the development will have to be demonstrated and generally clients and financiers will require an environmental impact assessment (EIA) to be undertaken. Provided that the planning has been thorough, the physical results and financial returns from the project should be comparable with those predicted in the studies.

Effective project planning will often proceed through a number of well defined stages:

- pre-feasibility study (project identification);
- feasibility study (project preparation); and
- detailed project planning.

Ideally the results of each stage will be evaluated and approved before the next stage is embarked upon. This approach will result in an impractical or inefficient project either being rejected at an early stage or being modified into a viable scheme.

The third of these stages can be quite costly and time consuming. Increasingly, therefore, that stage tends to proceed in parallel with the implementation of the project. The first two stages are, however, essential to the optimisation of a project and their omission or abbreviation can have disastrous consequences later on in the project cycle. Together they cover the range of activities needed to develop the initial ideas and aspirations of the client into a carefully researched and documented plan that can be successfully implemented.

It is not uncommon for clients to want to limit their expenditure during the early stages of investigation before the viability of the project concept has been proved. This can give rise to a paradoxical situation in which the client is impatient for quick, but definitive answers, which necessarily are based on inadequate data and study. Such pressure to truncate, or even skip, the pre-feasibility and feasibility stages must be resisted. The majority of the chapter is therefore devoted to these two key stages, largely as viewed from the perspective of the consultant.

THE ROLE OF THE CONSULTANT

The principal parties involved in a pre-feasibility or feasibility study include the client, relevant Government ministries of the country concerned, and the consultant. In a major new sugar project, the client in the 1960s and up to the 1980s was frequently the Government itself or a parastatal authority.

More recently, the continuing financial pressures on the developing countries (which are the main locations for sugarcane projects) have reduced almost to zero the number of such developments, which Governments can afford to finance as public sector projects. Moreover, it has become the conventional wisdom that Governments should not engage in the ownership or management of such enterprises that should operate within the private sector. New projects tend to require a major capital investment and greater financial returns may be offered by the rehabilitation of existing projects. The emphasis has therefore moved from new, Government-owned projects to rehabilitation programmes for existing projects or industries, generally in the private sector.

The participation by financing agencies in a project has always involved conditions. Provided that the project was technically feasible and financially viable, clear statements of support from the relevant Government (where the Government was the client) and satisfactory arrangements for management were often considered sufficient. Today, most of the major International Financing Institutions (IFI) require there to be a project sponsor who will typically be expected to provide 15–25% of the project equity, and to support the project in the case of unforeseen difficulties. The scope of the feasibility study may therefore need to be extended into the preparation of an outline financing plan and the identification of the project sponsor.

In this situation what is the role of the consultant? To the professional consultant it is self-evident that his services are essential. Many will have seen examples of sugar factories that have totally inadequate cane supplies, or even large areas of cane without a factory. Then there are examples of factories designed to be supplied with cane from irrigated estates for which the dam has not been built – or even an expensive dam and associated works without the associated agricultural development.

The consultant's role is essentially to ensure that the project plans do indeed constitute the best available plan, are internally consistent, are financially viable and meet fully the client's objectives (which may have been modified in the light of the consultant's advice). Yet many consultants will know some Minister of Agriculture who wants projects – but no more reports. No doubt the shelves of his Ministry are lined with reports that gather dust. The resources wasted in the preparation of these reports are indeed great, but much less serious than the waste involved in implementing even one inefficient or ill-conceived project.

What then may be done to ensure that the project achieves its primary objective, namely efficient and profitable use of resources? It is suggested that the key factors are as follows. First, it is necessary to use consultants of repute as, regrettably, it is always possible to find some group which will produce a report supporting even the most impractical scheme. In this context it should be noted that some consultants in the sugar sector have always been prepared to accept management responsibility for implementing any project they have studied. This greatly increases the likelihood that the advice and conclusions of the consultants will be realistic.

Secondly, it is important that the scope of the Terms of Reference (TOR) agreed between the client and the consultant is appropriate to the assignment. There is a danger that the client will expect too much from the consultant in a limited period of time by issuing TOR that are too broad. The consultant may accept these TOR assuming that various third parties have undertaken preparatory work to a satisfactory standard; if this advance work has not been done adequately, then the consultant's conclusions (and his reputation) may suffer. This risk is present at all stages of project planning, but is probably most serious at the feasibility study stage. Conversely, inflexible or restrictive TOR can also be a serious drawback. For example, detailed soil surveys may show that insufficient good land is available for the output originally envisaged and there may be no budget for re-casting the project objectives. Equally, the TOR may force the consultant to consider an area only for sugarcane when its characteristics might be more advantageous to cultivation of rice or other crops.

PRE-FEASIBILITY STUDY

Three principal types of pre-feasibility study may be distinguished. The first has the ultimate objective of establishing a specific enterprise, for example a new sugar project. In one example in Sri Lanka, the first study was commissioned by the Government and funded by the FAO/IBRD Co-operative Programme to ascertain whether existing areas of rainfed sugarcane, planted for jaggery (coarse sugar) production, could successfully be used in the development of a new sugar project. The main issues were the type of processing technology (conventional or 'appropriate'), project scale, location, and the proportions of cane to be supplied by the existing scattered private farmers, new settlers, and a nucleus estate. The findings of this study were positive, and the scheme proceeded through the feasibility study stage to implementation. The same consultancy company was involved in all stages, and was asked by the client – at that stage the Government – to take on the role of project promoter; the consultant associated with a leading merchant bank for the raising of project finance. The project company, the Pelwatte Sugar Company, was incorporated and became the official client.

A second type of pre-feasibility study is that relating to a country's total sugar industry; in economist's terminology – a sugar sector study. Here the objective is to develop a logical programme for the expansion, rehabilitation or rationalisation of the industry. The 1970 Indonesia Sugar Study was perhaps the most daunting task a consultant could undertake. Some 56 existing factories and cane growing areas on the island of Java were examined, and 20 potential new cane-growing areas were identified (and examined) in locations throughout the Indonesian archipelago. Cane was grown in rotation with the preferred crop (rice), the existing factories were small and old, and sugar production had declined to little more than one-third of the peak production level of 1940, to the extent that this traditional sugar exporter had become a major importer.

The report's main recommendations were:

- a programme of rationalisation of factory units, some being closed whilst others were rehabilitated and expanded; and
- expansion of the industry into new areas, mainly on the outer islands.

Some 30 years after its completion, the report still provides valuable source data and concepts that are used in current planning for the sugar industry.

The third type of pre-feasibility study is concerned with identifying the optimum use of natural resources in a defined area. Here the client may be a river valley development authority or a large, private sector landowner who is seeking professional advice in drawing up a development programme. Sugarcane would only be one of the options to be considered and the key issue to evaluate would be the comparative costs and returns of the various crops that could be grown in that agroclimatic environment.

In the first type of pre-feasibility study, the client may have (or may believe he has) a clear view of the overall project objective (say, increasing sugar production by 100 000 t/year within 5 years), but often requires assistance from the consultant in selecting, from a wide range of alternatives, the optimum means of achieving the objective. It is rarely possible (because of time and cost) or even desirable, to undertake a fully detailed evaluation of all the alternatives; thus a broad brush, preliminary evaluation of the major alternatives is required. This is the core function of the pre-feasibility or project identification study, which will:

- identify the major project options;
- undertake preliminary technical and financial analyses of the feasible options; and
- indicate the option recommended on technical and financial grounds.

During this stage it is particularly important that the consultant focuses upon (and endeavours to resolve) the key issues involved and does not become unduly distracted by details – however interesting these may be to him personally. For a sugar project the key issues may include:

- whether the land and the climate are suitable for sugarcane (or beet), or whether other crops would make better use of the available resources;
- the organisation of the agricultural production, i.e. estates, large farms or smallholders;
- whether irrigation is feasible and financially justified;
- the yields anticipated and, thus, the area and approximate location of the land required;
- the appropriate harvest season;
- major development costs which will vary from one site to another (e.g. flood protection measures, land clearing, provision of basic infrastructure);
- whether large-scale conventional processing technology or small-scale 'alternative' technology is appropriate;
- whether sugar can be produced at a price which will allow it to be sold profitably in the available markets; and
- political and social issues which may influence the choice of alternatives and the decision to implement the project.

In order that issues of this type may be dealt with thoroughly and yet speedily, it is particularly important that the individual consultants are all widely experienced in a number of interrelated disciplines. This is to ensure that the team for this stage is not too large; not more than four to six would be ideal, but occasionally the range of disciplines required necessitates a larger team. The project identification report would normally be produced within 3–4 months from commencement and might typically involve 200–300 man-days for a relatively straightforward assignment.

In the pre-feasibility study report the client should be presented with a clear project recommendation, on technical and financial grounds. Any alternatives should be ranked, as it is possible that the client may have overriding social (e.g. regional development), economic (e.g. changes in foreign exchange balances) or other reasons to select an apparently less attractive option. This ranking is chiefly applicable when the client is the Government. Even projects that are non-viable (in

the narrow financial sense) may be implemented because of special factors. In one example of this, the special factor was pacification of the project region that finally resulted in a major reduction in the national defence budget. The client, therefore, makes the final decision and defines (preferably with the assistance of the consultant) the TOR for the next stage – the feasibility study.

FEASIBILITY STUDY

The feasibility study ideally takes as its starting point the outline plan selected in the pre-feasibility study. In some cases, the client will have omitted this earlier stage and will simply define the feasibility study objectives and the major project parameters. These may, however, be inconsistent or would result in a non-feasible project. The best option in these circumstances is to revert to the scope and objectives of the feasibility study before proceeding further.

The feasibility study has to provide the definition of all the technical and financial aspects of the proposed project in sufficient detail to be confident that:

- the project is technically sound and can be implemented successfully on the timescale indicated in the report;
- the market forecasts demonstrate that the sale of the project's products are likely to be achieved at the projected prices; and
- the project financial returns are sufficiently attractive, and robust, to enable the project finance to be raised.

The study report has the overall purpose of providing all the information needed to advise the client whether or not the project is feasible and therefore should proceed to implementation. To achieve these objectives the study must, as a minimum:

- define all the project inputs in physical and financial terms;
- provide an implementation programme including staff recruitment, training and development;

- recommend a contract plan for the establishment of the processing plant and any necessary infrastructure;
- investigate fully the markets and prices for the products; and
- summarise all forecasts in terms of a phased project cash flow.

The physical aspects of the plan are generally the least difficult aspects of the feasibility study – though sometimes these aspects are given disproportionate attention by technical specialists. The contract plan will vary enormously according to local circumstances, and the spectrum of options will range from comprehensive turnkey contracts via engineering procurement and construction contracts (EPC), to establishing the facilities using the client's in-house resources. This is a very wide subject that is more properly the subject matter of project engineering and management texts than a cane sugar handbook. Those seeking further information on this area are referred to the relevant publications listed at the end of this chapter.

Total cost estimates must be reasonably accurate, even though firm quotations for each item cannot normally be obtained at this stage. Similarly, precise definitions of actual techniques and equipment are not required: during implementation management should be allowed flexibility within its total budget to modify specifications and techniques to suit the circumstances at that time. Some clients call for great detail in a feasibility study: this can be done but it requires considerable time, and is expensive; normally this detailed work is best done during the implementation phase.

Provided that the project objectives have been adequately defined either by the client or, preferably, in a pre-feasibility report, the task of the feasibility study is conceptually straightforward. In the real world, however, complications tend to arise from two main sources:

- incomplete and imprecise data; and
- the difficulties of simultaneously optimising all the various aspects of the project.

There is no perfect solution to the former problem; one must simply use the best data that can be found or developed. To take one specific example, the rainfall records in the project area may be incomplete. If there is a high correlation between data from the project area and from a nearby area with complete records, then the project area rainfall can be estimated, using appropriate statistical techniques, thereby allowing the calculation of crop yield potential and opportunity days for harvesting.

Even where complete and reliable data are available and used it is important, however, to recognise that forecasts (e.g. yields or rendement) inevitably involve an element of judgement. Best practice is generally considered to be:

- to present prudent mid-range forecasts; and
- to evaluate in the project financial analysis the consequences of changes in these key forecasts.

Frequently, therefore, in the so-called 'sensitivity analysis', the project returns will be re-calculated for cane yields, rendement, costs (operating and capital), sugar prices, etc., which are 10% better or worse than the base case assumptions.

This methodology will demonstrate which project variables have the greatest impact on project viability. As a general rule, sugar projects are more sensitive to changes in sugar prices and operating costs than to changes in capital costs. This does not imply that the level of the capital costs is not important: on the contrary, if it seems to be too high in relation to industry norms then it is unlikely that the necessary finance will be made available and the project will therefore not be able to proceed.

Sometimes, the quality of the planning data available allows a more sophisticated analysis of project risks to be undertaken. Where the frequency distribution of parameter values (e.g. expected yields, performance rates, prices, costs, etc.) can be stated with some confidence, appropriate computer software can be used to calculate the probability of achieving certain levels of project outcome (e.g. profit, rate of return on capital, etc.).

It is fairly obvious why sugar projects are highly sensitive to the sugar price. If the project's operating margin (revenue less costs) is 10%, then a 10% decrease in price will reduce the project's return to zero. Conversely, an increase of 10% will double the return. The practical problem is how to assess the probable level of sugar prices, not just in the next year or two but also for 10 or 20 years hence. For many years the standard approach to price forecasting involved:

- taking the long run price projections for raw sugar in constant US$, produced by the IBRD;
- adjusting for the quality of sugar actually being produced (mill white, refined, etc.); and
- adjusting for freight and insurance to give an import parity landed price in the project region.

This import parity price was generally recognised as providing a reasonable guide as to the probable level of sugar prices.

The IBRD forecasts were based on the concept that the world sugar price would tend to reflect the costs of bringing on line the new production capacity required to satisfy the expanding demand for sugar. Forecasts for raw sugar prices, in the 1980s, were typically in the region of US¢12–14/lb or US$285/tonne, fob, in terms of 1980 US$. At these prices, any reasonably efficient new sugar project would by definition, be shown to be viable as the sugar price estimate was itself based upon this type of calculation.

Regrettably there were and are huge distortions in the world sugar market because virtually all sugar-producing countries have erected barriers to free trade to protect their own domestic markets. In this situation, the IBRD model failed to forecast world sugar prices with any degree of accuracy. For example, prices in 2000–2003 averaged only about one-third of the earlier IBRD forecasts.

In the absence of a respected, independent source of price forecasts it has become extremely difficult to define a satisfactory methodology. For world market sales, one approach is to simply extrapolate the constant terms price trends that show a decline of around 1–2%, in line with the trends of many other agricultural commodities. Price forecasting within those markets presently enjoying tariff and quota protection is equally difficult as the future price level is entirely dependent upon the actions of national Governments and any changes to international trade agreements.

The other area of difficulty in a feasibility study is that of simultaneously optimising the various elements which constitute the total project. In principle, many of the interactions could be optimised using sophisticated techniques such a linear programming. In practice, the database is usually inadequate and the use of such techniques on inadequate data can lead to doubtful conclusions unless the consultant is very careful. Thus to a large extent one is forced to rely upon the experience of the individual consultants optimising their own sections of the plan and the team leader, reinforced by the financial analyst, synthesising the individual contributions into a logical total pattern.

One example of this problem occurred in a project that required the establishment of an extensive road network to provide access to the new areas on which outgrowers would grow cane to supply factory expansion. Complex analysis resulted in a least cost solution to the agricultural production options, taking account of not just the road development costs but also the anticipated cane yields in the different areas that had variable soils and rainfall. A simple analysis, however, showed that the level of cost optimisation was insignificant compared to the high marginal returns from processing extra cane in the factory, hence the priority was to maximise the production of cane, almost irrespective of roads, soils and rainfall.

A further example occurred in a study of a national sugar industry comprising a number of separate factories and estates supplying the cane. The complication here was that the size and efficiencies of the factories varied significantly, as did the productivity of the estates, some of which could supply cane to more than one factory. After several false starts the eventual and highly successful solution was to take a modular approach to the factory development options and combine these in different ways with a comparable range of the estate development options. This fairly complex approach resulted in new insights into the options for optimising the industry. However, the political dimensions of the technically most attractive options have so far prevented their implementation.

Typically a project preparation report for a new sugar project would require 4–6 months for completion and would involve 600–800 man-days. If extensive soil surveys were required, the total man-days would be much greater. The costs are therefore significant and consequently it is important that the client is satisfied, before committing this expenditure, that the project has a reasonable prospect of proceeding to implementation. This re-emphasises the need for a prior (and less costly) pre-feasibility study.

With the present emphasis on the role of the private sector in owning and operating business enterprises it is evident that the market and financial analyses are of critical importance. Unless the analysis here is complete, and convincing, the project is unlikely to proceed. In this context it may be queried whether there is still a role for the economic (as opposed to financial) evaluation of a project. By this is meant the re-evaluation of the project cash flows in terms of the economic (as opposed to financial) cost of the inputs and value of the outputs.

Sugar projects are often on a scale such that they have a significant impact on large numbers of people, occupy large areas of land, use large volumes of irrigation water (for which there may be alternative uses), and they can significantly influence the economy as a whole by their role in developing new regions of the country. Even a private sector project will usually require the active or passive support of the Government for its plan, and this support will be more readily available if the economic benefits are clearly demonstrated. Hence both financial and economic analyses will usually be required for a major new or rehabilitation project.

A frequently stated requirement of a feasibility study report is that it is a 'bankable document'. In other words, it is a document of sufficient stature that it may be used in the raising of project finance. To achieve this objective the report must be clear, complete, and, most important, authoritative. In the context of sugar projects the consultant should have sufficient confidence in the conclusions of the study to offer to implement and subsequently manage the project on the basis of the plans and forecasts presented in the report. This approach is adopted by several of the major consultants in this field who operate and manage sugar projects throughout the world. This is perhaps the most important contribution that can be made by a consultant, in effect providing a corporate guarantee that the project is practical and realistic.

FURTHER READING

Project planning can be viewed as a subsection of development economics, a subject that is extensively covered in the literature. Some useful texts that concentrate on project planning and project appraisal are listed below, together with a selection of the literature dealing with engineering project management.

Baum, W.C. & Tolbert, S.M. (1985) *Investing in Development.* World Bank, Washington, DC.

British Standards Institution (2000) *BS 6079 Project Management, Part 1: Guide to Project Management.* BSI, London.

British Standards Institution (2001) *BS 6079 Project Management, Part 2: Vocabulary.* BSI, London.

British Standards Institution (2002) *BS 6079 Project Management, Part 3: Guide to the Management of Business Related Project Risk.* BSI, London.

British Standards Institution (2001) *BS IEC 62198 Project Risk Management – Application Guidelines.* BSI, London.

Chartered Institute of Building (1996) *Code of Practice for Project Management for Construction and Development,* 2nd edn. Longman, Harlow.

Curry, S. & Weiss, J. (2000) *Project Analysis in Developing Countries,* 2nd edn. Macmillan, New York.

Gittinger, J.P. (1982) *Economic Analysis of Agricultural Projects.* World Bank/Johns Hopkins University Press, Baltimore, MD.

Hamilton, A. (2001) *Managing Projects For Success, A Trilogy.* Thomas Telford, London.

Harrison, F.L. (1985) *Advanced Project Management.* Gower, ISBN 0 566 02475 6.

Haslam, J. M. (2002) *Writing Engineering Specifications.* E. & F. N. Spon, London.

Holroyd, T. (1999) *Site Management for Engineers.* Thomas Telford, London.

Holroyd, T. (2000) *Principles of Estimating.* Thomas Telford, ISBN 0 7277 2763 X.

Horgan, M. O'C. (1984) *Competitive Tendering for Engineering Contracts.* E. & F. N. Spon, London.

Horgan, M.O'C. & Roulston, F.R. (1989) *The Foundations of Engineering Contracts.* E. & F. N. Spon, London.

Horgan, M. O'C. & Roulston, F.R. *Project Control of Engineering Contracts.* E. & F. N. Spon, London.

Landon, J.R. (ed.) (1991) *Booker Tropical Soil Manual: A Handbook for Soil Survey and Agricultural Land Evaluation in the Tropics and Sub-Tropics.* Longman, Harlow.

Lansley, P.R. & Harlow, P.A. (eds) (1989) *Managing Construc-*

tion Worldwide, Vol. 3. Routledge, London.

Loftus, J. (ed.) (1999) *Project Management of Multiple Projects and Contracts*. Thomas Telford, London.

Morris, P. (1997) *The Management of Projects*. Thomas Telford, London.

Thompson, P. & Perry, J. (1992) *Engineering Construction Risks*. Thomas Telford, London.

UN Food and Agriculture Organisation (1986) *Guide for Training in the Formulation of Agricultural and Rural Investment Projects*. FAO, Rome.

UN Food and Agriculture Organisation (1997) *Formulation of Agricultural and Rural Investment Projects: planning tools, case studies and exercises*. FAO, Rome.

Author Index

Abbott, E.V., Zummo, N. & Tippett, R.L., testing for disease resistance 60
Abbott, E.W., sugarcane diseases in Louisiana, USA 55
Albert, H.H. & Schenk, S., molecular markers 44
Al-Janabi, S.M., Honeycutt, R.J., McClelland, M. & Sobral, B.W.S., genetic linkage map of *S. spontaneum* 25
Allam, A.I. & Abou Dooh, A.M., varietal resistance to borers 83
Allsopp, P.G.
locust spp. in Australia and damage caused 94, 95
sap feeder spp. in Australia 93
Allsopp, P.G. & McGhie, T.K., control of white grubs 87
Allsopp, P.G., Bull, R.M. & McGill, N.G., reduction in sugar yield through soil pest damage 85
Allsopp, P.G., Chandler, K.J., Samson, P.R. & Story, P.G.
control of cicadas and earth pearls 89
control of white grubs 87
life cycles of white grubs 86
locust spp. in Australia and damage caused 94, 95
rodent spp. and control 95–6
Allsopp, P.G., McGhie, T.K., Cox, M.C. & Smith, G.R., breeding for canegrub resistance 39
Allsopp, P.G., McGhie, T.K., Smith, G.R., Ford, R. & Cox, M.C., varietal resistance and control of white grubs 87
Allsopp, P.G., McGill, N.G., Licastro, K.A. & Milner, R.J., biological control of white grubs 88
Anderson, D.L., Snyder, G.H. & Martin, F.G., necessary nutrient elements for growth 151
Antoine, R. & Ricaud, C., screening for

resistance to leaf scald disease 66
Arencibia, A., Molina, P.R., de la Riva, G. & Selman-Housein, G., production of transgenic sugarcane 45
Arencibia, A., Vazquez, R.I., Prieto, D. *et al.*, biotechnological methods for borer control 83
Artschwager, E., characteristics of clones 8
Artschwager, E. & Brandes, E.W., ancestry of sugarcane 2
Artschwager, E., Brandes, E.W. & Starrett, R.C., anatomy of inflorescence 14
Ashref, M. & Fatima, B., breeding for resistance to Chilo spp 39
Autrey, L.J.C., Dhayan & Sullivan, S., losses due to gumming disease 64
Autrey, L.J.C., Saumtally, S., Dookun, A., Sullivan, S. & Dhayan, S., virulence of leaf scald disease strains 65
Awadallah, W.H., El-Metwally, E.F. & Aly, F.A., effect of insecticides against borer damage 82
Azab, Y.E. & Chilton, S.J.P., inheritance of red rot resistance 60

Bailey, R.A., hot water treatment (HWT) control of smut disease 63
Bailey, R.A. & Bechet, G.R., crop loss due to RSD 68
Bailey, R.A. & Fox, P.H.
control scheme against RSD 68
time of planting and mosaic disease incidence 72
yield losses due to mosaic disease 70
Bailey, R.A. & McFarlane, S.A.
crop loss due to RSD 68
spread of RSD 69
Bailey, R.A. & Tough, S.A., spread of RSD 69

Baird, D., use of nearest neighbour models 35
Baker, R.E.D., Martyn, E.B. & Stevenson, G.C., sugarcane diseases in the Caribbean 55
Bakker, H.
control of flowering 152
field design 145
replanting gaps in rows 155
Balance, M.C., Milanes Ramos, N. & Mesa Lopez, J.M., smut disease resistance testing 38
Balu, R., production of commercial sugar in India *c.* 2500 years ago 1
Barber, C.A.
origin of sugarcane 3
study of cane morphology 8
types of Saccharum barberi 7
Barnes, A.C.
field design 145
importance of surface drainage 146
India home of sugarcane 3
Mexican sugar industry 4
Bartlett, M.S. nearest neighbour models in field experiment analyses 35
Benath, L.L. & Monteith, M.H. soil oxygen deficiency and root growth 144, 151
Berding, N.
improved flowering 28
selection for weevil borer resistance 40
Berding, N. & Koike, H., germplasm conservation 21
Berding, N. & Moore, P.H., optimised flowering induction 28
Berding, N. & Roach, B.T., germplasm collection, maintenance and use 21, 22, 23, 26
Berding, N. & Skinner, J.C.
improvement in sugarcane fertility 21, 28
increase in productivity 36, 42
Berding, N., Brotherton, G.A. le Brocq,

D.G. & Skinner, J.C., near infrared reflectance (NIR) spectroscopy for sugarcane analysis 36

Berding, N., Moore, P.H. & Smith, G.H., review of sugarcane breeding 20

Berding, N., Skinner, J.C. & Ledger, P.E., screening clone resistance against common rust disease 38

Besse, P., McIntyre, C.L. & Berding, N., rDNA variations in *Erianthus* and characterisation of its germplasm 23, 25, 29

Bessin, R.T., Moser, E.B. & Reagan, T.E., use of insecticides against borers 81

Bessin, R.T., Reagan, T.E. & Martin, F.A., measurement of varietal resistance to borers 83

Birch, R.G., Bower, R., Elliott, A., Potier, B., Franks, T. & Cordeiro, G., expression of foreign genes in sugarcane 45

Blackburn, F., soil temperatures and root growth 13

Bleszynski, S., taxonomy of *Diatraea* 80

Boedijono, W.A., cost for insecticidal control of borers 82

Bond, R.S.
observations on family selection 32
selecting for eldana resistance 39
use of controlled environments 27

Bower, R. & Birch, R.G., transgenic sugarcane plants via micro projectile bombardment 27, 45

Brandes, E.W.
distinguishing *Saccharum* species 4
qualities of noble canes 6
swidden agriculture 2
transfer of sugarcane 3

Breaux, R.D., breeding to improve sugar production, breeding strategies by bi-parental means and polycrosses 37

Breaux, R.D. & Miller, J.D., seed handling, germination and seedling propagation 30

Brett, P.G.C., flowering and pollen fertility 27

Brotherton, G.A., Cross, K.W.V. & Stewart, P.N., breeding for low starch content 40

Brown, A.H.D., Daniels, J. & Latter, B.D.H., quantitative genetics: analysis of variation and correlation analysis 31, 40

Buckle, A.P., rodent control 96

Bull, J.K., Basford, K.E., Cooper, M. & DeLacy, I.H., enhanced interpretation of pattern analyses of environments 34

Bull, J.K., Basford, K.E., DeLacy, I.H. & Cooper, M., classifying genotypic data, determining appropriate group numbers and composition for data sets 34

Bull, J.K., Bull, T.A. & Cooper, M., importance of water and nitrogen in generating clone x environment interaction 34

Bull, J.K., Hogarth, D.M. & Basford, K.E., impact of genotype x environment interaction 32, 33

Bull, J.K., Mungomery, V.E. & Hogarth, D.M., improvements in commercial production 36, 37, 42

Bull, T.A., row spacing and potential productivity 41

Bull, T.A. & Bull, J.K., increasing yields through higher planting density 41

Bull, T.A. & Glasziou, K.T., floral initiation 14

Burner, D.M., cytogenetic analysis 24

Burner, D.M. & Legendre, B.L.
chromosome transmission and meiotic stability 24
genome amplification for the subtropics 23

Burner, D.M. & Webster, R.D., cytological studies on North American species of Saccharum 21

Burner, D.M., Grisham, M.P. & Legendre, B.L., resistance of sugarcane relatives to smut disease 24

Burnett, J., social history of diet in England 1

Burnquist, W.L., development and application of restriction length polymorphism 22

Cackett, K.E., damage by earth pearls and control 89

Carnegie, A.J.M., management practices for borer control 84, sap feeder spp. 91, 92

Carnegie, A.J.M. & Leslie, G.W., pheromone trapping 85

Carnegie, A.J.M., Dick, J. & Harris, R.H.G., leaf feeder spp. in South Africa 94

Celarier, R.R., evolution of sugarcane 2

Chang, Y-S., pedigree analysis of genetic relationships and implications of inbreeding 26

Chang, Y-S. & Lo, C.C., genetic relationships among cultivars 26

Chang, Y-S. & Milligan, S.B., univariate and bivariate cross prediction methods 29, 32

Chapman, L.S., increase in sugar yield potential from breeding 37

Charernsom, K. & Suasa-ard, D.W.
control of white grubs in Thailand 87
white grub species in Asia 86

Chen, C.T. & Kusalwong, A., white leaf disease 73

Chen, Y.C. & Lo, C.C., disease resistance and sugar content 23

Cheng, W.Y., loss in sugar because of damage by borers 79

Chona, B.L., sugarcane diseases in India 55

Chowdhury, M.K.U. & Vasil, L.K., molecular analysis of plants regenerated from embryonic cultures 45

Clarke, M., Reece, N.E. & Elcock, H.L., outbreak of smut disease in Barbados 55

Clowes, M. StJ. & Breakwell, W.L.
control of termites 88
fungicide treatment against smut disease 114

Coleman, R.E., floral initiation 14

Comstock, J.C. & Lockhart, B.E.L., sugarcane bacilliform virus (SCBV) 73

Comstock, J.C., Irvine, J.E. & Miller, J.D., yellow leaf syndrome (YLS) 39

Comstock, J.C., Miller, J.D., Shine, J.M. & Tai, P.Y.P., tissue culture-generated disease-free plants for controlling RSD 69

Comstock, J.C., Shine, J.M. & Raid R.N., effect of common rust disease on growth and yield 61

Comstock, J.C., Shine, J.M., Tai, P.Y.P. & Miller, J.D., breeding for RSD resistance 57

Conlong, D.E.
biological control of *Eldana saccharina* 81
borers in Far East and Australasian region, biological control of white grubs 88

Cooper, M., Woodruff, D.R., Eisemann,

R.L., Brennan, P.S. & DeLacy, I.H., genotype x environment interaction in wheat 34

Cox, M.C. & Croft, B.J., selection for smut disease resistance in Australia 57

Cox, M.C. & Hansen, P.B., improved productivity trends with introduction of new varieties 36, 37, 38, 42

Cox, M.C. & Hogarth, D.M., effectiveness of family selection 32

Cox, M.C., Hogarth, D.M. & Hansen, P.B., breeding for high early season sugar content 37

Cox, M.C., Hogarth, D.M. & Mullins R.T., potential gain in sugar production related to time of harvest 37

Cox, M.C., McRae, T.A., Bull, J.K. & Hogarth, D.M., family selection improves efficiency and effectiveness 32

Croft, B.J. & Margarey, R.C., *Pachymetra* root rot 39

Croft, B.J., Greet, A.D., Leaman, T.M. & Teakle, D.S., detection of RSD pathogen 68

Cromarty, A.S., Ellis, R.H. & Roberts, E.H., design of seed storage facilities 30

Cronje, C.P.R. & Bailey, R.A., world distribution of yellow leaf syndrome (YLS) 72

Cronje, C.P.R., Tymon, A.M., Jones, P. & Bailey, R.A., yellow leaf syndrome (YLS) caused by a phytoplasma 39, 72

Cuenya, M.I. & Mariotti, J.A., potential gain in sugar production related to time of harvest 37

Cullis, B.R., Gleeson, A.C. & Thomson, F.M., analysis of early generation variety trials 35

Da Silva, J., Honeycutt, R.J., Burnquist, W., Al-Janabi, S.M., Sorrells, M.E., Tanksley, S. D. & Sobral, B.W.S., genetic linkage map 25

Da Silva, J., Sorrells, M.E., Burnquist, W.L. & Tanksley, S.D., RFLP linkage map and genome analysis 25

Daniels, J. & Roach, B.T., taxonomy and evolution of sugarcane 21

Daniels, J., Husain, A.A. & Hutchinson,

P.B., sugarcane diseases in Fiji 55

Daniels, J., Smith, P., Paton, N. & Williams, C.A., origin of genus Saccharum 20

Dasart, L. & Victorine, C., rodent spp. in Guyana and control 95, 96

Davis, M.J. & Bailey, R.A., description of RSD pathogen 67

Davis, M.J., Dean, J.L., Miller, J.D. & Shine, J.M., detection of RSD pathogen 68, screening for resistance to RSD 69

Davis, M.J., Gillaspie, A.G., Harris, R.W. & Lawson, R.H., causal agent of ratoon stunting disease 67

Davis, M.J., Gillaspie, A.G., Vidaver, A.K. & Harris, R.H., causal agent of ratoon stunting disease 67

Davis, M.J., Rott, P., Warmuth, C.J., Chatenet, M. & Baudin, P., new strain of leaf scald disease in Florida 65

Dean, J.L., screening for resistance to mosaic disease 71

Dean, J.L. & Davis, M.J., crop loss due to RSD 68

Deerr, N., early production of sugar 5, discovery of sugarcane fertility 6, incorrect source of Saccharum sinense 7

Dent, D.

   host plant resistance against borers 82

   monitoring methods 84

D'Hont, A., Grivet, L., Feldmann, P., Rao, S., Berding, N. & Glaszmann, J.C., characterisation of double genomic structure 25

D'Hont, A., Grivet, L., Lu, Y.H. *et al.*, genome of modern sugarcane varieties 25

D'Hont, A., Lu, Y-H., Feldmann, P. & Glaszmann, J.C., cytoplasmic diversity in sugarcane 26

D'Hont, A., Rao, P.S., Feldmann, P. *et al.*, identification and characterisation of sugarcane intergeneric hybrids 23, 25, 29

Dodds, J.H., in vitro methods for conservation 27

Duke, N.H. & Eastwood, D., management practices for borer control 84

Earle, F.S., disease resistance 7

Edgerton, C.W., red rot disease and poor cane stands in ratoons 60

Egan, B.T., Ryan, C.C. & Franki, R.I.B., Fiji disease 73

Eksomtramage, T., Paulet, F., Noyer, J.L., Feldmann, P. & Glaszmann, J.C., utility of isozymes 22

El-Amin, E.M., management practices for borer control 84

Ellis, R.D., Wilson, J.H. & Spies, P.M., irrigation policy to optimise yield 113

Embaby, M.M., end of noble cane era 5–6

Espinoza, R. & Galvez, G., genotype x environment interaction 33

Evans, H., categories of roots 13

Evtushenko, Ll., Dorofeeva, L.V., Subbotin, S.A., Cole, J.R. & Tiedje, J.M., causal agent of ratoon stunting disease (RSD) 67

Fauconnier, R. & Bassereau, D., mechanisation guidelines for cane harvesting 164

Fegan, M., Croft, B.J., Teakle, D.S., Hayward, A.C. & Smith, G.R., detection of RSD pathogen 68

Fehr, W.R., yield gain linked to genetic improvement 42

Ferreira, S.A. & Comstock, J.C., smut disease 61, 63

Ferrer, W.F.

   control of white grubs 86, 87

   first use as sweetener in China and India 1, 2

Florez, S., biological control of borers 81

Frison, E.A. & Putter, C.A.J.

   hot water treatment to control leaf scald disease in quarantine 66

   safe movement of sugarcane through quarantine 55

Fukui, K., Ohimido, N., Ha, S. & Moore, P.H., identification of wild sugarcane chromosomes 25

Gallacher, D.J., development of minimum descriptor set 22

Gallacher, D.J., Lee, D.J. & Berding, N., use of isozymes 22

Galloway, J.H.

   description of clones 8

   sugar as staple food 1, 2

Ganeshan, S. & Rajabalee, A.

   control of leaf feeders 95

   leaf feeder spp. in Mauritius and

damage caused 94

Gillaspie, A.G. & Teakle, D.S., heat-treatment to control RSD 67, 69

Gillaspie, A.G., Davis, R.E. & Worley, J.E., causal agent of ratoon stunting disease (RSD) 67

Glimelius, K., evolution and crop improvement 27

Gomez, L.A., artificial rearing of parasitoids and effective control of Diatraea sp. 80

Grassl, C.O., *Saccharum edule* not an authentic *Saccharum* species 4

Greathead, D.J.
locust spp. in Africa and control 94, 95
sap feeder damage 90

Grimes, N. & des Vignes, W.G., monitoring froghopper populations 85

Grisham, M.P.
mosaic and yield losses due to disease 70
strains of smut disease 63

Grisham, M.P., Pan, Y.B., Legendre, B.L., Godshall, M.A. & Eggleston, G., effect of yellow leaf syndrome (YLS) on yield and juice quality 73

Grivet, L., D'Hont, A., Dufour, P., Hamon, P., Roques, D. & Glaszmann, J.C., comparative genome mapping 25

Gupta, S.C. & Gupta, A.P., sap feeder damage 90

Hall, D.G., sap feeder spp. 92

Harborne, K.M., mosaic disease vector 70

Harlan, J.R. & De Wet, J.M.J., classification of cultivated plants 21

Harvey, M., Huckett, B.I. & Botha, F.C., use of PCR and RAPDs 22, 26

Hassanien, M.H. & El-Naggar, M.Z., insecticidal control of borers 82

Heinz, D.J.
chromosome engineering 24, 26
yield improvement 41, 42

Heinz, D.J. & Tew, T.L., hybridisation procedures 28

Hemaprabha, G. & Ram, B., genetic variability in nobilisation stages of *S. robustum* 23

Hernandez, J., biological control of sap feeders 92

Herrera, G., Snyman, S.J. & Thomson, J.A., biotechnological methods for borer control 83

Hill, D.S., sap feeder spp. and biology 94

Hogarth, D.M.
gains from plant breeding 36, 37
sugarcane genetics, quantitative inheritance studies: estimation of variance components, correlations and predicted responses, effect of competition 31, 32

Hogarth, D.M. & Bull, J.K., implications of genotype x environment interactions 32

Hogarth, D.M. & Cross, K.W.V., inheritance of fibre characteristics 40

Hogarth, D.M. & Kingston, G., breeding for low ash content 40

Hogarth, D.M. & Mullins, R.T., mobile weighing equipment 31, 32, 35

Hogarth, D.M. & Skinner, J.C., computerisation of breeding records 29–30

Hogarth, D.M., Braithwaite, M.J. & Skinner, J.C., selection of families 32

Hogarth, D.M., Cox, M.C. & Bull, J.K., review of sugarcane breeding 20

Hogarth, D.M., Reimers, J.F., Ryan, C.C. & Taylor, P.W.J., quantitative inheritance of resistance to rust and Fiji disease 38

Hogarth, D.M., Wu, K.K. & Heinz, D.J., estimation of genetic variance 31, 37

Holden, J.R., principles and theory of irrigation and drainage 123

Hong, H.L., resistance to pokkah boeng disease 59

Horau, M., sugarcane diseases in Réunion 55

Hoy, J.W. & Flynn, J.L., tissue culture-generated disease-free plants as an RSD control method 69

Hsu, S-Y., Lin, C-J. & Lo, C.C., computer as an aid in crossing programme 29

Hsu, S-Y., Lo, C.C. & Shih, S.C., use of *Saccharum spontaneum* derivative in sugarcane crossing 23

Hughes, C.G., crop loss due to RSD 68

Hughes, C.G., Abbott, E.V. & Wismer, C.A., sugarcane diseases and world distribution 54, 55

Humbert, R.P., field design 145

Hunsigi, G., field design 145

Ingram, K.T. & Hilton, H.W., fertiliser, soil moisture and their effects on growth 150

Irvine, J.E. & Benda, G.T.A., row spacing and plant population 145

Irvine, J.E., Richard, C.A., Garrison, D.D. *et al.*
row spacing and crop yield potential 41
row spacing, plant population and yields in ratoon 146

Jackson, P.A.
genetic relationships between clones closely related to Saccharum spontaneum 24
use of performance in plant cane 33

Jackson, P.A. & Hogarth, D.M., patterns of response across sites and crop years 33

Jackson, P.A. & Roach, B.T., performance of sugarcane progeny 24

Jackson, P.A., Bull, J.K. & McRae, T.A., role of family selection 32

Jackson, P.A., Horsley, D., Foreman, J., Hogarth, D.M. & Wood, A.W., genotype x environment interactions in variety trials 33

Jackson, P.A., McRae, T.A. & Hogarth, D.M., selection across variable environments 32, 34

Jenkin, M.J., Reader, S.M., Purdie, K.A. & Miller, T.E., detection of rDNA sites 25

Jeswiet, J.
Pansahi clones included with Saccharum sinense 7
Saccharum barberi & *S. sinense* indigenous to India 3
study of sugarcane morphology in Java 8

Jones, P.N., Ferraris, R. & Chapman, L.S., minimizing confounding of genotype x year and genotype x crop type effects 33

Kandisami, P.A., Sreenivasan, T.V., Ramana Rao, T.C. *et al.*, catalogue on sugarcane genetic resources 21

Kang, M.S. & Miller, J.D., genotype x environment interactions 33

Kang, M.S., Miller, J.D. & Tai, P.Y.P., genetic and phenotypic path analyses and heritability 31

Kar, K., Gupta, S.C. & Kureel, D.C., testing for red rot disease resistance 60

Kay, M., synopsis of sprinkler systems and equipment 128

Keeping, M.G. & Leslie, G.W., varietal resistance to borers 83

Khan, Z.R., Chiliswa, P., Ampong-Nyarko, K. *et al.*, management practices for borer control 84

King, A.G., loss of recoverable sucrose because of *Eldana saccharina* damage 79

King, N.G., Mungomery, R.W. & Hughes, C.G., field design 145

Kira, M.T. & El-Sherif, H., loss in sugar because of damage by Chilo agamemnon 79

Koike, H. & Gillaspie, A.G., mosaic and yield losses due the disease 70

Kuniata, L.S.
insecticide use against borers in Papua New Guinea 82
lepidopterous and coleopterous spp. in the Far East, weevil borer in Papua New Guinea 80
loss in sugar because of damage by *Chilo sacchariphagus*, separation damage caused by *Sesamia grisescens* and *C. terrenellus* 79
management practices for borer control 84

Kuppen, J.P. & Leslie, G.W., use of methyl bromide for borer control 84

Law, C.N., genetic modification 44–5

Lee, D.J., enhancement of *Saccharum* spp. hybrid material 23, 29

Lee, M., DNA markers and plant breeding 43

Lefebvre, L.W., Ingram, C.R. & Yang, C., rodent spp. in Florida and damage 95

Legendre, B.L., increasing sucrose content 37

Leiva, E. & Barrantes, A., tramline system of drip irrigation in Venezuela 135

Leslie, G.W.
Eldana saccharina lays eggs on dead leaf material attached to stalks 78
pest status, biology and control measures 80

Leslie, G.W. & Keeping, M.G.
assessing cane borer resistance 83

rodent spp. in Mexico and damage 95

Linedale, A.L. & Ridge, D.R., control of harvesting losses 36, 40

Lingle, S.E. Wiedenfield, R.P. & Irvine, J.E., response to saline irrigation water 144

Liu, L.J., Rosario, T. & Roig, F.M., sugarcane diseases in Puerto Rico 55

Liu, M.C., plant generation from suspension culture protoplasts 44

Liu, Z.C., Sun, Y.R., Warg, Z.Y. & Lie, G.F., management practices and biological control of sap feeders 93

Lockhart, B.E. & Cronje, C.P.R., yellow leaf syndrome (YLS) 72

Lu, Y-H., D'Hont, A., Paulet, F., Grivet, L., Arnaud, M. & Glaszmann, J.C., molecular diversity and genome structure 26

Lu, Y-H., D'Hont, A., Walker, D.I.T., Rao, P.S., Feldmann, P. & Glaszmann, J.C., relationships between ancient species of sugarcane 25

MacDonald, D.W & Fenn, M.G.P., rodent biology 96

McDonald, L.M. & Milligan, S.B., field evaluation of check plot adjustments 35

McFarlane, S.A., Bailey, R.A. & Subramoney, D.S., detection of RSD pathogen 68

McGlinchey, M.G. & King, A.G., the CANEGROW climatic model 108

MacKinnon, A., use of polyacrylamide soil conditioners 127

McPhee, J., introduction of lemonade in Paris (1630) 1

McRae, T.A. & Jackson, P.A., selection of families 32, competition effects in selection trials 35

McRae, T.A., Bull, J.K. & Robotham, B.G., billet samples for measuring sugar content 35, 36

McRae, T.A., Bull, J.K., Robotham, B.G. & Sweetnam, R.C., measuring sugar content in variety trials 35–6

McRae, T.A., Hogarth, D.M., Foreman, J.W. & Braithwaite, M.J., selection of seedling families 32

Madan, Y.P., Maan Singh & Singh, M.,

control of termites 89

Magar, S.S., drip irrigation in India 133

Magarey, R.C., Croft, B.J. & Willcox, T.G. orange rust disease 55, 60, 61

Mangelsdorf, A.J., positive correlation between number of sunshine hours and sugar yield 16

Mariotti, J. A., potential gain related to time of harvest 37

Martin, J.P., sugarcane diseases and world distribution 54

Martin, J.P., Abbott, E.V. & Hughes, C.G., sugarcane diseases and world distribution 54–5

Martinez, A., use of insecticides against borers 82

Matassa, V.J., Basford, K.E. & Jackson, P.A., intergenotypic competition in single-row plots 35

Matassa, V.J., Basford, K.E., Stringer, J.K. & Hogarth, D.M., application of spatial analysis 35

Meager, R.L. Jun., Smith, J.W. Jun. & Johnson, K.J.R.
loss in sugar: damage by *Eldana loftini* 79
use of insecticides against borers 81

Merry, R.E., tramline system of drip irrigation in Swaziland 135

Milligan, S.B., Gravois, K.A., Bischoff, K.P. & Martin, F.A., crop effects on broad-sense heritabilities and genetic variances 33

Milner, R.J., Lomer, C.J. & Prior, C., control of termites 89

Mintz, S.W., comment on sugar consumers 2

Mirkov, T.E., Barrilleaux, A., Paterson, A.H., Yang, M. & Irvine, J.E., transgenic sugarcane 157

Mirzawan, P.D.N., Cooper, M. & Hogarth, D.M., magnitude and impact of genotype x environment interactions 33

Mirzawan, P.D.N., Cooper, M., De Lacy, I.H. & Hogarth, D.M., retrospective analysis of the relationships among test environments 33

Mohyuddin, A.I.
reduction in sugar yield through termite damage 85
sap feeder spp. in India and control 92
white grub species in India and Pakistan 86–7

Moore, P.H.,
    molecular biology 25
    physiology and control of flowering
        27, 152
    varietal improvement 41–2, 44
Moore, P.H. & Nuss, K.J., flowering
    synchronisation 27
Mrig, K.K. & Chaudhary, J.P., control of
    termites 88
Msomi, N. & Botha, F.C., molecular
    markers 44
Muchow, R.C., Hammer, G.L. &
    Kingston, G., achievements in
    sugarcane production 41
Mukherjee, S.K., extensive hybridisation
    to produce cultivated canes 4, 20
Murphy, D.J., production of industrial
    oils 45
Musikavanhu, F., use of ultra-violet light
    traps 85

Nawrath, C., Poirier, Y. & Somerville, C.,
    biodegradable plastic production
    45
North, D.S., losses due to gumming
    disease 64
Nuss, K.J., screening for eldana
    resistance 39

Orian, G., possible origin of gumming
    disease 64
Ortiz, R & Caballero, A., effectiveness of
    early selection procedures 32

Pandey, K.P., Singh, R.G. & Singh,
    S.B., insecticidal and biological
    control of borers 82
Panje, R.R., classification of *Saccharum
    spontaneum* on vegetative
    characters 7
Pantoja, N., biological control 81
Parthasarathy, N., definition of groups of
    sugarcanes 3
    Saccharum officinarum & *S.
    spontaneum* ancestral stock to N
    Indian canes 7
Pathak, R.S., Chilo resistance 39
Paulet, F., Engelmann, F. & Glaszmann,
    J.C., cryopreservation of apices
    27
Pearse, T.L., roguing to control smut
    disease 63
Pemberton, C.E. & Williams, J.R., source
    of sugarcane pests 78
Plucknett, D.L. & Smith, N.J.H.,
    improvement in production 36,
    42–3

Pollock, J.S., variety x environment
    interaction and selection 33

Raid, R.N. & Comstock, J.C., common
    rust disease 60
Rajabalee, A.
    Chilo spp. associated with sugarcane
        worldwide 80
    control of cicadas 89
    control of white grubs 87–8
    loss in sugar because of damage by
        Chilo sacchariphagus 79
    pheromone trapping by pheromones
        85
    white grub species in Africa 86
Ramana Rao, T.C., Sreenivasan, T.V., &
    Palanichami, K., catalogue on
    sugarcane genetic resources 21
Ramdoyal, K., Badaloo, G. & Mangar,
    M., effect of potassium
    metabisulphite on flowering 28
Reagan, T.E., sap feeder damage 90
Ricaud, C. & Autrey, L.J.C., description
    of gumming disease 64
Ricaud, C. & Ryan, C.C., description of
    leaf scald disease 64
Ricaud, C., Egan, B.T., Gillaspie, A.G.
    & Hughes, C.G., sugarcane
    diseases and world distribution
    54
Ridge, D.R. & Dick, R.G., green cane
    harvesting and dirt rejection by
    mechanical harvesters 40
Ridge, D.R. & Hurney, A.P., row spacing
    41
Riley, I.T., Jubb, T.F., Egan, B.T. & Croft,
    B.J., outbreak of smut disease in
    Australia 55
Roach, B.T. & Daniels, J., progress made
    by breeding 37
Rosenfield, A.H., parentage of H109 5
Ross, C.W., Sojka, R.E. & Lentz, R.D.,
    use of polyacrylamide soil
    conditioners 127
Rott, P. & Davis, M.J., description of leaf
    scald disease 64
Rott, P., Bailey, R.A., Comstock, J.C.,
    Croft, B.J. & Saumtally, S.,
    sugarcane diseases and world
    distribution 55, 63
Ryan, C.C. & Egan, B.T., common rust
    disease 60

Salazar, J., effective biological control 80
Salisbury, F.B. & Ross, C.W., necessary
    nutrient elements for growth 151
Samoedi, D., most serious cane borer sp.

in Java 80
Samoedi, D., Allsopp, P.G. & Kuniata,
    L., white grub species in
    Australia 86
Sandhu, S.S., Mehan, V.K. & Singh, K.,
    losses caused by red rot disease
    60
Saumtally, S., races of gumming disease
    64
Saumtally, S. & Dookun, A., possible
    origin of gumming disease 64
Scagliusi, S.M. & Lockhart, B.E.L.,
    viral pathogen of yellow leaf
    syndrome (YLS) 72
Shaver, T.N., Brown, H.E. & Hendricks,
    D.E., monitoring borer
    populations 85
Shaw, D.V. & Hood, J.V., use of clonal
    replicates 35
Sheridan, R.B., consumption of sugar 1
Simmonds, N.W.
    dispersal of smut disease 36, 38
    family selection 32
    principles of crop improvements 34
    strategies for the use of crop genetic
        resources 22
Simpson, G.B. & Reinders, F.B., tests on
    floppy system of irrigation 132
Singh, K. & Singh, R.P., red rot disease
    59
Singh, R.P. & Lal, S., red rot disease 59
Singla, M.L. & Duhra, M.S., insecticidal
    control of borers 82
Skinner, J.C.
    competition between varieties in
        trials 34
    controlled pollination and selection
        28
Skinner, J.C. & Hogarth, D.M., use of
    border rows 34
Skinner, J.C., Hogarth, D.M. & Wu,
    K.K., selection methods, criteria
    and indices 30, 31, 33, 34
Smedema, L.K. & Rycroft, D.W.,
    principles and theory of
    irrigation and drainage 123
Smith, G.R., Fraser, T.A., Braithwaite,
    K.S. & Harding, R.M., yellow
    leaf zzsyndrome caused by a
    luteovirus 39
Smith, J.S.C. & Smith, O.S.,
    differentiation of US maize
    hybrids 22
Smith, R.H., rodent control 96
Smith, R.H. & Hood, E.E.,
    transformation of
    monocotyledons 44

Sojka, R.E. & Lentz, R.D., use of polyacrylamide soil conditioners 127

Soopramanien, G.C. & Batchelor, C.H., drip irrigation design, operation and maintenance 135

Sorrells, M.E., development and application of RFLPs in polyploids 25

Sosa, O.
host plant resistance against borers 82–3
reduction in cane and sugar yield through soil pest damage 85
sap feeder spp. 92

Sosa, O., Cherry, R.H. & Nguyen, R., sap feeder spp. 92

Sosa, O., Shine, J.M. & Tai, P.Y.P., weevil borer pest in Florida 80

Southwood, T.R.E., sampling programmes 85

Spaull, V.W., biological control by nematodes 81

Spaull, V.W. & Cadet, P., reduction in cane yield through nematode damage 85, 90

Specht, J.E. & Williams, J.H., contribution of genetic technology to soybean productivity 42

Sreenivasan, T.V. & Nair, N.V., catalogue on sugarcane genetic resources 21

Sreenivasan, T.V., Ahloowalia, B.S. & Heinz, D.J., improvement through breeding 24, 26

Steindl, D.R.L.
control of RSD 67
heat-treatment to control RSD 69

Stevenson, G.C.
close relationship of *S. robustum* with *S. officinarum* 7
floral initiation 13
history 2

Stevenson, R.A. & Rands, R.D., first list of fungal and bacterial diseases 54

Stringer, J.K., McRae, T.A. & Cox, M.C., use of linear unbiased prediction for estimating breeding value 30

Suasa-ard, D.W. & Charernsom, K.
host plant resistance 93
sap feeder spp. in Asia and biological control 92

Sugihara, R.T., Tobin, M.E. & Koehler, A.E., rodent control in Hawaii 96

Sukarso, G., assessment of family selection 32

Suma, S. & Jones, P., Ramu stunt disease 74

Sweby, D.L., Huckett, B.I. & Botha, F.C., minimizing somaclonal in tissue cultures 45

Sweet, C.P.M. & Patel, R., expression of yield in terms of age 108

Tai, P.Y.P., long-term storage of pollen 29

Tai, P.Y.P. & Miller, J.D.
family performance at early stages of selection 32
genotype x environment interaction for cold tolerance 33

Tai, P.Y.P., Gan, H., He, H. & Miller, J.D.
characteristics and flowering hybrids from commercial sugarcane and *Saccharum spontaneum* crosses 23
variation for juice quality and fibre content 24

Tai, P.Y.P., Miller, J.D. & Legendre, B.L.
evaluation of World Collection 21
preservation of germplasm 27

Tamanikaiyaroi, R., Johnson, S.S. & Wood, R.A., losses caused by weevil borers 79–80

Taylor, P.W.J. & Dukic, S., development of an *in vitro* culture technique 27

Teakle, D.S., causal agent of ratoon stunting disease 67

Thompson, G.D., water use by sugarcane 122

Trenor, K.L. & Bailey, R.A., red rot disease infection incited by borer damage 60

Troyer, A.F., improvement in maize yields 42

van Dillewijn, C.
anatomy of inflorescence 14
botany 8
boundaries of sugarcane cultivation 16
curling of leaves 11
replanting gaps in rows 155
soil oxygen deficiency and root growth 151

van Dillewijn, J., pokkah boeng and losses caused by this disease in Java 58

van Leerdam, M.B., Johnson, K.J.R. & Smith, J.W. Jun., *Eoreuma loftini* lays eggs on dead leaf material

attached to stalks 78

Viswanathan, R., grassy shoot disease 73

Walker, D.I.T.
breeding for disease resistance 23, 37–9
number of fertile florets 16

Walker, D.I.T. & Simmonds, N.W., performance of new varieties in trials and commercially 36, 42

Walker, W.R., procedures for designing and appraising surface irrigation systems 123

Wang, T-H., Hour, A-L., Hsu, S-Y. & Lo, C.C., database management 29

Ward, A.L. & Cook, I.M., control of white grubs 87

Weber, W.E. & Stam, P., optimum grid size in unreplicated trials 35

Webster, R.D. & Shaw, R.B., taxonomy of the native North American species of *Saccharum* 21, 23

Went, F.A.F.C., red rot disease in Java 59

White, W.H., assessing cane borer resistance 39, 82

White, W.H. & Reagan, T.E., biological control of borers 81

Whiteing, G.C., Norris, C.P. & Paton, D.C., extraneous matter (EM) levels and cane loss related to speed of chopper harvester's speed 169, 170

Whittle, A.M., yield loss due to smut disease 38

Wiedenfield, R.P., fertiliser, soil moisture and their effects on growth 150

Wiehe, P.O., sugarcane diseases in Mauritius 55

Wilbrink, G., leaf scald disease in Indonesia 65

Wilkinson, G.N., Eckert, S.R., Hancock, T.W., Mayo, O., Rathgen, A.J. & Sparrow, D.H.D., statistical methodology for design and analysis 35

Williams, J.R.
biological control in Mauritius 81
sap feeder spp. 91–2
use of trap crop for borer control 84

Wismer, C.A., benomyl as a fungicide 58

Wismer, C.A. & Bailey, R.A., pineapple disease 57

Withers, B. & Vipond, S., principles and theory of irrigation and drainage 123

Wood, B.J., rodent spp. in Taiwan 95

Wu, K.K. & Tew, T.L., evaluation of
    crosses by family yields 32
Wu, K.K., Burnquist, W., Sorrells,
    M.E., Tew, T.L., Moore, T.L.
    & Tanksley, S.D., detection
    and estimation of linkage in
    polyploids 25
Wu, K.K., Heinz, D.J. & Hogarth, D.M.,
    heritability against smut disease
    races A and B 38

Wu, K.K., Heinz, D.J. & Meyer, H.K.,
    heritability of smut disease
    resistance and yield potential 38

Yadav, R.L., stalk population density
    management 41
Yang, Z.N. & Mirkov, T.E., mosaic
    disease strain detection70
Yates, R.A. & Taylor, R.D., fertiliser, soil
    moisture and their effects on

growth 150

Zhang, L. & Birch, R.G., genetic
    engineering for resistance to leaf
    scald disease 45
Zheng, D., Lin, Y. & Tai, P.Y.P.,
    electrophoretic analysis of
    intergeneric hybrids 29
Zummo, N. & Charpentier, L.J., mosaic
    disease vector 70

Subject Index

Acidovorax avenae 69
agriculture *see* cultivation systems
Alexander the Great, sugarcane from
    India 4
altitude, and inflorescence 13
ancestry of sugarcane 2–4
anther 15
anthocyanins 10
apical meristem 13
armyworms, leaf feeders 93–5
arrow (inflorescence) 14
Australia, disease control 55
Australian Sugar Milling Council, sugar
    quality 185

bacterial diseases 63–9
    first listing 54
Badilla, source of Q813 & HQ409 5
Barbados, rodents 96
biodegradable plastic production 45
biological controls 80–1
    artificial rearing of parasitoids 80
    Diatraea sp. 80
    Eldana saccharina 81
    Mauritius 81
    sap feeders 92–3
    soil pests 88
    stalk borers by nematodes 81–2
    white grubs 88
biotechnological methods, borer control
    83
Bipolaris sacchari 63
Bligh (Capt.) transfer of sugarcane from
    Santo Domingo to Jamaica 5
bounties/subsidies 18
Bourbon cane, noble cane variety 4
Brazil
    biological control of *Diatraea* sp. 80
    control of termites 88
    locusts 95
    sugarcane diseases 55
British Guiana, early noble cane varieties,
    D109, D145 & D625 5
brown spot *(Cercospora longipes)* 63

bulliform cells 11
Bureau of Sugar Experiment Stations
    drip irrigation 135
    sugarcane diseases in Australia 55

Caledonian cane, noble cane variety 5
cane borers *see* stalk borers
Ceratocystis paradoxa 57–8
Cercospora longipes 63
Cheribon cane, noble cane variety 5
Chilo sacchariphagus 39, 79
China canes, *S. sinense* 4
chocolate 1
chromosome engineering 24, 26
chromosomes, identification, wild
    sugarcane 25
cicadas, and earth pearls, control 89
classification
    cultivated plants 21
    genotypic data, determining group
        numbers and composition for
        data sets 34
    S. spontaneum, vegetative characters 7
clonal replicates 35
clones 8
    closely related to *S. spontaneum* 24
coffee 1
coleopteran borers *see* stalk borers
Columbus
    transfer of sugarcane to New World 1
    transfer of sugarcane on second voyage
        from Spain 4
confounding, genotype x year and
        genotype x crop type effects 33
controlled environments 27
Creole cane, noble cane variety 4
crop management and control 151–2
crop yield 16
    cane varieties 17
    damage by borers 79
    evaluation of crosses by family yields
        32
    evolution and crop improvement 2, 27
    expression of yield in terms of age 108

gain, linked to genetic improvement
    42
higher planting density 41
improvement 41, 42
irrigation policy to optimise yield 113
light, positive correlation between
        hours and yield 16
row spacing and yield potential 41, 146
selective breeding 37
    and smut disease resistance 38
soil conservation and field layout 103
soil pest damage 85
sugarcane varieties 17
termite damage 85
cross-pollination 27–30
Crusaders, news of cane sugar to Europe
    1
cultivation systems 16–18, 101–8, 143–58
    monoculture of sugarcane 156
curling of leaves 11
cytological studies, North American
        species of *Saccharum* 21

D109, D145 & D625, early noble cane
        varieties in British Guiana 5
database management 29
dewlaps 11
Diatraea sp.
    biological control 80
        artificial rearing of parasitoids 80
    taxonomy 80
disease control 54–77
disease resistance *see* host plant resistance
Dominican Republic, and Haiti 4
downy mildew *(Peronosclerospora
        sacchari)* 63
drainage *see* irrigation and drainage
drip irrigation 135
    India 133
    tramline system, Swaziland, Venezuela
        135

early commercial cane varieties 4–5
earth pearls, and control 89

East India Company 1
economics 158
    see also payment systems
Eldana saccharina
    biological controls 81
    loss of recoverable sucrose 79
    screening for eldana resistance 39
electrophoretic analysis, intergeneric
        hybrids 29
environment, ideal 16
Eoreuma loftini, eggs on dead leaf
        material attached to stalks 78
Erianthus germplasm, rDNA variations
        23, 25, 29
evolution of sugarcane 2, 27
eye spot *(Bipolaris sacchari)* 63

fertiliser, and soil moisture, effects on
        growth 150
fertility, improvement 21, 28
field design/layout 145–6
    clearing 145
    fallow 17
    field factors and cane quality 153–5
    herbicide recommendations 118
    intergenotypic competition in single-
        row plots 35
    management problems 143–4
    row spacing and crop yield 103
    row spacing and yield potential 41, 146
    and soil conservation 101
field evaluation of check plot adjustments
        35
Fiji disease virus (FDV) 73
fires, control measures and equipment
        163–4
first use as sweetener in China and India
        1, 2
Florida, rodents 95
flower *see* inflorescence
flowering
    floral initiation 13–14
    improvement 28
    induction 28
    physiology and control 27, 152
    and pollen fertility 27
    synchronisation 27
flowering period 14
froghopper populations, monitoring 85
fungal diseases 57–63
    first list 54
fungicides 58
    smut disease 114
Fusarium, sett and stem rot 63

genetic engineering 44–5
    chromosome engineering 24, 26

resistance to leaf scald disease 45
and soybean productivity 42
genetic linkage map 25
    S. spontaneum 25
genetic and phenotypic path analyses and
        heritability 31
genetic relationships
    among cultivars 26
    between clones closely related to *S.
        spontaneum* 24
genetic resources 21
genetics
    characterisation of double genomic
        structure 25
    characteristics of clones 8
    chromosome transmission and meiotic
        stability 24
    confounding, genotype x year and
        genotype x crop type effects 33
    evaluation of crosses by family yields 32
    expression of foreign genes in
        sugarcane 45
    variability in nobilisation stages of *S.
        robustum* 23
    see also heritability
genome
    comparative genome mapping 25
    modern sugarcane varieties 25
genome amplification, subtropics 23
genome structure, and molecular
        diversity 26
genotype x environment interaction 32,
        33
    cold tolerance 33
    variety trials 33
    wheat 34
genotypic data, determining group
        numbers and composition for
        data sets 34
germination 16
germplasm 20–27
    collection, maintenance and use 21,
        22, 23, 26
    preservation 27
Gibberella fujikuroi 58–9
Gibberella spp. 63
Gibberella subglutinans 58–9
Glagah cane, Javanese variety of *S.
        spontaneum* 7
Glomerella tucumanensis 59–60
glume 14
grassy shoot (phytoplasmal disease) 73–4
green cane harvesting 161–2
gumming disease *(Xanthomonas
        axonopodis pv vasculorum)* 63–4
    resistance of Tanna cane in Mauritius
        5

Guyana, rodents 95, 96

H109, early noble cane variety in Hawaii
        · 5
Haiti 4
harvesting 152–3, 160–80
    coordination and control 179
    green cane harvesting 161–2
    mechanisation guidelines 164
    potential gain related to time of
        harvest 37
Hawaii
    early noble cane variety (H109) 5
    rodents 96
heat treatment
    HWT control of RSD 67, 69
    HWT control of smut disease 63
    leaf scald disease 66
Herbaspirillum rubrisubalbicans 69
heritability
    crop effects on broad-sense
        heritabilities and genetic
        variances 33
    fibre characteristics 40
    red rot resistance 60
    resistance to rust and Fiji disease 38
    smut disease resistance 38
Hispaniola, (Dominican Republic and
        Haiti) 4
history of sugarcane 1–2
host plant resistance 7, 93
    Eldana saccharina 39
    leaf scald disease 45, 66
    mosaic (SCMV/SrMV) 71
    ratoon stunting disease (RSD) 57, 69
    red rot disease 60
    rust and Fiji disease 38
    smut disease 38
        sugarcane relatives 24
    and sugar content 23
    weevil borer 40
    see also selective breeding
hot water treatment (HWT)
    control of leaf scald disease in
        quarantine 66
    control of smut disease 63
HQ409, early noble cane variety in
        Australia 5
hybrid cane 5–6
hybrid enhancement 23, 29
hybridisation procedures 28

improvement *see* germplasm; selective
        breeding
in vitro methods for conservation 27
inbreeding, pedigree analysis 26
incentives

expanding production 187–90
improving technical performance 183
socioeconomic groups 190–1
India
commercial sugar, *c.* 2500 years ago 1
home of sugarcane 3
Indian Sugar, field performance in Uttar
Pradesh 192
industrial oils 45
Industrial Revolution, effect on
population and diet 1
Industry Sources Ltd, cane payment
systems in eleven countries 186
inflorescence 13–15
and altitude 13
number of fertile florets 16
intercalary meristem 13
intercropping 157–8
intergeneric hybrids, electrophoretic
analysis 29
intergenotypic competition in single-row
plots 35
internode 9
Irian Jaya 2
irrigation and drainage 122–41
appraising surface irrigation systems
123
drip irrigation 133, 135
principles and theory 123
response to saline irrigation water 144
tests on floppy system of irrigation 132
isozymes 22

Java
noble cane variety SW11 5
pokkah boeng disease 58
red rot disease 59
study of sugarcane morphology 8
sugarcane fertility 5
'wonder cane', POJ2878 5–6

labour, indentured, Indian subcontinent
and China 17–18
labour incentives
different socio-economic groups
190–91
expanding production 187–90
Lahaina cane, variety of noble cane 5
leaf 10–12
bulliform cells 11
dewlaps 11
lamina 10
leaf sheath 10
midrib 11
screening for resistance to leaf scald
disease 66
siliceous cells 11

stomata 11
leaf feeders 93–5
armyworms 93–5
biology and control 94–5
control 95
damage 94
distribution 94
Mauritius 94
South Africa 94
leaf scald disease 64–6
Florida 65
hot water treatment 66
Indonesia 65
quarantine 66
resistance, genetic engineering 45
leaf trash 12
Leifsonia xyli subsp. *xyli* 66–9
lemma 14
lemonade 1
lepidoteran borers *see* stalk borers
Leptosphaeria sacchari 63
light 13
hours, and sugar yield 16
ligules 11
linear unbiased prediction for estimating
breeding value 30
linkage in polyploids, detection and
estimation 25
LMC International Ltd., cane sucrose
content and season length 188
locust spp.
Africa, and control 94, 95
Australia, and damage caused 94, 95
Brazil 95
lodicule 15
Lousier cane, variety of noble cane 5
luteovirus *see* yellow leaf syndrome
(YLS)

maize, US maize hybrids 22
maize yields 42
markers, and plant breeding 43
Mauritius Sugar Industry Research
Institute, races of gumming
disease 64
mechanisation, cane harvesting 164
meristem
apical 13
intercalary 13
methyl bromide, borer control 84
Mexican sugar industry 4
Mexico
cane quality 184
rodents 95
minimum descriptor set 22
molecular analysis, plants regenerated
from embryonic cultures 45

molecular biology 25
molecular diversity, and genome
structure 26
molecular markers 44
monitoring
froghopper populations 85
stalk borer populations 84–5
monitoring methods 84
morphology 8–17
mosaic (SCMV/SrMV) 70–2
and crop yield 70
screening for resistance 71
strain detection 70
time of planting and disease incidence
72
vector 70
mottled stripe *(Herbaspirillum
rubrisubalbicans)* 69
Mungo cane *S. barberi* 7

Nargori cane *S. barberi* 7
near infrared reflectance (NIR)
spectroscopy, sugarcane analysis
36
nearest neighbour models 35
NEGROW climatic model 108
nematodes 89–90
New World 1
nitrogen, clone x environment interaction
34
nobilisation 6
noble cane, end of era 5–6
noble cane varieties 4–5
qualities 6
nutrients necessary for growth 151
nutrition 13

ontogenetic analysis 24
orange rust disease *(Puccinia kuehnii)*
60–61
organic sugar 157
origin of genus *Saccharum* 3, 20
Otaheite cane, noble cane variety 4
ovary 15

Pachymetra root rot 39, 63
palea 15
Palm Tree Line 16
Panache Indian Canes, *S. sinense* 4
panicle 14
Pansahi cane
clones included with *S. sinense* 7
S. barberi 7
Papua New Guinea 2
Paris, introduction of lemonade in Paris
(1630) 1
payment systems 181–94

based on average quality 184–7
   performance 191–3
   sugar products 183
pedigree analysis, implications of
      inbreeding 26
Peronosclerospora sacchari 63
pests 78–100
   rodents 95–6
   soil pests 85–90
   status, biology and control measures
      80
Phaeocytostroma sacchari 5
pheromone trapping 85
physiology and control of flowering 27,
      152
phytoplasmal diseases, Ramu stunt and
      white leaf 73–4
pineapple disease (*Ceratocystis paradoxa*)
      57–8
pistil 15
Planalsuçar
   planning and layout 145
   sap feeder spp. and control 92
   sugarcane diseases in Brazil 55
plant generation from suspension culture
      protoplasts 44
plot technique
   border rows 34
   row spacing and yield potential 41, 146
POJ2878, Java's 'wonder cane' 5–6
pokkah boeng disease (*Gibberella
      fujikuroi* & *G. subglutinans* 58–9
pollen
   long-term storage of pollen 29
   viability 15
pollination, and selective breeding 28
polyacrylamide soil conditioners 127
pre-feasibility study 197–8
project planning 195–200
   feasibility studies 198–201
Puccinia kuehnii 60–61
Puccinia melanocephala 60–61

Q813, early noble cane variety in
      Australia 5
qualities, ennobilised canes 6
quantitative genetics: analysis of variation
      and correlation analysis 31, 40

Ramu stunt (phytoplasmal) disease 73–4
ratoon, row spacing, plant population and
      yields 146
ratoon stunting disease (RSD) 67–9
   control, tissue culture-generated
      disease-free plants 69
   control scheme 68
   crop loss due to RSD 68

heat-treatment 67, 69
Leifsonia xyli subsp *xyli* 66–9
   resistance
      screening for 69
      selective breeding 57
   spread 69
rDNA variations, *Erianthus* germplasm
      23, 25, 29
red rot disease (*Glomerella tucumanensis*)
      59–60
   incited by borer damage 60
   Java 59
   losses 60
red stripe (*Acidovorax avenae*) 69
restriction length polymorphisms 22, 25
   RFLP linkage map and genome
      analysis 25
retrospective analysis, relationships
      among test environments 33
ring spot (*Leptosphaeria sacchari*) 63
ripening 16–17
rodents
   Barbados 96
   biology and control 95–6
   Florida 95
   Guyana 95, 96
   Hawaii 96
   Mexico 95
   Taiwan 95
root rot (*Pachymetra chaunorhia*) 39, 63
roots 13
   categories of roots 13
   rooting depth 13
   sett root 12
   shoot root 12
rose bamboo cane, male parent of H109 5
row spacing
   and plant population 145
   and potential productivity 41
   replanting gaps in rows 155
RSD *see* ratoon stunting disease (RSD)
rust disease *(Puccinia melanocephala)*
      60–61
   screening clone resistance 38

Saccharum barberi
   characteristics 3
   major types, Mungo, Nargori,
      Pansahi, Saretia and Sunnabile
      7
Saccharum edule
   characteristics 7
   not an authentic *Saccharum* species 4
Saccharum officinarum 4
Saccharum robustum 7
   origins of *S. officinarum* 4
   and *S. officinarum* 7

Saccharum sinense 3–4
   Uba cane 7
Saccharum spontaneum
   derivative in sugarcane crossing 23
   origins 3
   subsp. *aegyptiacum* 7
   subsp. *indicum* 7
   use in 'nobilisation' process 6
   vegetative characters 7
saline irrigation water 144
sampling programmes 85
sampling and testing of cane 183–4
sap feeders 90–3
   in Asia and biological control 92
   in Australia 93
   biological control 93
   biology 90
   and biology 94
   damage 90
   distribution and control 91–3
   in India and control 92
   populations 92
   West Indies 93
Saretia cane, *S. barberi* 7
Sclerostachya, common ancestry with
      Saccharum 4
seed storage facilities 30
seed/caryopsis 15–16
seedlings
   production 16
   selection 32, 35
   varieties, noble canes 5
selection 30–36
   seedling families 32
      competition effects in selection
         trials 35
   varieties for disease resistance 57
   weevil borer resistance 40
selective breeding 20–53
   analysis
      application of spatial analysis 35
      early generation variety trials 35
   bi-parental means and polycrosses 37
   canegrub resistance 39
   competition, varieties in trials 34
   computerisation
      breeding records 29–30
      crossing programme 29
   crop effects on broad-sense
      heritabilities and genetic
      variances 33
   crop yield 37
   cross-pollination 27–30
   detection of rDNA sites 25
   development of *in vitro* culture
      technique 27
   disease resistance 23, 37–9

early selection procedures 32
eldana resistance 39
enhanced interpretation of pattern
    analyses of environments 34
family performance at early stages of
    selection 32
gains from plant breeding 36, 37
high early season sugar content 37
hybrid enhancement 23, 29
hybridisation procedures 28
improvements in commercial
    production 36, 37, 42
increasing sucrose content 37
linkage in polyploids 25
low ash content 40
low starch content 40
optimum grid size in unreplicated
    trials 35
performance in plant cane 33
performance of progeny 24
principles of crop improvement 34
relationships between ancient species
    of sugarcane 25
resistance to *Chilo* 39
restriction length polymorphism 22
    development and application of
    RFLPs in polyploids 25
RSD resistance 57
variation for juice quality and fibre
    content 24
see also host plant resistance
sett 8
site selection 144
slave trade, and sugarcane cultivation 17
smallholders 191
smut disease 38, 61–3
    Barbados 55
    dispersal 36, 38
    fungicides 114
    heat treatment 63
    resistance
        Australia 57
        heritability and yield potential 38
    resistance of sugarcane relatives 24
    resistance testing 38
social history of diet in England 1
soil conservation and field layout 101
    herbicide recommendations 118
    management problems 143–4
    row spacing and crop yield 103
soil oxygen deficiency and root growth
    151
soil pests 85–90
    control
        agronomic practices 87
        biological control 88
        insecticides 87

and crop yield 85
    distribution 86
soil temperatures, and root growth 13
Soltwedel (SW), cane fertility 5
somaclonal variation, tissue cultures 45
sorghum mosaic virus (SrMV) 70–2
South Africa, field performance 193
South African Sugar Association
        Experiment Station, sugarcane
        diseases 5
speciality crops 157
spikelet 14
stalk borers 78–85
    biology 78
    control 80–4
        biological control 81–2
        biotechnological methods 83
        insecticides 81–2
        methyl bromide 84
    damage caused 79
    distribution 80
    host plant resistance 82–3
        measurement of varietal resistance
        83
    management practices 83–4
    monitoring 84–5
    Papua New Guinea 82
    trap crop for borer control 84
    weevil borers, Fiji 79–80
stalk population density management 41
statistical methodology for design and
        analysis 35
stem
    anatomy 8
    anthocyanins 10
    chlorophyll 10
    epidermis 8
    internode 9
    parenchyma 8
    seed-piece 8
    sett 8
stem rot *(Gibberella* spp.) 63
stigma 15
stomata 11
Sugar Association of Caribbean
    losses in yield, rodents 96
    sap feeders, West Indies 93
sugar beet 1
sugar consumption 1
sugar content, variety trials 35–6
Sugar Milling Research Institute, field
        performance in N/NE Thailand
        and South Africa 193
sugar yield *see* crop yield
sugarcane
    bounties/subsidies 18
    cytoplasmic diversity 26

developments in 20th C 18
early commercial cane varieties 4–5
future/improvement 43–6
green cane harvesting 161–2
growing cycle 17
history 1–2
monoculture 156
payment systems 181–3
production achievements 41
reaping and transport 164–78
safe movement through quarantine 55
sampling and testing 183–4
and slave trade 17
storage 178–9
water use 122
see also payment systems
sugarcane bacilliform virus (SCBV) 73
sugarcane diseases
    Caribbean 55
    Fiji 55
    India 55
    Louisiana 55
    Mauritius 55
    Puerto Rico 55
    Reunion 55
    world distribution 54–5
sugarcane mosaic virus (SCMV) 70–2
sugarcane yellow leaf virus (SCYLV) 39,
    72–3
sugarcane yellows phytoplasma (SCYP)
    39, 72–3
Sunnabile cane, *S. barberi* 7
surface drainage 146
SW11, noble cane variety in Java 5
Swaziland, tramline system of drip
        irrigation 135
swidden agriculture 2
synopsis of sprinkler systems and
        equipment 128

Taiwan, rodents 95
Tanna cane, noble cane variety 5
taxonomy
    Diatraea 80
    and evolution of sugarcane 21
    native North American species of
    Saccharum 21, 23
tea 1
termites 85
    biology 88
    control 88, 89
    genera and species 88
Thailand, field performance 193
time of harvest 37
tissue culture-generated disease-free
        plants, controlling RSD 69
tissue cultures, somaclonal variation 45

tramline system, drip irrigation 135
transfer of sugarcane 3
transformation of monocotyledons 44
transgenic sugarcane 45, 157
    via micro projectile bombardment 27, 45
Transparent cane, noble cane variety 5

Uba cane, *S. sinense* 7
ultra-violet light traps 85
univariate and bivariate cross prediction methods 29, 32
Ustilago scitaminea 61–3
    see also smut disease

varietal improvement 41–2, 44
varietal resistance
    borers 83
    control of white grubs 87
varieties of sugarcane, crop yield 17

variety trials, genotype x environment interaction 33
Vellai cane, noble cane variety 5
Venezuela, tramline system of drip irrigation 135
vertebrate pests 95–6
viral diseases 69–73
virulence of leaf scald disease strains 65

water, and nitrogen, clone x environment interaction 34
water use, sugarcane 122
weed control 116–22
weevil borers
    Fiji, losses 79–80
    Florida 80
    Papua New Guinea 80
white grubs 87
    Asia 86
    Australia 86

biological controls 88
control 86, 87
genera and species 86
India 86–7
insecticides 87
life cycles 86
Pakistan 86–7
Thailand 87
white leaf disease 73–4
World Collection, evaluation 21

Xanthomonas albilineans 5, 64–6
Xanthomonas axonopodis pv *vasculorum* 5, 63–4

yellow leaf syndrome (YLS) 39, 72–3
    caused by a luteovirus 39
    caused by a phytoplasma 39, 72
    world distribution 72
    yield and juice quality 73